The Institute of Mathematics
and its Applications
Conference Series

The Institute of Mathematics
and its Applications
Conference Series

Previous volumes in this series were published by
Academic Press to whom all enquiries should be addressed.
Forthcoming volumes will be published by
Oxford University Press throughout the world.

NEW SERIES

1. *Supercomputers and parallel computation* Edited by D. J. Paddon
2. *The mathematical basis of finite element methods* Edited by David F. Griffiths
3. *Multigrid methods for integral and differential equations* Edited by D. J. Paddon and H. Holstein
4. *Turbulence and diffusion in stable environments* Edited by J. C. R. Hunt
5. *Wave propagation and scattering* Edited by B. J. Uscinski
6. *The mathematics of surfaces* Edited by J. A. Gregory

Wave propagation and scattering

Based on lectures at a conference
organized by the Institute of Mathematics and its Applications
and held at the University of Cambridge, 5–6 April 1984

Edited by

B. J. USCINSKI
University of Cambridge

CLARENDON PRESS · OXFORD · 1986

Oxford University Press, Walton Street, Oxford OX2 6DP

Oxford New York Toronto
Delhi Bombay Calcutta Madras Karachi
Kuala Lumpur Singapore Hong Kong Tokyo
Nairobi Dar es Salaam Cape Town
Melbourne Auckland

and associated companies in
Beirut Berlin Ibadan Nicosia

Oxford is a trademark of Oxford University Press

Published in the United States
by Oxford University Press, New York

British Library Cataloguing in Publication Data

Uscinski, B. J.
Wave propagation and scattering – (The Institute of Mathematics and its Applications conference series; 5)
1. Scattering (Physics) 2. Waves
I. Title II. Series
531'.1133 QC20.7.53

ISBN 0 19 853607 0

Printed in Great Britain by
St Edmundsbury Press, Bury St Edmunds, Suffolk

PREFACE

The subject of wave propagation and scattering in random media has seen many fundamental developments during the last decade. There has been extensive progress in both experiment and theory. New instrumentation such as lasers, fibre optics, and digital photon counters as well as advances in computer software and microprocessors have allowed accurate and reliable experiments to be carried out. These experiments which test theory and investigate the random media themselves include light propagation in the atmosphere, radio-wave scattering in the interstellar and interplanetary media, and ocean acoustic experiments. At the same time theoretical work has moved forward steadily, so that there is now a much improved understanding of propagation through the deeply modulated phase screen and in extended random media.

As a result of all these developments there has been a feeling among workers in random wave propagation that some review of the field and comparison of progress in the separate areas is needed. As a result of this a two day symposium on the subject was organised by the Institute of Mathematics and its Applications in Cambridge on 5th - 6th April, 1984. The aims of the symposium were to allow workers in the field to meet and compare notes and also to provide potential users, such as industry and Government Establishments, with an opportunity to assess the expertise available in the UK. The audience thus consisted of representatives from both science and industry.

This book is based on the presentations made at the IMA symposium on Wave Propagation and Scattering in Random Media, Cambridge, 5th - 6th April, 1984. It contains a selection of topics written by prominent workers in the field covering many important areas of progress. These will provide the reader with an overview of how the subject stands at present and constitute a useful source of specialised information on individual topics.

B.J. Uscinski
University of Cambridge

ACKNOWLEDGEMENTS

The Institute thanks the authors of the papers, the editor, Dr. B. Uscinski (University of Cambridge) and also Mrs. Janet Parsons, Mrs. Denise Bumpus, Miss Pamela Irving and Miss Karen Jenkins for typing the papers.

CONTENTS

CONTRIBUTORS

M.V. BERRY; *H.H. Wills Physics Laboratory, University of Bristol, Royal Fort, Tyndall Avenue, Bristol, BS8 1TL.*

R. BUCKLEY; *Institute of Sound and Vibration Research, University of Southampton, Southampton, SO9 5HN.*

R.H. CLARKE; *Department of Electrical Engineering, Imperial College, Exhibition Road, London, SW7 2BT.*

J.C. DAINTY; *Blackett Laboratory, Imperial College, Prince Consort Road, London, SW7 2BZ.*

P.J. DUFFETT-SMITH; *Cavendish Laboratory, University of Cambridge, Madingley Road, Cambridge, CB3 OHE.*

T.E. EWART; *Applied Physics Laboratory and School of Oceanography, University of Washington, Seattle, Washington, USA.*

J.H. HANNAY; *H.H. Wills Physics Laboratory, University of Bristol, Royal Fort, Tyndall Avenue, Bristol, BS8 1TL.*

A. HEWISH; *Cavendish Laboratory, University of Cambridge, Madingley Road, Cambridge, CB3 OHE.*

E. JAKEMAN; *Theoretical Physics Section, Royal Signals and Radar Establishment, St. Andrews Road, Great Malvern, Worcs, WR14 3PS.*

C. MACASKILL; *(University of Cambridge) now at Department of Applied Mathematics, University of Sydney, NSW 2006, Australia.*

J.T. MACKLIN; *MSDS Laboratory, Marconi Research Centre, West Hanningfield Road, Great Baddow, Chelmsford, Essex, CM2 8HN.*

D. NEWMAN; *Blackett Laboratory, Imperial College, Prince Consort Road, London, SW7 2BZ.*

C.J. OLIVER; *Royal Signals and Radar Establishment, St. Andrews Road, Great Malvern, Worcs, WR14 3PS.*

K. OUCHI; *(Queen Elizabeth College) now at Department of Physics, King's College, Strand, London, WC2R 2LS.*

B.J. RICKETT; *(University of Cambridge) now at Department of Electrical Engineering and Computer Science, University of California, San Diego, La Jolla, California 92093, USA.*

D.L. ROBERTS; *(University of Cambridge) now at the Meteorological Office, London Road, Bracknell, Berks, RG12 2SZ.*

S. ROTHERAM; *GEC Research Laboratories, Marconi Research Centre, West Hanningfield Road, Great Baddow, Chelmsford, Essex, CM2 8HN.*

F.G. SMITH; *Jodrell Bank, Macclesfield, Cheshire, SK11 9DL.*

B.J. USCINSKI; *Department of Applied Mathematics and Theoretical Physics, University of Cambridge, Silver Street, Cambridge, CB3 9EW.*

M.E.R. WALFORD; *H.H. Wills Physics Laboratory, University of Bristol, Royal Fort, Tyndall Avenue, Bristol, BS8 1TL.*

K.D. WARD; *Royal Signals and Radar Establishment, St. Andrews Road, Great Malvern, Worcs, WR14 3PS.*

INTRODUCTORY ADDRESS

B.J. Uscinski
(Department of Applied Mathematics and Theoretical Physics, University of Cambridge)

ABSTRACT

In recent years important experimental and theoretical developments have occurred in many areas of random propagation and scattering. Topics like the statistical structure of the scattered wave field, including the probability density of intensity fluctuations, the nature and influence of caustics and speckle, to name but a few, are now much better understood. In the course of the first day some of the more important advances in the mathematical theory will be discussed.

Equal progress has been made in associated experimental work, partly as a result of improvements in instrumentation and computational techniques. Thus the whole subject is now entering a stage of rapid growth since the new theoretical results can be carefully checked by controlled experiments. On the second day some typical new experimental results will be described and the application of recent advances for practical purposes will be discussed.

When this conference on random wave propagation and scattering was first planned we had in mind to gather together a relatively small number of workers in the field to discuss the most recent areas of progress. The subsequent wide interest in the conference has been most gratifying, and we have present today, not only those who have specialized to a greater or lesser extent in problems of random wave propagation, but also many who may be less familiar with the technical side of the subject. The effects of randomness in wave propagation are important in many branches of physics and engineering and so it is not surprising that a knowledge of contemporary scattering theory, and the practical answers it can offer, has come to be regarded as a desirable, if not essential tool, for both theoretician as well as experimenter.

Because many of the audience may be unfamiliar with some of the concepts frequently discussed in scattering theory, or may even be encountering the subject for the first time, I would like to provide a brief introduction to a few of the more important ideas and problems that we will be discussing during the next two days. Probably the most convenient way to do this is to give a short history of random propagation theory and the accompanying experimental scene. In this way a certain order, historical if not logical, will be imposed, and the ideas and problems will arise as part of the story.

I should like to begin, not at the beginning of scattering theory, which goes back a long way, certainly past Lord Rayleigh to Isaac Newton, and possibly even further, but some twenty years ago. It was then as an experimenter that I was first faced with a practical problem in the random scattering of radio waves in the ionosphere and forced to look to existing theory for answers. The next twenty years have been spent still looking for answers, fortunately not to the same question. All history is to some extent subjective and so this will be one man's view of random propagation theory. It is my own view of how I saw the scene twenty years ago, of what developments have since occurred and appear to me to be important, and finally where we stand today.

TWENTY YEARS AGO

The random propagation problems encountered by the experimenter twenty years ago were not so very different from those that are of interest today - except that then they could not be satisfactorily explained. Radio waves passing through the earth's ionosphere exhibited irregular fading of both amplitude and phase, while similar fluctuations of light intensity had long been of interest. Lasers may have been a little thinner on the ground then than now but the twinkling of optical stars due to the atmosphere had long been the object of wonder, and measurement. It was about twenty years ago that radio stars were also found to twinkle in the same way due to irregular diffraction in the solar corona. Radar returns from rough surfaces such as ocean waves, rocky terrain, and even from the moon's surface, were all found to exhibit intensity fluctuations, and the problem of reflection from a rough surface ranked alongside transmission through a random medium as a cardinal scattering problem.

Many of the situations mentioned above have a common feature. In the transmission problem the scattering is through small angles in the cases of most interest, ie. very nearly forward scattering. Another common element, soon apparent to the experimenter, was that the amount of scatter in almost all real situations was large. This phenomenon was known as "multiple scatter". It was also soon clear to the experimenter looking for theory to compare with data, that no satisfactory theory of multiple scatter existed.

There was one notable expection. Radiative transport theory could deal with the multiple scatter problem of the directional distribution of the radiation flux. However, the experimenter in random propagation was generally interested in other properties of the scattered field. He wanted to know not just the directional or angular spectrum of the field, but, principally, what were the properties of the intensity fluctuations produced in the wave, such as their spectrum and variance. Similar properties of the phase fluctuations were also of interest. These phenomena, coherent intensity and phase fluctuations, are important when the scattering angles are very small. The scattered waves then retain a large degree of correlation with each other and exhibit marked interference effects.

THEORY

Twenty years ago a survey of the theory of scattering in an extended medium revealed two main theories for dealing with intensity and phase fluctuations. These were the Born approximation and Rytov's method of smooth perturbations, and the best known authors were Tatarski and Chernov in the Soviet Union. The Born approximation is based on first order perturbation theory and can describe only very small fluctuations of phase and intensity. Rytov's method was initially thought to be valid for large fluctuations, but as early as 1965 it was realised that while it could deal with larger fluctuations than the Born method it was valid only for small intensity fluctuations. Unfortunately the belief that Rytov's method is valid for large intensity fluctuations has persisted in some quarters even to the present day and continues to lead to error, confusion, and much waste of valuable time.

The other important theoretical concept much in use twenty years ago is that of the random phase screen. This is a model of scattering situations where the wave acquires a phase modulation on passing through a layer of medium, and then developes amplitude modulation on propagating away from the layer. The importance of this model cannot be overestimated.

Its use allowed the basic concepts of random propagation to be understood and explained. It served as a satisfactory model for many experimental situations and ultimately proved to be the key that has unlocked the problems of multiple scatter in the extended medium.

In its earliest and simplest form in which the screen induces small phase fluctuations it was associated with the names of Ratcliffe, Booker, Clemmow, Hewish and others (Ratcliffe, 1956). A more interesting extension was that of the screen producing large phase fluctuations. This was required to model many real situations, but the intensity fluctuations due to such a screen proved difficult to study. Mercier in 1962 gave a mathematical formulation of the problem and pointed the way to some important properties of the intensity fluctuations, such as the focussing effect.

The situation twenty years ago could be summed up as follows: most random wave propagation experiments at the time involved multiple scatter, but no satisfactory theory of multiple scatter existed.

PROGRESS IN THE 1970'S

For the reasons just mentioned the 1970's were marked by a search for methods to deal with multiple scatter. Two important steps were made when treating the extended medium:

(a) Firstly the fact that the scattering angles are small led to the adoption of the parabolic form of the wave equation as a basis for more theories, and

(b) Secondly, attention was concentrated on the various moments of the random wave field.

These two ideas were combined to lead to the now well-known parabolic partial differential equations for the moments. The best known of these is the equation for the fourth moment first proposed by Shishov in 1967. The fourth moment is important because it gives one of the most useful measures of the intensity fluctuations, ie. its spectrum and variance. The parabolic equation for the fourth moment has a simple clear physical meaning. It represents repeated application of narrow angle Born scattering to the wave-field and hence offers good prospects for obtaining useful expressions for the fourth moment that reliably reflect the physical process of multiple scatter.

The second moment of the complex amplitude was a familiar quantity twenty years ago, being the Fourier transform of the directional spectrum. Another form of the second moment, a product of complex amplitudes at different wave frequencies began to assume importance. Its Fourier transform gives the shape of a radiation pulse broadened by random scattering, and the appropriate second moment equation was successfully solved to give this quantity. This enabled the pulses from the newly discovered pulsars, broadened by scattering in the interstellar medium, to be described theoretically.

THE DEEP PHASE SCREEN

Although the parabolic equation for the fourth moment was well established in the 1970's it proved difficult to solve. In the absence of analytical expressions for the fourth moment the experimenter tended to turn his attention to the deeply modulated phase screen to model multiple scatter situations.

The spectrum of intensity fluctuations produced by such a screen has a convenient integral form. This integral was first evaluated analytically in 1971 by Buckley and Shishov independently. The resulting expressions for the spectra and normalised variance of intensity fluctuations, sometimes called the scintillation index, revealed some of the most interesting and important features of multiple scatter. These included the growth of the scintillation index with distance from the screen to attain a maximum value that could be well in excess of unity with a subsequent fall off to unity at large distances from the screen. The form of the intensity fluctuation spectrum could also differ radically from that predicted by the Rytov method.

It is interesting to note that modelling an extended scattering medium by a strong phase screen concentrated in one plane seems like a contradiction. However, strangely enough, this model had considerable success in explaining experimental observations. It is only quite recently, with the development of analytical solutions of the fourth moment equation that the reason for this has become clear.

DESCRIPTION OF THE MEDIUM

Another significant development that occurred during the 1970's concerned the description of the random medium itself. Tatarski had favoured the use of the Kolmogorov turbulence spectrum. This implies a wide spread of scale sizes and spatial frequencies. Chernov, on the other hand used a Gaussian auto-correlation function to describe the medium in his theoretical work, and this implies the presence of really only one scattering scale in the medium. Much of the random propagation theory done outside the Soviet Union tended to use the Gaussian auto-correlation function basically because of its mathematical convenience. However, there was an increasing awareness in the 1970's that most real scattering media contain not just one scattering scale, but rather a whole range of scale sizes.

The adoption of power-law spectra of various types not only provided a better description of the scattering medium but also revealed a number of interesting phenomena in the random propagation itself. It became clear that the spectrum and variance of intensity fluctuations could behave quite differently depending on the range of scattering scales present in the screen or medium. Introduction of the concept of fractals allowed screens to be treated in which there is no outer or inner bound to the range of scales and waves passing through such a screen can exhibit very interesting properties indeed.

MOST RECENT ADVANCES

Progress in understanding intensity fluctuations in an extended medium in the 1970's was fairly slow. Apart from the items mentioned above the most notable contributions were probably those from the Soviet Union. Many papers appeared by Shishov and Gochelashvili dealing with the case when the intensity fluctuations are saturated and one approaches the very far field where the scintillation index is almost unity and the statistics are almost Gaussian.

Among these papers was one by Shishov in 1971 treating whole range of scatter and not just the asymptotic far field case. This paper is certainly one of the most significant to be written about intensity fluctuations. It contains the seeds of many powerful ideas and pointers to the solutions we now have. Unfortunately it was a difficult paper to understand and was almost completely passed over by workers in the field.

Another notable development was the introduction of path integrals. Tatarsky and Zavorotuyi in the Soviet Union, Hannay in the United Kingdom and Dashen in the United States all made use of this approach to obtain additional intuition about multiple scatter. It proved particularly useful for describing the second moment, ie. the complex coherence including the mean intensity when the presence of a mean refractive index profile leads to the appearance of curved deterministic rays.

THE 1980'S

In the early years of the 1980's an analytical solution was found in Cambridge for the parabolic fourth moment equation. The method followed in the steps of Shishov but the resulting expressions are uncomplicated and have a simple physical meaning (Uscinski, 1982, Macaskill 1983). The same results can now be reproduced by a variety of methods. The theoretical spectra and variances of intensity fluctuations that they yield have been checked by comparison with numerical solutions of the equation and scattering simulations. The agreement obtained is very good and we are now confident in their accuracy.

Attention is now being directed at finding the higher order statistics. In particular the full probability distribution of intensity fluctuations both for the extended medium as well as for the deep phase screen is being sought. This will certainly prove a difficult question to resolve and as yet only partial answers are available.

EXPERIMENTS, REAL AND NUMERICAL

Just as the theory of scattering and random propagation had its origins in the investigation of natural phenomena, so its progress has been helped very substantially in recent times by controlled experiments both real and numerical. The advent of lasers has enabled exact measurements to be made of the propagation of light in randomly fluctuating media, while the availability of sophisticated detectors and microprocessors allows the required spectra and other statistics to be readily extracted from what would once have been overwhelming amounts of experimental data.

The main drawback of many random propagation experiments lies in the fact that for various reasons the irregular scattering medium is not adequately measured. Now if the true space-time autocorrelation function of the medium is not known it is not possible to make proper comparisons of theory and experiment. For this reason it is worth while making special mention of two acoustic experiments carried out by Ewart in the United States (Ewart, 1976; Ewart and Reynolds, 1984). These are almost unique in that very thorough measurements were made of the irregular medium over the entire propagation path, at the same time that the fluctuations of both phase and amplitude of the scattered sound field were recorded. Ewart's acoustic experiments provide an opportunity to test many aspects of current scattering theory in a way that was not previously possible.

Mention should also be made of the numerical scattering experiments that can be carried out in the much larger and faster computers that are now available. In these experiments a stochastic medium with a specified autocorrelation function can be synthesized in the computer and a wave front propagated through it in accordance with the parabolic approximation. This process can be repeated many times for different realizations of the medium and any desired statistical quantity of the wave field can be determined. Numerical simulations, if carefully carried out, provide an excellent method of testing theoretical results and obtaining intuition into the random propagation problem.

SUMMARY

Compared with twenty years ago today's random propagation theory can offer a wide range of reliable answers to many of the questions that arise in multiple scattering situations. Indeed it is our hope that this conference may serve to acquaint the user with what the theory now has to offer that may be employed in practice. This can cover applications as diverse as the development of technical equipment or the interpretation of scientific experiments.

Moreover the prospects for further progress in the near future are very good. We now have a strong body of reliable theory that has opened up the area of multiple scattering and given us much insight. Advances in computer hard and software mean that it is now possible to solve some of the important equations numerically, as well as to conduct large scale numerical simulations that will provide answers and insight where theory fails. Progress in electronics and microprocessors means that we can look forward to accurately controlled experiments that will help to resolve many of the outstanding questions in this branch of science.

The present conference should help to lend perspective to the developments in scattering theory that have taken place over the last two decades and allow the participants to present highlights of their recent work in this broader context. It is hoped that the interest generated by the conference will lend impetus to the excellent work in these fields already in progress in the UK.

REFERENCES

Buckley, R. 1971, "Diffraction by a random phase screen with very large R.M.S. phase deviation", *Aust. J. Phys*. **24**, 351.

Ewart, T.E. 1976, "Acoustic fluctuations in the open ocean - a measurement using a fixed refractive path", *J. Acoust. Soc. Am.* **60**, 46

Ewart, T.E. and Reynolds, S.A. 1984, "The Mid-Ocean Acoustic Transmission Experiment - MATE", *J. Acoust. Soc. Am.* **75**, 785.

Macaskill, C. 1983, "An improved solution to the fourth moment equation for intensity fluctuations", *Proc. R. Soc. Lond.*, **A386**, 461.

Mercier, R.P. 1962, "Diffraction by a screen causing large random phase fluctuations", *Camb. Phil. Soc. Proc.* **58**, 382.

Ratcliffe, J.A. 1956, "Some aspects of diffraction theory and their application to the ionosphere", *Reps. Proc. Phys.* **19**, 188.

Shishov, V.I. 1967, *Tr. Fiz. Akad. Nauk SSSR.*, **38**, 171.

Shishov, V.I. 1971, "Diffraction of waves by a strongly refracting phase screen", *Izv. Vyssh. Ucheb. Zaved. Radiofiz.* **14**, 85.

Shishov, V.I. 1971, "Strong intensity fluctuations of a plane wave propagating in a random refractive medium" *Zh. Eksp. Theor. Fiz* **61**, 1399.

Uscinski, B.J. 1982, "Intensity fluctuations in a multiple scattering medium. Solution of the fourth moment equation", *Proc. Roy. Soc. Lond.*, **A380**, 137.

TWINKLING EXPONENTS IN THE CATASTROPHE THEORY OF RANDOM SHORT WAVES

M.V. Berry
(H.H. Wills Physics Laboratory, Bristol)

ABSTRACT

The difficulty in obtaining a theoretical description of the statistics of random short waves arises from the presence of ray caustics, which cause intensity moments to diverge. The divergences are determined by the result of a competition between singularities, which contribute exponents calculated using catastrophe theory.

1. INTRODUCTION

The applications and illustrations of catastrophe theory in physics are now many and varied, and have been recently and extensively reviewed in several books (Poston and Stewart 1978, Gilmore 1981) and articles (Stewart 1981). In this paper I will concentrate on one group of topics, centred on the origin and applications of scaling laws in the theory of short waves and explain how these form the basis of a theory of the statistics of the waves when they encounter randomness.

There are several reasons why it is appropriate to write about this subject. The first reason is that although the theory is subtle it nevertheless provides an explanation of a familiar phenomenon: the intense twinkling of starlight.

The second reason is that the theory embodies in a novel way two themes with which theoretical physics has been much preoccupied, namely <u>scaling</u> and <u>nongaussian fluctuations</u>. Scaling is an expression of the fact that some physical quantities can depend nonanalytically on others. This is most familiar in phase transitions (Stanley 1971) where thermodynamic properties are nonanalytic functions of temperature at critical points. In this paper the scaling laws will concern the short-wave limit, in which the wavenumber $k(= 2\pi/\text{wavelength})$ becomes infinite. Nongaussian fluctuations occur in random variables or functions governed by processes with a high degree of mutual correlation, so that the central limit theorem cannot be applied. Familiar examples are the stable distributions of probability theory (Jona-Lasinio 1975), and certain fractal noises (Mandelbrot 1982). My example is unfamiliar in this context and involves large fluctuations for which statistical quantities (moments of wave intensity) may scale in ways dominated by competition amongst singularities (i.e. the catastrophes).

The third reason is that the theory makes essential use of concepts characteristic of catastrophe theory. This essential use is to be contrasted with the use 'in principle' which has been so common when catastrophe theory is invoked outside physics (and which has given rise to controversy - see Sussmann and Zahler (1975), Zahler and Sussmann (1977) and subsequent correspondence in Nature **270** (1977) 381-384, 658). The characteristic concepts in question are firstly, the structural stability of certain universality classes of singularity, and secondly the hierarchy of normal forms representing the singularities in each class.

The fourth reason is that this group of subjects is one of the less well known applications of catastrophe theory. The subjects are still in their infancy, and it is likely that much remains to be discovered by imaginative investigators.

The plan of this paper is as follows. Section 2 contains a brief account of the catastrophe theory of waves. Section 3 is devoted to the scaling laws expressing the nonanalyticity of short waves. These are essential preliminaries to section 4, where at last randomness is incorporated into the asymptotics, thus introducing the central idea of singularity-dominated strong fluctuations by deriving scaling laws for moments of wave intensity.

2. SHORT WAVES AS DIFFRACTION CATASTROPHES

Consider a monochromatic wave with wave number k, represented by a scalar wavefunction $\psi(C;k)$, where C is an abbreviated notation for any quantities $C_1, C_2 \ldots$ on which the waves depend, such as time, position coordinates, or parameters describing diffracting objects or refracting media. In the language of catastrophe theory, C are control parameters. ψ will be assumed to satisfy a linear wave equation with boundary conditions. This framework is very broad, including optics, quantum mechanics, acoustics, elasticity and small-amplitude water waves.

The short-wave limit is $k\to\infty$. It is nontrivial because ψ is a nonanalytic function of $1/k$, with an essential singularity at $1/k=0$. In optics, for example, it is not possible to express monochromatic electromagnetic wave fields as Taylor series in the wavelength with geometrical optics as the leading term. Thus the connection between wave and ray optics (and, similarly, between quantum and classical mechanics) is much more complicated than the connection between special relativity and Newtonian mechanics, which is simply a matter of expanding in powers of v/c (relative velocity of coordinate frames divided by speed of light).

Large-k asymptotics must give the correct mathematical description of three physically obvious facts. Firstly, $\psi(C;k)$ must be constructed in terms of the trajectories of the corresponding Hamiltonian problem (for example rays of light). Secondly, on the caustic or focal set, that is on the envelope of the rays representing the wave, ψ must rise to high values, diverging as $k\to\infty$. And thirdly, the scale of diffraction fringes in C space must vanish as $k\to\infty$.

A crucial element in formulating asymptotics in accordance with these three criteria is the recognition that a wave corresponds not to a trajectory but to a family of trajectories; one is reminded of Dirac's (1951) remark: "presumably the family has some deep significance in nature, not yet properly understood". Different trajectories may pass through different points C, and more than one trajectory may pass through a given point C. To label the trajectories in a family we employ variables $s= s_1,s_2\ldots$. In the terminology of catastrophe theory, s are state variables; they may represent, for example, points at which trajectories intersect an initial wavefront, or directions of trajectories at an instant of time. The different trajectories through a point will be denoted by $s^\mu(C)$ $(\mu=1,2\ldots)$.

The trajectories $s^\mu(C)$ are determined by ray dynamics. These are governed by equations derived from a Hamiltonian function (whose operator generalization generates the wave equation satisfied by $\psi(C;k)$). Here it will be necessary only to invoke the fact that ray dynamics may also be derived from a variational principle, which may be expressed as follows: there exists an optical distance function (or action function) $\phi(s;C)$, whose stationary values $s^\mu(C)$ are the rays. Thus

$$\frac{\partial\phi}{\partial s_i}(s;C)=0 \quad \forall\, i \quad \text{if} \quad s=s^\mu(C). \qquad (1)$$

An elementary example of this way of formulating ray dynamics is the evolution of rays from a curved wavefront W in a plane filled with homogeneous isotropic medium (figure 1).

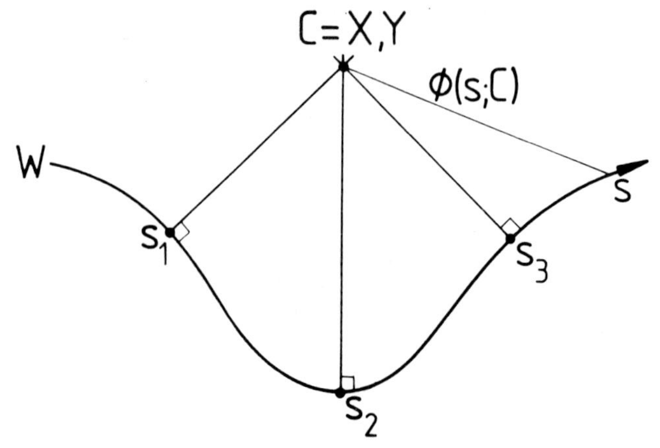

Fig. 1. Optical distance $\phi(s;C)$ from s on a wavefront W to control point C; also shown are three trajectories, for which ϕ is stationary.

C corresponds to position X,Y in the plane, s is a coordinate on the wavefront, and ϕ is the distance to C from s on W. It is clear that (1) simply expresses the condition that in these circumstances rays are straight lines normal to W.

In catastrophe terminology, (1) states that rays are determined by a <u>gradient map</u> (figure 2) from s to C.

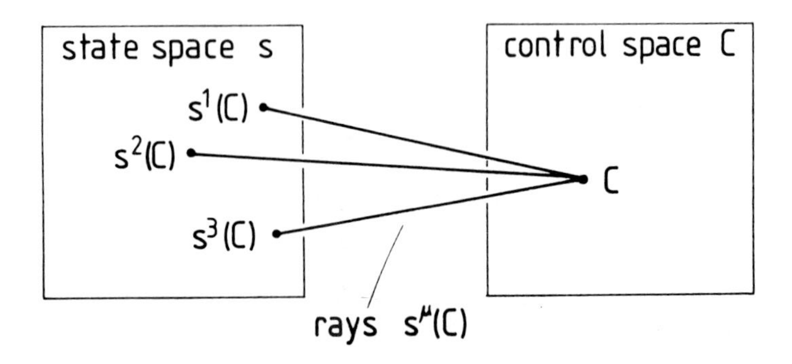

Fig. 2. Multivalued gradient mapping from state space to control space, induced by rays $s^\mu(C)$.

The caustics are envelopes of the ray family described by $\phi(s,C)$ and are defined as singularities in C space of the gradient map (1), that is hypersurfaces across which the number of rays suddenly changes. The condition for this is that ϕ is stationary to higher order, so that in addition to (1) the equation

$$\det\left\{\frac{\partial^2\phi}{\partial s_i \partial s_j}\right\} = 0 \tag{2}$$

must hold. This is illustrated for our simple example in figure 3, which shows the family of rays normal to W. At points such as A, three rays pass through each point; at points such as B, one ray passes through each point. The separator set is the caustic, in this case a cusped curve whose points satisfy the focal condition, which follows from (2), of lying on the locus of centres of curvature of W.

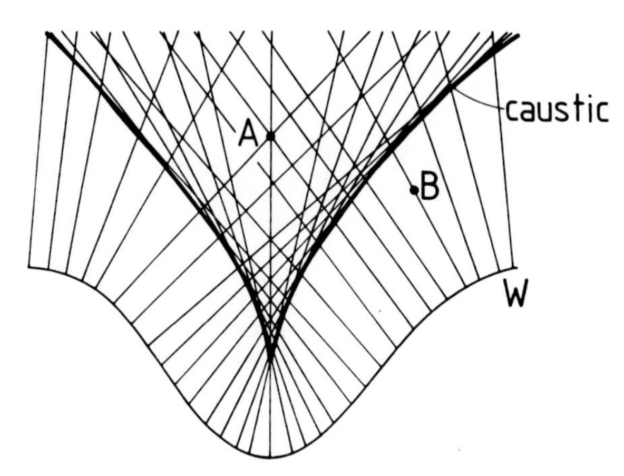

Fig. 3. Cusped caustic formed as envelope of trajectories (normals) from wave-front W; three trajectories reach A, one reaches B.

Caustics organize the multivaluedness of the ray family. In the space sXC the different solutions $s^{\mu}(C)$ join to form a smooth surface, called the critical manifold, whose foldings over the control space C correspond to the rays. This is illustrated for our example by figure 4.

The reason for introducing the function $\phi(s;C)$ is that as well as generating rays and caustics it also generates wave-functions $\psi(C;k)$ in the short-wave limit $k\to\infty$. A substantial body of asymptotic analysis leads to the following integral representation:

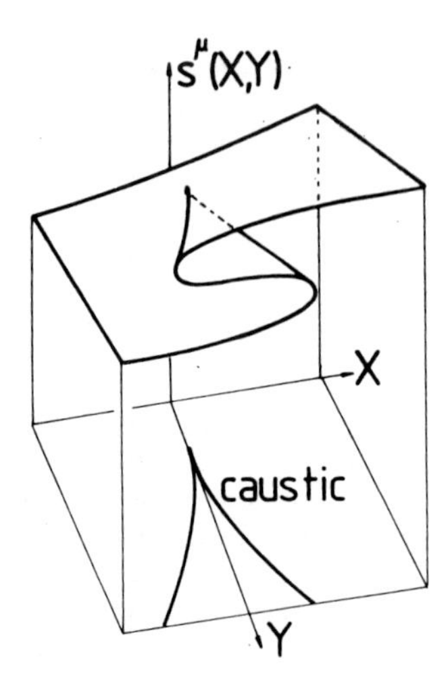

Fig. 4. Caustic of figure 3 formed by projection of critical manifold in s,C space.

$$\psi(C;k) \simeq \left(\frac{k}{2\pi}\right)^{n/2} \int d^n s\ a(s;C)\, e^{ik\phi(s;C)} \qquad (3)$$

In this equation, n is the number of state variables and a is a smooth function of s and C dependent on the ray family. The integral (3) generalizes a number of mode expansions, Kirchhoff diffraction integrals and Fourier-transform representations (whose variety is a consequence of arbitrariness in interpreting s), which all become equivalent in the short-wave limit. The derivation of such appoximations and their correction terms was put on a firm basis by Maslov, in work reviewed by Kravtsov (1968), Duistermaat (1974) and in the book by Maslov and Fedoriuk (1981). Intuitive presentations have been given by Berry (1976, 1981) and Berry and Upstill (1980). Physicists will find (3) reminiscent of Feynman-path integrals, which are exact representations of ψ as a 'super-superposition' in which s is infinite-dimensional and ϕ a functional; this point of view is well presented by Shulman 1981.

For large k the integrand in (3) is a rapidly-oscillating function of s and direct evaluation of the integral is impractical. It is natural to attempt an approximate evaluation by the method of stationary phase, which consists in expanding to second order about its stationary points, which are precisely the rays $s^\mu(C)$ defined by (1). This gives

$$\psi(C;k) \approx \sum_\mu \left[\frac{a(s;C)\exp\{ik\phi(s;C)+i\alpha_\mu \pi/4\}}{\left|\det \dfrac{\partial^2\phi(s;C)}{\partial s_i \partial s_j}\right|^{\frac{1}{2}}} \right]_{s=s^\mu(C)} \tag{4}$$

where a is an amplitude and α_μ is the signature of the matrix $\partial^2\phi/\partial s_i \partial s_j$. Thus ψ appears as a superposition in which each ray contributes a wave whose phase is precisely the stationary value of the optical distance function.

Although (4) gives a useful description of many interference effects, it suffers the fatal defect of failing just where we shall be most interested in ψ, namely on caustics. This is clear from (2), which implies that the approximation (4) diverges to infinity as C moves onto a caustic because of the vanishing of the denominators of its terms. Now when $k = \infty$ this divergence must surely occur: it simply expresses the infinite concentration of rays at a caustic. But we are interested in large-k asymptotics rather than the trajectory limit, and so seek to determine exactly <u>how</u> the divergence occurs as k increases. Thus near caustics (4) is too crude an approximation to the integral (3).

It is at this point that catastrophe theory enters to provide two remarkable simplifications of the problem for the important case (which occurs 'almost always') when the caustic has the property of <u>structural stability</u>. This means that a small smooth deformation of ϕ (i.e. a diffeomorphism such as would be produced by changing the nature or positions of diffracting objects) will cause a smooth deformation of the caustic. In figure 3, for example, the cusp will remain a cusp under small changes in the form of W. The central theorem of catastrophe theory (Poston and Stewart 1978, Gilmore 1981) is that <u>the world of caustics is partitioned into universality classes</u>. Any two caustics in the same class can be transformed into each other by diffeomorphism of their ϕ's. It is these universality classes (or in mathematical terminology, equivalence classes) that constitute the <u>catastrophes</u>. The classification of catastrophes, begun by Thom (1975), has been carried much further by Arnol'd (1975). Figure 5 shows the forms of the caustics for the catastrophes whose <u>codimension</u> K (essential number of control parameters C) satisfies $K \leq 3$.

The first way in which this classification enables the diffraction integral (3) to be simplified follows from replacing all $\phi(s;C)$ in a given universality class (labelled j) by one <u>normal form</u> $\Phi_j(s;C)$. The transformation of ϕ into Φ_j is achieved by diffeomorphism of s and C. From this point of view, catastrophe theory is the classification of normal forms; table I is a list of normal forms for the first few catastrophes (Arnol'd 1975 lists many more). Making the same transformation in (3), and replacing the factor a and the transformation Jacobian by unity (because these are smooth functions and near a caustic the contributing s values lie close together), we obtain, instead of the infinitely many integrals corresponding to all possible ϕ, the following finite set of diffraction catastrophes

$$\Psi_j(C;k) = \left(\frac{k}{2\pi}\right)^{n/2} \int d^n s\ e^{ik\Phi_j(s;C)} \quad . \qquad (5)$$

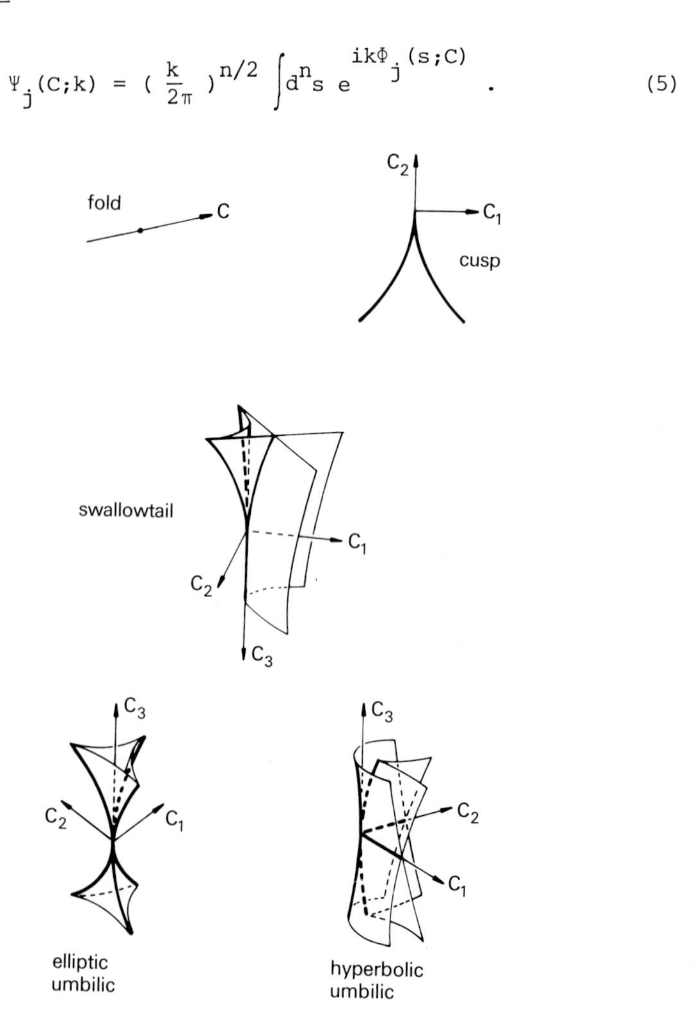

Fig. 5. The elementary catastrophes of codimension $K \leqslant 3$.

The polynomial exponents in the integrands represent the irreducible topological complexity of the collisions of stationary points $s^{\mu}(C)$ of Φ_j (i.e. of caustics) as the parameters C vary. Diffraction catastrophes can be very complicated functions of C. Consider, for example the cusp. From table I, the quartic polynomial Φ_{cusp} (which on using (1) can easily be seen to correspond to the cubic critical manifold of figure 4) gives

$$\Psi_{cusp}(C_1,C_2) = (k/2\pi)^{1/2} \int_{-\infty}^{\infty} ds \exp\{ik(s^4/4+C_2 s^2/2+C_1 s)\} \qquad (6)$$

Table I

Name	Symbol	K	$\phi(s:C)$
fold	A_2	1	$s^3/3+Cs$
cusp	A_3	2	$s^4/4+C_2s^2/2+C_1s$
swallowtail	A_4	3	$s^5/5+C_3s^3/3+C_2s^2/2+C_1s$
elliptic umbilic	D_4^-	3	$s_1^3-3s_1s_2^2-C_3(s_1^2+s_2^2)-C_2s_2-C_1s_1$
hyperbolic umbilic	D_4^+	3	$s_1^3+s_2^3-C_3s_1s_2-C_2s_2-C_1s_1$
butterfly	A_5	4	$s^6/6+C_4s^4/4+C_3s^3/3+C_2s^2/2+C_1s_1$
parabolic umbilic	D_5	4	$s_1^4+s_1s_2^2+C_4s_2^2+C_3s_1^2+C_2s_2+C_1s_1$

Standard polynomials Φ for the elementary catastrophes with codimension $K \leqslant 4$

Photographs of the intensity pattern of $|\Psi_{cusp}|^2$, and a computer simulation based on computing contours of (6), are shown in figure 6. This function was first studied by Pearcey (1946). The simplest diffraction catastrophe - the fold - was studied by Airy (1838). The more complicated elliptic umbilic diffraction catastrophe was studied in detail by Berry, Nye and Wright (1979). The computation of the oscillatory integrals (5) is growing into a small industry (see, for example, Connor and Farrelly 1981, Connor and Curtis 1982, and Upstill et al 1982).

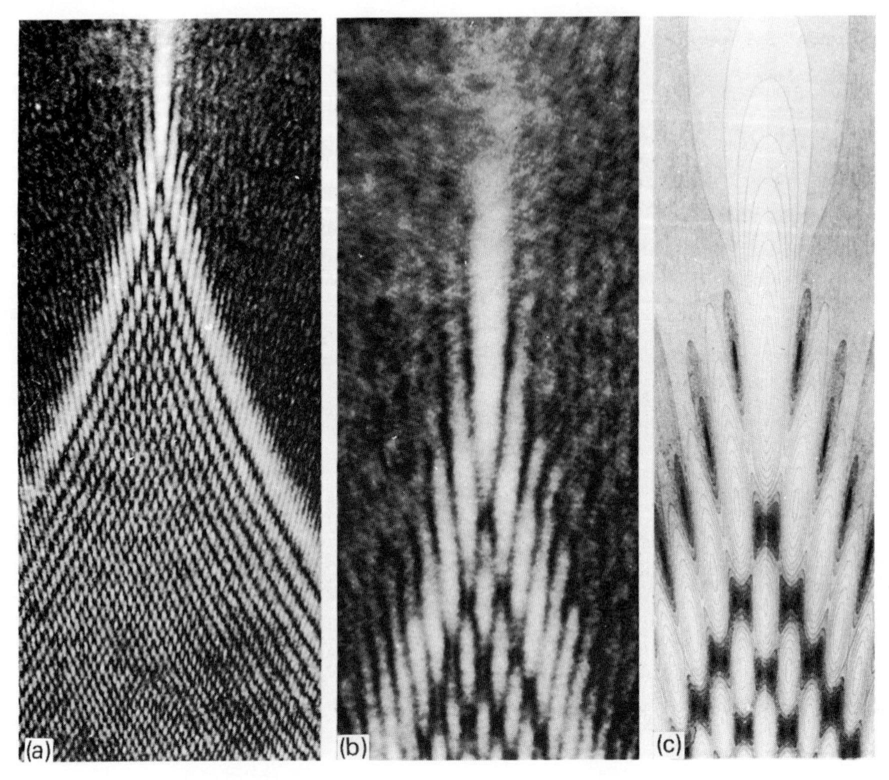

Fig. 6. Cusp diffraction catastrophe; a) experiment; b) magnification of a); c) computer simulation made by shading contours of $|\Psi_{cusp}(C_1,C_2)|^2$ calculated from (6).

3. SCALING

The second way in which catastrophe theory can simplify the asymptotic diffraction integrals follows from the fact that the normal forms $\Phi_j(s,C)$ are quasi-homogeneous polynomials in the variables s with coefficients linear in the parameters C. This has the consequence that the asymptotic parameter k can be scaled out of (5), so that the diffraction catastrophe Ψ_j for any value of k can be expressed in terms of Ψ_j for any other value (e.g. k=1). The explicit scaling law is (apart from possible logarithmic modifications to be mentioned later)

$$\Psi_j(C_i;k)=k^{\beta_j}\Psi_j(C_i k^{\sigma_{ij}};1), \tag{7}$$

where the separate control parameters have been denoted by C_i.

Table II lists the values of the exponents β_j and σ_{ij}, and also the additional exponent

$$\gamma_j \equiv \sum_{i=1}^{K} \sigma_{ij} \tag{8}$$

for catastrophes with codimension $K \leqslant 4$. The method for obtaining the exponents is simply to first rescale s to eliminate k from the C-independent terms in the exponent of (5) and to then rescale the C_i to eliminate k from the other terms. It is easy to confirm that when applied to (6) this procedure yields the correct cusp exponents in table II.

Table II

catastrophe	β	σ_i	γ
fold	1/6	$\sigma_1=2/3$	2/3
cusp	1/4	$\sigma_1=3/4, \sigma_2=1/2$	5/4
swallowtail	3/10	$\sigma_1=4/5, \sigma_2=3/5, \sigma_3=2/5$	9/5
elliptic umbilic	1/3	$\sigma_1=2/3, \sigma_2=2/3, \sigma_3=1/3$	5/3
hyperbolic umbilic	1/3	$\sigma_1=2/3, \sigma_2=2/3, \sigma_3=1/3$	5/3
butterfly	1/3	$\sigma_1=5/6, \sigma_2=2/3, \sigma_3=1/2, \sigma_4=1/3$	7/3
parabolic umbilic	3/8	$\sigma_1=5/8, \sigma_2=3/4, \sigma_3=1/2, \sigma_4=1/4$	17/8

Exponents governing scaling of wave amplitudes and fringe spacing as $k \to \infty$

In mathematical terms, the scaling (7) is a precise expression of the nonanalyticity of wave functions near caustics, as $k \to \infty$. In physical terms, the exponent β_j describes the short-wave divergence of wave intensity $|\Psi_j|^2$ at the caustic singularity ($C_i=0$): the intensity scales as $k^{2\beta_j}$. The exponents σ_{ij} describe the shrinking of the diffraction fringes in the C_i control-space direction: the fringe spacings scale as $k^{-\sigma_{ij}}$. The exponent γ_j describes the shrinking of the K-dimensional hypervolume of the main diffraction maximum: this scales as $k^{-\gamma_j}$. β_j is the 'singularity index' introduced by Arnol'd (1975) and computed for a large number of cases by Varchenko (1976); σ_{ij} and γ_j (the 'fringe index') were introduced by Berry (1977).

To illustrate with a quantum-mechanical example the way in which the scaling laws can quickly lead to interesting physics, consider an isotropic source emitting particles with mass m and speed v in a gravitational field with acceleration $\underset{\sim}{g}$ (figure 7).

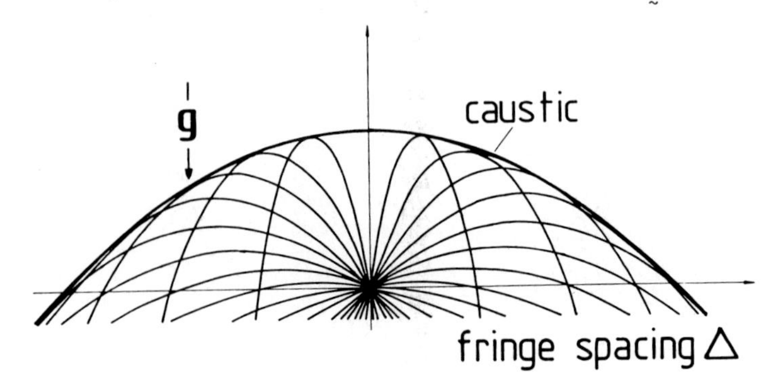

Fig. 7. Section of paraboloidal caustic formed by particles issuing from a source under uniform gravity.

The corresponding parabolic Newtonian trajectories envelop a caustic which is a fold catastrophe in the form of a paraboloid (this is the 'bounding paraboloid' of elementary gunnery). Within the paraboloid, two rays pass through each point, and their interference gives rise to diffraction fringes in space. What is the semiclassical separation Δ (figure 7) of the bright fringes nearest the caustic? The answer can be obtained by realizing that Δ may depend on m,v,g and Planck's constant $\hbar$. Since $\hbar$ corresponds to k^{-1} in the scaling laws, table II gives $\Delta \sim \hbar^{2/3}$ for the fold. Elementary dimensional analysis now leads to

$$\Delta = a(\hbar^2/m^2g)^{1/3}, \tag{9}$$

where a is dimensionless. This does not involve v and so the spacing of these quantal fringes is, curiously, unaffected by altering the de Broglie wavelength h/mv of the particles (this is a case where the semiclassical limit is not quite the same as the short-wave limit). A more refined analysis (Berry 1982a) gives, for the constant, a=3.53897, which for neutrons in the earth's gravity gives Δ=0.026mm - an almost-macroscopic quantum effect.

Chillingworth and Romero-Fuster (1983) have proved an interesting relation between γ_j and β_j, namely

$$\gamma_j = (K_j+1)(1-\beta_j) - 1 \tag{10}$$

where K_j is the codimension of the j'th catastrophe. For the catastrophes in table II the relation can be checked directly. For some higher catastrophes, the γ_j and β_j as calculated by Berry (1977) do not satisfy (10) and this is connected with the phenomenon of *modality* which must now be briefly discussed.

A remarkable feature of Thom's early classification of catastrophes was that the list of universality classes of given codimension was finite. If the number of state variables s-the *corank* - is unity, there is indeed only one singularity for each value of K. But if the corank is two then a new phenomenon appears when K=7: the finiteness of classification breaks down, and universality classes are parameterized by one or more continuous moduli, denoted a. Singularities with different a are not equivalent under diffeomorphism, although they may be topologically equivalent.

If the moduli a are treated as additional control parameters and their indices σ_{ij} are computed and included in the computation of γ_j from equation (8), then the relation (10) is satisfied. But this procedure although mathematically impeccable, is physically unsatisfactory, because the σ_{ij} for moduli are always found to be negative or zero, so that as $k\to\infty$ *diffraction fringes do not shrink along modal directions in parameter space.*

An example is the singularity Z_{11}, whose normal form is

$$\Phi\,(s_1s_2;a,C_1\ldots C_9)=s_1^3s_2+s_2^5+as_1s_2^4+C_1s_1+C_2s_1^2+C_3s_1s_2$$

$$+C_4s_1s_2^2+C_5s_1s_2^3+C_6s_2+C_7s_2^2+C_8s_2^3+C_9s_s^4. \qquad (11)$$

Removing k in (5) from the first two terms by rescaling s_1, s_2 gives $\beta=8/15$. The rescaling C_1----C_9 to eliminate k from the last nine terms gives a series of positive σ_i whose sum is $=21/5=63/15$, which with K=9 does not satisfy (10). If now the modulus a is rescaled, the corresponding index $\sigma_a=-1/15$, so that diffraction fringes expand along the a direction as $k\to\infty$. When this is included to give $\gamma=62/15$, and K increased to 10, then (10) is satisfied.

Another example is X_9, whose normal form may be written

$$\Phi\ (s_1,s_2;a,C_1\ldots C_7)=s_1^4+s_2^4+as_1^2s_2^2+C_1s_1+C_2s_2+C_3s_1^2+C_4s_2^2$$
$$+C_5s_1s_2+C_6s_1s_2^2+C_7s_1^2s_2. \quad (12)$$

Rescaling the first two terms gives β =1/2, and rescaling the last seven terms gives γ=7/2 which with K=7 fails to satisfy (10). If now a is rescaled, its index is σ_a=0, so that diffraction fringes neither expand nor shrink in the a-direction as $k\to\infty$. Of course this zero index cannot alter the value of γ, but if a is counted as a control parameter, thus increasing K to 8, then (10) is again satisfied.

These arguments show that although (10) can be confidently employed to relate β_j and γ_j for nonmodal catastrophes, it should not be relied on for modal ones if γ_j is taken to represent the index governing fringe shrinking. This is an indication of the fact that in diffraction physics moduli a have a significance different from that of control parameters C. In some cases, modality requires (7) to be modified by logarithmic factors, as discussed by Varchenko (1976). I emphasise, however, that in the great majority of applications of shortwave scaling the catastrophes involved are nonmodal, so that both (7) and (10) may be employed without regard to the complications introduced by modality.

Diffraction catastrophe scaling laws and exponents are strongly reminiscent of those occurring in the theory of phase transitions (Stanley 1971, Fisher 1967), with $k\to\infty$ corresponding to $T \to T_c$. There too one starts with integrals (for partition functions rather than wave functions) whose quadratic approximation yields the wrong behaviour close to singularities (thermodynamic critical points rather than caustics). Moreover there exist exponent-equalities analogous to (10), and variables may be relevant (analogous to control parameters) or irrelevant or marginal (analogous to moduli).

But there are differences between the two sorts of scaling. Diffraction exponents are always rational numbers, whilst thermodynamic exponents need not be. Diffraction integrals can be reduced to low-dimensional integrals, and scaling accomplished in a finite number of steps, whilst partition functions involve infinite dimensions and it is their recursive transformation, via the renormalization group (Wilson 1975 -

see also Pfeuty and Toulouse 1977), which generates the scaling laws.

4. TWINKLING OF RANDOM SHORT WAVES

When refracted by atmospheric turbulence, the steady light from a star acquires a fluctuating intensity that we perceive as twinkling. When reflected or refracted by irregular undulations on a water surface, sunlight forms fantastic patterns of moving bright lines on the sides of boats, under bridges and on the bottoms of swimming pools. Because of the unpredictability of the air turbulence or the undulating water, one may speak of the fluctuating light intensity as a random function and seek to comprehend its statistics. The wavelength of light is small in comparison with the smallest cells of atmospheric turbulence or the smallest ripples on water, so that these problems lie in the domain of random short waves.

Wave propagation in random media has given rise to an extensive literature (as illustrated by the other papers in this volume, and reviewed by Uscinski 1977 and in a more general context by Ziman 1979) as has wave reflection from irregular surfaces (Beckmann and Spizzichino 1963, Bass and Fuks 1979). In spite of intense study, these problems have proved remarkably resistant to analytical solution. This is the case even for the simple phase screen model Mercier 1962, Bramley & Young 1967, Salpeter 1967) in which a random spatial phase modulation is imparted to the wavefronts of an initially plane wave which then propagates freely; the phase fluctuations are converted by diffraction into intensity fluctuations which are the object of study. Until recently the only tractable case was that of weak scattering, which could be treated by perturbation techniques such as the Born approximation.

Now, however, a clear picture is emerging of the short-wave limit, i.e. the limit $k\to\infty$, based on the realization that random waves must be dominated by random caustics, near which wave functions $\psi(C;k)$ take the form of one of the diffraction catastrophes discussed in sections 2 and 3. The caustics give rise to intense nongaussian fluctuations in the intensity $|\psi|^2$. In the ray limit $k=\infty$, the caustics are singularities of ψ (not softened by diffraction), and the intensity fluctuations are infinitely strong. It is important to realise that these singularity-dominated strong fluctuations are produced by natural focusing of the rays when the fluctuations in atmospheric refractive index, or in the height of the water surface, are themselves gentle (or even Gaussian-distributed).

The large fluctuations are described by the <u>moments</u> I_m of the probability distribution of the wave intensity, defined by

$$I_m \equiv \langle|\psi|^{2m}\rangle. \tag{13}$$

where $\langle\ \rangle$ denotes averaging over an ensemble of random media or undulating surfaces. I_1 is the average wave intensity and is not large when $k\to\infty$ because the caustic singularities are integrable. $I_m \geqslant 2$ diverge as $k\to\infty$, behaviour to be contrasted with that of a Gaussian random wave (Reψ and Imψ independent random variables with zero mean), for which $I_m = m!$ and which is independent of k. Nongaussian fluctuations in the twinkling light from Sirius were measured by Jakeman, Pike and Pusey (1976). In precatastrophic studies of the second moment, Shishov (1971) and Buckley (1971), established that $I_2 \sim \ell nk$ as $k\to\infty$. In what follows I shall outline the leading k-asymptotics of I_m for general m; a fuller treatment is given in the original paper (Berry 1977), and an important development has been made by Hannay (1982, 1983, 1985).

What catastrophe theory provides is the understanding of the nature of the divergence of I_m as $k\to\infty$. A crucial step is realising that the ensemble (of phase screens, or atmospheres, for example), over which the average in (13) is taken, can be <u>smoothly parameterised</u> by a large number of variables which can be considered as extra control parameters C. Each choice of C thus gives an atmosphere, or an irregular surface. For example, the deviation of an irregular surface from a plane, or the variations in refractive index of air, may be described by Gaussian random functions, which are superpositions of infinitely many sinusoids whose phases are the extra controls (giving a control space in the form of an infinite-dimensional torus). Averaging consists of integrating over these C with smooth probability density P(C) of realisations of members of the ensemble, so that

$$I_m = \int dC P(C)\ |\psi(C;k)|^{2m}. \tag{14}$$

Now, for large k this enormously augmented control space is dominated by caustics on which ψ is large; because of the high dimensionality of C, catastrophes of very high codimension can occur. The integral (14) is dominated by the caustics, and it is natural to assess the separate contributions of each universality class of singularity, that is to discover the k-dependence of the contribution I_{mj} of the j'th catastrophe to the m'th moment. The assumption here is that contributions of different catastophes to (14) can be separated.

To estimate I_{mj} the diffraction catastrophe scaling laws are employed as follows. The localized control-space regions of high intensity corresponding to the j'th catastrophe give contributions whose 'strength' is $|\psi|^{2m} \sim k^{2m\beta_j}$ and whose 'extent' is $k^{-\gamma_j}$ where γ_j and β_j are the exponents of section 3. Thus

$$I_{nj} \sim k^{2m\beta_j - \gamma_j} \qquad (15)$$

This estimate is confirmed by a careful scaling of the integral (14).

Thus each catastrophe contributes a power-law divergence to the n'th moment, provided the exponent $2m\beta_j - \gamma_j$ is positive. Obviously I_n is dominated by the catastrophe(s) for which this exponent is largest, so that the asymptotic behaviour is

$$I_m \to A_m k^{\nu_m} \quad \text{as } k \to \infty, \qquad (16)$$

where

$$\nu_m \equiv \max_j (2m\beta_j - \gamma_j) \qquad (17)$$

will be called the twinkling exponents. In the competition to dominate I_m, which catastrophe wins? This is fully discussed by Berry (1977). The main result is that the codimension K(m) of the winning catastrophe increases with m: higher moments are dominated by higher catastrophes. This is physically reasonable, because high moments of $|\psi|^2$ are dominated by large rare fluctuations at the light detector, and these correspond to the close passage of high-codimension diffraction catastrophes.

The value of the twinkling exponent ν_m depends on which catastrophes are permitted to enter the competition, and this in turn depends on physical circumstances. For waves propagating in two space dimensions, wavefronts are one-dimensional and so only catastrophes with corank unity (i.e. one state variable s- cf.section 2) can compete. These are the cuspoid singularities, and (17) gives, for the twinkling exponents,

$$\nu_m = \max_K \frac{K(2m-K-3)}{2(K+2)} \tag{18}$$

The first few exponents are listed in table III. The value $\nu_2=0$ reflects the fact, already mentioned, that I_2 grows as ℓnk rather than as a power. As $n\to\infty$, $\nu_m\to m$ and the codimension $K_{max}(m)$ of the dominating catastrophe grows as $K_{max}(m)\to 2\sqrt{m}$.

Table III

m	2	3	4	5	6	7	8	9	10	11	12	13
K	1	1	2	2	3	3	3 and 4	4	4	4 and 5	5	5
ν_m	0	$\frac{1}{3}$	$\frac{3}{4}$	$\frac{5}{4}$	$\frac{9}{5}$	$\frac{12}{5}$	3	$\frac{11}{3}$	$\frac{13}{3}$	5	$\frac{40}{7}$	$\frac{45}{7}$

Twinkling exponents ν_m and codimension K for catastrophes of corank unity winning competition to dominate intensity moments I_m.

For waves in three space dimensions, wavefronts are two-dimensional and so catastrophes with corank unity and two may compete. The difficulty now is that many corank-two singularities exhibit modality (discussed near the end of section 3) and the classification of modal catastrophes is incomplete. Nevertheless, a study of the completely-classified singularities with $K \leqslant 11$ shows that for given m the quantity $2m\beta_j-\gamma_j$ increases with K and then decreases, so that there is still one (or sometimes two) dominating catastrophe(s). The resulting twinkling exponents and the dominating catastrophe(s) are listed in table IV. It is clear that the corank-two singularities soon dominate those of corank unity.

An interesting unsolved problem for wave propagation in three space dimensions is: how does ν_m behave as $m\to\infty$? The answer would require knowledge of the asymptotics of β_j and γ_j as $K\to\infty$.

For coranks unity and two, the existence of a maximum in $2m\beta_j-\gamma_j$ for finite K, on which the whole concept of dominating singularity depends, came as a pleasant surprise. It is not at all clear whether this feature would persist for singularities with corank $\geqslant 3$, that is for wave propagation in spaces of higher dimensionality. Perhaps there is a critical dimensionality, above which $2m\beta_j-\gamma_j$ does not have a maximum. Progress on this problem is frustrated by lack of systematic classification of high-corank catastrophes.

Table IV

m	2	3	4	5	6	7	8	9	10	11	12	13
Symbol	A	A_2 and D_4	D_4	D_4 and E_6	E_6 and X	X	X	X	X and W_{12}	W_{12}	W_{12}	W_{13}
K	1	1 and 3	3	3 and 5	5 and $\gtrsim 7$	$\gtrsim 7$	$\gtrsim 7$	$\gtrsim 7$	$\gtrsim 7$ and 10	10	10	11
ν_m	0	$\frac{1}{3}$	1	$\frac{5}{3}$	$\frac{5}{2}$	$\frac{7}{2}$	$\frac{9}{2}$	$\frac{11}{2}$	$\frac{13}{2}$	$\frac{38}{5}$	$\frac{87}{10}$	$\frac{157}{16}$

Twinkling exponents ν_m, codimension K and symbolic notation for catastrophes of corank unity or two winning competition to dominate intensity moments I_m.

The central result (16) of this theory is that intensity moments diverge as $k\to\infty$, with twinkling exponents ν_m which are universal for a given class of competing catastrophes. This universality means that ν_m are independent of the statistics of the random medium or undulating surface. But the coefficients A_m in (16) are not universal and do depend on the nature of the randomness. Hannay (1982, 1983, 1985), in a powerful analysis, has calculated the A_m for the corank-unity catastrophes involved in diffraction from a one-dimensional random phase screen. He finds that for certain exceptional moments I_m (where 16m-7 is the square of an odd number, i.e. m = m = 2,3,5,5.5,8....) namely those at which the dominating catastrophe changes (see table III) the power-law divergence is multiplied by a factor $\ell n k$. For the particular case where the phase screen has Gaussian statistics and m=2, he obtains results in agreement with the earlier calculations of Shishov (1971) and Buckley (1971).

In optics, the twinkling exponents predicted by (16) can be tested by measuring the moments using light of different wavelengths, because

$$\nu_m = \lim_{k\to\infty} \frac{d\,(\ell n I_m)}{d\,(\ell n\ k)}\ . \qquad (19)$$

Such a test has been carried out by Walker, Berry and Upstill (1983) with just two wavelengths of laser light refracted by randomly rippling water. The experiment was very difficult because the high moments depended on rare events and so their values took a long time to stabilise. Figure 8 shows the measured exponents compared with the predictions given by various classes of catastrophes. The best fit is given by cusps with γ_j calculated using only one of the control directions (across the cusps), and is consistent with the visual observation that in the experiments the detector plane was dominated by lines where it was almost touched by the cusped edges of caustic surfaces in space. This question of a 'partial asymptotics', for cases where caustics are so asymmetrically deformed (e.g. elongated by paraxiality) that not all control parameters contribute to the γ_j involved in scaling the intensity moments, is very subtle and needs further study.

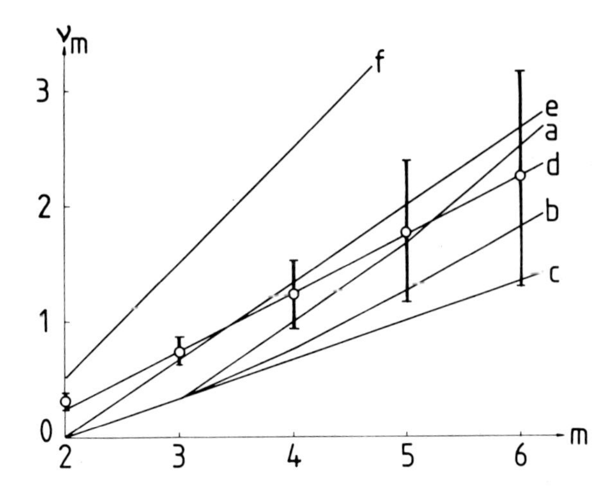

Fig. 8. Twinkling exponents ν_m (after Walker, Berry and Upstill 1983). Circles and error bars: experimental; curve a: unrestricted competition amongst catastrophes of coranks unity and two (table IV); curve b: competition restricted to corank-unity (cuspoid) catastrophes (table III); curve c: exponents from fold catastrophes; curve d: exponents from nongeneric transverse sections of cusp catastrophes; curve e: exponents from nongeneric transverse sections of elliptic and hyperbolic umbilic catastrophes; curve f: exponents from nongeneric transverse sections of X_9 catastrophes.

Finally, I wish to point out that there is an important class of problems involving waves and randomness for which the shortwave asymptotics is not characterised by singularity-dominated strong fluctuations. This occurs whenever the randomness has a self-similar, i.e. fractal structure (Mandelbrot 1982) extending to infinitesimal length scales. Then the limit $k\to\infty$ probes ever finer scales and the randomness never appears smooth on a wavelength scale. Caustics do not form and instead it is expected and found that the wave statistics as $k\to\infty$ will depend on a fractal dimension D describing the self-similarity of the randomness.

Two such cases have been studied so far, both concerning the phase-screen model for waves propagating in two space dimensions. Such a phase screen is characterised by the deviation h(x) it produces it initially rectilinear wavefronts normal to the incident beam. In the first case ('diffractals'), Berry (1979) and Berry and Blackwell (1981) studied the propagation of monochromatic waves and quasimonochromatic pulses when the wavefront and thus the graph of the function h(x) is a D-dimensional fractal curve (for example one of the Weierstrass-Mandelbrot functions studied by Berry and Lewis 1980); such curves are continuous but not differentiable, so that rays (normals) do not exist and a fortiori caustics do not exist. In the second case ('subfractals'), Jakeman (1982ab) studied monochromatic waves evolving from wavefronts for which h(x) was smooth but its derivative dh/dx is a D-fractal curve; thus rays exist but caustics do not (because the curvature does not exist).

REFERENCES

Airy, G.B., 1838, "On the intensity of light in the neighbourhood of a caustic" *Trans.Camb.Phil.Soc.* **6** 379-403

Arnol'd, V.I., 1975, "Critical points of smooth functions and their normal forms", *Usp.Mat.Nauk.* **30** No.5 3-65 (1975 Russ.-Math.Surv. **30** No.5 1-75

Bass, F.G. and Fuks, I.M., 1979, "Wave Scattering by Statistically Rough Surfaces, Pergamon:Oxford

Beckmann, P., and Spizzichino, A., 1963, "The Scattering of Electromagnetic Waves from Rough Surfaces, Pergamon:Oxford

Berry, M.V., 1976, "Waves & Thom's theorem", *Adv.in Phys.* **25** 1-26

Berry, M.V., 1977, "Focusing and twinkling: critical exponents from catastrophes in non-Gaussian random short waves", *J.Phys.A.* **10** 2061-2081

Berry, M.V., 1979, "Diffractals", *J.Phys.A.***12** 781-797

Berry, M.V., 1981, "Singularities in Waves and Rays" in Physics of Defects (Les Houches Lectures XXXIV ed. R. Balian, M. Kleman and J-P. Poirier) (North-Holland; Amsterdam) pp 453-543

Berry, M.V., 1982a, "Wavelength-independent fringe spacing in rainbows from falling neutrons", *J.Phys.A* **15** L385-L388

Berry, M.V. and Blackwell, T.M., 1981, "Diffractal echoes", *J.Phys.A* **14** 3101-3110

Berry, M.V. and Lewis, Z.V., 1980,: On the Weierstrass-Mandelbrot fractal function", *Proc.Roy.Soc.* **A370** 459-484

Berry, M.V., Nye, J.F., and Wright, F.J., 1979, "The elliptic umbilic diffraction catastrophe", *Phil. Trans.Roy.Soc.Lond.* **A291** 453-484

Berry, M.V. and Upstill, C,1980, "Catastrophe optics: morphologies of caustics and their diffraction patterns", *Prog.Opt.* **18** 257-346

Bramley, E.N., and Young, M., 1967, "Diffraction by a deeply modulated random phase screen", *Proc.IEE* **114** 533-556

Buckley, R., 1971, "Diffraction by a random phase screen with very large R.M.S. phase deviation I. One-dimensional screen. II Two-dimensional screen" *Aust. J.Phys.* **24** 351-71, 373-96

Chillingworth, D., and Romero-Fuster, C., 1983, "Remarks on the singularity indices of Arnol'd & Berry", *Proc.Roy.Soc.Edin.* **94A** 339-350

Connor, J.N.L., and Curtis, P.R., 1982, "A method for the numerical evaluation of the oscillatory integrals associated with the cuspoid catastrophes: application to Pearcey's integral and its derivatives", *J.Phys.A* **15** 1179-1190

Connor, J.N.L., and Farrelly, D., 1981, "Theory of cusped rainbows in elastic scattering: uniform semiclassical calculations using Pearcey's integral", *J.Chem.Phys.* **75** 2831-2846

Dirac, P.A.M., 1951, "The Hamiltonian form of field dynamics" *Can.J.Math.* **3** 1-23

Duistermaat, J.J., 1974, "Oscillatory integrals, Lagrange immersions and unfolding of singularities", *Commun.Pure App. Maths* **27** 207-281

Fisher, M.E., 1967, "The theory of equilibrium critical phenomena", *Reps.Prog.Phys.* **30** 615-730

Gilmore, R., 1981, "Catastrophe Theory for Scientists and Engineers" Wiley: New York

Hannay, J.H., 1982, "Intensity fluctuations beyond a one dimensional random refracting screen in the short-wavelength limit", *Optica Acta* **29** 1631-1649

Hannay, J.H., 1983, "Intensity fluctuations from a one-dimensional random wavefront", *J.Phys.A* **16** L61-L66

Hannay, J.H., 1985, "Intensity fluctuations from a random wavefront". This volume.

Jakeman, E., 1982a, "Scattering by a corrugated random surface with fractal slope", *J.Phys.A* **15** L55-L59

Jakeman, E., 1982b, "Fresnel scattering by a corrugated random surface with fractal slope", *J.Opt.Soc.Amer.* **72** 1034-1041

Jakeman, E., Pike, E.R., and Pusey, P.N., 1976, "Photon correlation study of stellar scintillation", *Nature* **263** 215-217

Jona-Lasinio, G., 1975, "The renormalization group: a probalistic view(')", *Nuovo.Cim.* **B26** 99-119

Kravtsov, Yu.A. 1968, "Two new asymptotic methods in the theory of wave propagation in inhomogeneous media (Review)" *Sov. Phys.Acoust.* **14** 1-17

Maslov, V.P., and Fedoriuk, M.V., 1981, "Semiclassical Approximation in Quantum Mechanics", D.Reidel:Dordrecht

Mandelbrot, B.B., 1982, "The Fractal Geometry of Nature", Freeman: San Francisco

Mercier, R.P., 1962, "Diffraction by a screen causing large random phase fluctuations", *Proc.Camb.Phil.Soc.* **58** 382-400

Pearcey, T., 1946, "The structure of electromagnetic field in the neighbourhood of a cusp of a caustic", *Phil.Mag.* **37** 311-317

Pfeuty, P., and Toulouse, G., 1977, "Introduction to the Renormalization Group and to Critical Phenomena" Wiley: New York

Poston, T., and Stewart, I.N., 1978, "Catastrophe Theory and its Applications", Pitman: London

Salpeter, E.E., 1967, "Interplanetary scintillations. I. Theory" *Astrophys. J.* **147** 433-448

Schulman, L.S., 1981, "Techniques and Applications of Path Integration", Wiley-Interscience: New York

Shishov, V.I., 1971, "Diffraction of waves by a strongly refracting random phase screen", *Izv.Vuz.Radiofiz.(USSR)* **14** 85-92

Stanley, H.E., 1971,"Phase Transitions and Critical Phenomena", Oxford University Press:London

Stewart, I.N., 1981, "Applications of catastrophe theory to the physical sciences", *Physica 2D* 245-305

Sussmann, H.J., 1975, "Catastrophe theory as applied to the social and biological sciences: a critique". *Synthese* **37** (1978) 117-216

Thom, R., 1975,"Structural Stability and Morphogenesis", Benjamin: Reading, Mass.

Upstill, C., Wright, F.J., Hajnal, J.V. and Templer, R.H., 1982, "The double-cusp unfolding of the $^{o}X_{9}$ diffraction catastrophe", *Optica Acta* **29** 1651-1676

Uscinski, B.J., 1977,"The Elements of Wave Propagation in Random Media", McGraw-Hill: New York

Varchenko, A.N., 1976, "Newton Polyhedra and Estimation of Oscillating Integrals", *Funkt.Anal.i Prilozhen (Moscow)* **10** No.3 13-38, Funt.Anal.Appl. 1976 10 175-196

Walker, J.G., Berry, M.V., and Upstill, C., 1983, Measurement of twinkling exponents of light focused by randomly rippling water", *Optica.Acta* **30** 1001-1010

Wilson, K.G., 1975, "The renormalization group: critical phenomena and the Kondo problem" *Rev.Mod.Phys.* **47** 773-840

Zahler, R.S., and Sussmann, H.J., 1977, "Claims and accomplishments of applied catastrophe theory", *Nature* **269** 759-763

Ziman, J.M., 1979,"Models of Disorder",Cambridge University Press

INTENSITY FLUCTUATIONS FROM A ONE-DIMENSIONAL RANDOM WAVEFRONT

J.H. Hannay
(H.H. Wills Physics Laboratory, University of Bristol)

ABSTRACT

A wave with an irregular wavefront develops intensity fluctuations through natural focusing as it propagates in free space. The moments $\langle I^m \rangle$ of the intensity distribution are calculated in terms of the wavefront randomness for a one dimensional wavefront, in short wavelength limit.

A curved wavefront produces at its centre of curvature, a focus, that is, an enhanced intensity. So the question arises, in principle and in practice, what are the fluctuations of intensity produced at an observation point by a randomly curving wavefront - an irregular wavefront drawn from a specified statistical ensemble? How violently, in other words, does the source of this wavefront appear to twinkle? This contribution describes how these questions are answered in the simplified circumstance of a one-dimensional wavefront in the short wavelength, geometrical optics limit which the term wavefront already implies. The answer can and should be considered as an application of catastrophe theory which was recognized and exploited as the central ingredient in this problem by M.V. Berry (see his contribution in this volume). In fact, however, the one-dimensional (corank 1) catastrophes are so simple in structure that one can actually proceed unaware that one is using the theory at all.

Figure 1 shows an irregular wavefront and its wave-normals or rays. The wavefront might for example have been produced by the propagation of a plane wave through a thin layer of refracting medium (a bathroom window pane, perhaps) with irregular thickness. The wave intensity is almost unchanged by the passage if the layer is thin so the associated density of rays along the wavefront is uniform. Some distance away from the wavefront though, the situation has altered radically. The rays, after first developing mild density fluctuations, 'have folded over themselves' to produce cusp-shaped regions where there are three 'layers' of rays with consequent enhanced intensity. Most dramatically though, as indicated by the graph on the right of the Figure, the density of rays has risen to infinity on the envelope - the caustic or fold lines,- which are the locus of the centre of curvature of the wavefront. These are the (extended) foci of this process of <u>natural focusing</u>.

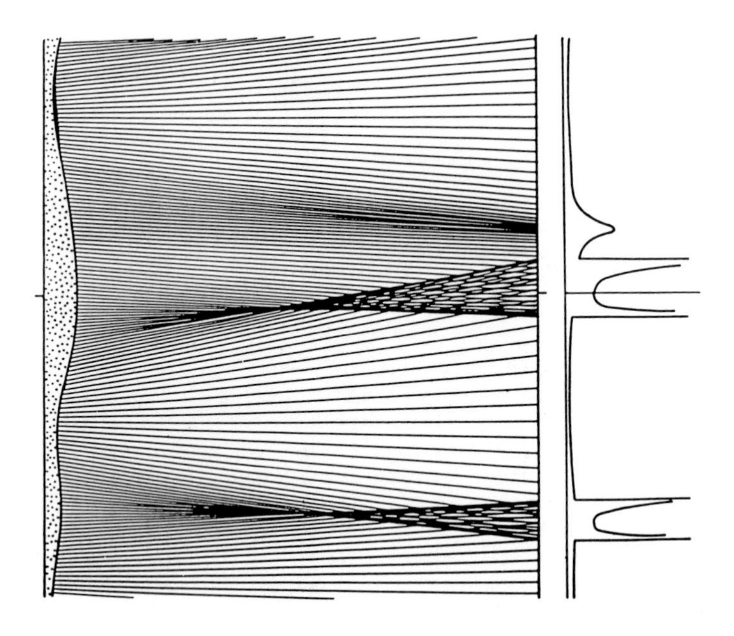

Fig. 1

Figure 2 shows that there are ~ $\sqrt{\ell}$ rays crossing a small perpendicular element of length ℓ and hence that the density of rays rises inversely as the square root of the distance away from these lines. The infinity is obviously unphysical and this square root form of divergence is particularly devastating because it means that even the simplest measure of intensity fluctuation, the <u>square</u> of the intensity averaged, say, over the observation screen, is infinite. The same is true of the more convenient <u>ensemble</u> average $\langle I^2 \rangle$ mentioned earlier; indeed all the intensity moments $\langle I^m \rangle$ diverge in geometrical optics. It is essential to take into account the wave nature of light -

the intrinsic diffraction that accompanies any variation in the density of rays.

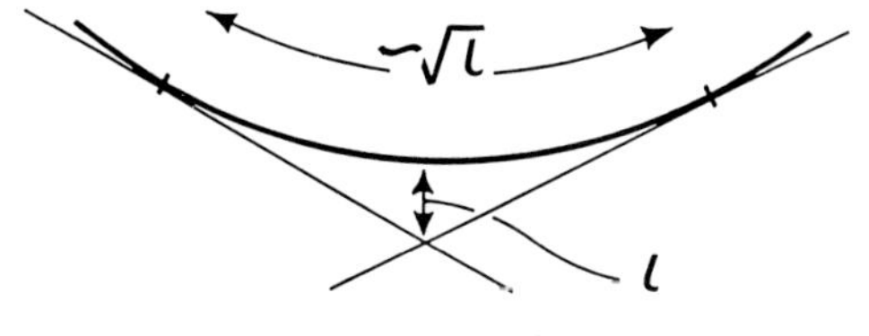

Fig. 2

Before tackling the wave theory it is worth pausing to appreciate the phenomenon from another viewpoint, namely what an observer moving across the fold would actually see. Each ray through the observation point correspond to an image of the (point) source, so, starting inside the three-layered region, three distinct images are seen. Moving towards the fold, two of the images approach each other increasing in brightness and finally coalesce and annihilate as the fold is crossed. The third image which remains is substantially undisturbed throughout this process (it participates instead in crossing the other fold). The fold line therefore is associated with the coalescence and annihilation of <u>two</u> rays, the third being incidental. It is quite possible to involve all three rays jointly, however, by moving to the cusp point in the caustic where a three-ray coalescence takes place. As will emerge below, the fold (A_2) and the cusp (A_3) are the first two members of a hierarchy of <u>catastrophes</u> (A_n) in which the nth member is a coalescence of n rays. For the analysis now though, we set aside the picture of rays and work entirely in terms of the wavefront.

Let the wavefront be described by its distance function S(x) (Fig.3) from a fixed observation point. Contrived focusing, as produced by a lens for example, would aim to yield (in two dimensions) a wavefront which was a perfect circular arc centred on the observation point. The corresponding function S(x) would be a constant over a range of x (the arc). In natural focusing from an irregular wave-front, on the other hand, there is zero chance that the corresponding irregular function S(x) has any finite range of x for which it is a constant; it can only be expected to be a constant over an infinitesimal range of x, i.e., to have zeros in its derivatives with respect to x. This is the essential difference between contrived and natural focusing: although the former is a definite sense 'stonger', the local constancy of natural focusing is already sufficient to produce the infinite geometrical intensity, which qualifies it as 'focusing'.

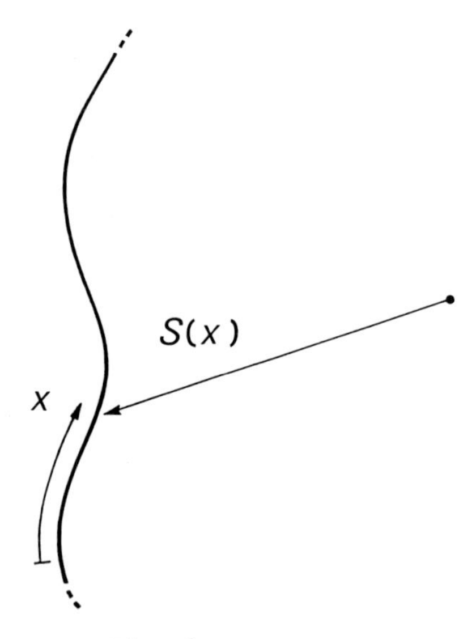

Fig. 3

The analysis of natural focusing then, is to be based on the Taylor expansion of $S(x)$. It could not be more straightforward: there is a hierarchy of focusings and an n-th order focus or 'catastrophe A_n' is produced by a point x on a wavefront where the first n derivatives dS/dx, d^2S/dx^2 ... d^nS/dx^n are simultaneously zero but the next is nonzero. It is instructive to examine the first few orders explicitly.

(a) A_1: first-order focusing $dS/dx=0$. This is not actually focusing at all; it is simply the condition that a ray (wavenormal) passes through the observation point. Locally, $S(X) \sim X^2$, quadratic (where the origin of X is at x).

(b) A_2: second-order focusing

$$\frac{dS}{dx} = \frac{d^2S}{dx^2} = 0, \text{i.e.} S(X) \sim X^3, \text{ cubic.}$$

This lowest true focusing occurs when the observation point lies at the centre of curvature of the patch of wavefront at x. It is therefore the order of focusing found on folds (since the fold was the locus of centres of curvature).

(c) A_3: third-order focusing

$$\frac{dS}{dx} = \frac{d^2S}{dx^2} = \frac{d^3S}{dx^3} = 0, \text{i.e.} S(X) \sim X^4. \text{ quartic.}$$

The order of focusing found at cusps, since cusps occur at maxima and minima of the curvature d^2S/dx^2 of the wave-front.

(d) A_4: fourth-order focusing

$$\frac{dS}{dx} = \frac{d^2S}{dx^2} = \frac{d^3S}{dx^3} = \frac{d^4S}{dx^4} = 0, \text{ i.e. } S(X) \sim X^5. \text{quintic.}$$

The order of focusing found at swallowtails.

This is all the background we need for the desired analysis of twinkling. The idea is to combine two ingredients to yield directly the intensity moments $\langle I^m \rangle$. The first is the probability (density) $p_x^{(n)}$ of finding an n-focus at (i.e. arising from) a point x on the wavefront. And the second is the wavefield $\psi^{(n)}$ it produces at the observation point. Then, schematically at least, we will have a contribution

$$\int p_x^{(n)} \; |\psi^{(n)}|^{2m} dx \tag{1}$$

of n-focusing to the intensity moment $\langle I^m \rangle$. Obvious objections to this assertion, for example that it does not account for any interference there would be between contributions due to separate foci from the same wavefront, are overruled by virtue of the short wavelength limit which will be taken. This eliminates interference, for instance, because there is, in this limit, a vanishingly small probability of more than one focus occurring from the same wavefront. In fact the schematic formula needs refinement in only one respect, that one must admit not only perfect n-foci but slightly imperfect or 'broken' or 'unfolded' foci. They too have substantial wave fields at the observation point and the parameters describing them must be integrated over, along with x.

If the wavefront producing an n-focus is imagined slightly disturbed in a general way (i.e. by adding a small arbitrary function to S(x)), the focus at the observation point which we are considering to be fixed can be called 'broken'. Pictorially the initially rather flat function S(x) has developed ripples, and its stationary points (i.e. maxima and minima) correspond to distinct rays through the observation point into which the original n-fold coalescence of rays has been split by the breaking. Now comes the crucial theorem. Provided the disturbance is sufficiently slight there are exactly $n - 1$ basic ways (or abstract 'directions') in which the focus can break up: a general (sufficiently small) break up is a linear combination of these ways with different components in the different 'directions'. In other words, a broken focus is associated with a point in a standard $(n - 1)$-dimensional 'control space' whose co-ordinates are these components.

The prescription for finding the components is again based on the local Taylor expansion of the disturbed $S(x)$. It involves the first n derivatives of $S(x)$ whose vanishing defined the n-focus, but which are now slightly nonzero. All of these serve to break the n-focus, but not all independently: one combination of them merely shifts the n-focus and therefore does not count. More precisely there is necessarily a new point where the nth derivative d^nS/dx_n, which we will write $S_n(x)$ for short, is zero. (This follows because $S_{n+1}(x)$ was, and still is, nonzero throughout the neighbourhood). The new point can be taken as the natural centre of the disturbed n-focus and the Taylor expansion about it reads

$$S = S_{n+1}\frac{X^{n+1}}{(n+1)!} + S_{n-1}\frac{X^{n-1}}{(n-1)!} + \ldots + S_1X \tag{2}$$

to order n+1. The n-1 'control' co-ordinates are then proportional to S_{n-1}, S_{n-2} ... S_1 (each with a constant of proportionality depending on S_{n+1} to some power). These are the variables which break the n-focus.

The wave field at the observation due to a broken n-focus in a unit intensity wavefront is in the short wavelength limit given by the Kirchhoff diffraction integral

$$\psi^{(n)}_{k\to\infty} = \sqrt{\frac{k}{2\pi i S_o}}\, e^{ikS_o} \int \exp\{ik \left(S_{n+1}\frac{X^{n+1}}{(n+1)!} + S_{n-1}\frac{X^{n-1}}{(n-1)!} + \ldots S_1X\right)\}dX \tag{3}$$

The square-root prefactor is such as to make $\psi = \exp(ikS_o)$ for a flat wavefront ($n = 1$, $S_2 = S_o^{-1}$). In the exponent appears k times the local form of the distance function. This can safely be extended out to infinity because as $k \to \infty$ the contributions to ψ from regions where X is large negligible for both the approximated and true wavefronts, so it does not matter which is used. For the same reason no obliquity factor in the integrand is required: for X small the wavefront is nearly normal to the line to the observation point. The lowest non-trivial diffraction integral for example is that of the fold A_2 which is recognized as the integral representation of the Airy function

$$\psi = \sqrt{\frac{k}{2\pi i S_o}}\, e^{ikS_o} \int \exp\{ik(S_3 \frac{X^3}{3!} + S_1 X)\}dX \tag{4}$$

$$= \sqrt{\frac{k}{2\pi i S_o}}\, e^{ikS_o} \left(\frac{2}{9kS_3}\right)^{1/3} Ai\left(S_1 \left(\frac{2k^2}{9S_3}\right)^{1/3}\right) \tag{5}$$

The higher ones are not standard functions and are complex rather than real as Ai is.

It is now straightforward to write down the contribution $\langle I_m \rangle_n$ of n-focusing to $\langle I^m \rangle$.

$$\langle I^m \rangle_n = \int \ldots \int P(S_{n+1}, 0, S_{n-1}, \ldots, S_1, S_o)$$

$$|\psi^{(n)}|^{2m} dS_o dS_1 dS_2 \ldots dS_{n-1} \left|\frac{dS_n}{dx}\right| dx dS_{n+1} \tag{6}$$

Here P is the joint probability density at x of the first n+1 derivatives of S (including the zeroth S_o). With zero inserted as its second argument it is the probability density of a (broken) n-focus at x. The $|\psi^{(n)}|^{2m}$ is the mth power of intensity due to this broken n-focus, where ψ is given by eqn.3.

In the small wavelength limit eqn(6) reduces directly by scaling to a simpler form in which all the control variables are zero:

$$\langle I^m \rangle_n = \int \ldots \int P(S_{n+1}, 0, 0, \ldots, 0, S_o)$$

$$|\psi^{(n)}|^{2m} dS_o dS_1 \ldots dS_{n-1} \left|\frac{dS_n}{dx}\right| dx\, dS_{n+1} \tag{7}$$

Physically this can be understood because in this limit even a very slightly imperfect focus reduces the wavefield dramatically - the variation of ψ with the control variables $S_1 \ldots S_{n-1}$ is (infinitely) much faster than the variation of P. Having now no dependence on the wavefront statistics, the integrals over these n-1 variables can be performed once and for all to yield a pure number N times a power of k, S_o, S_{n+1}. A mere triple integral is left

$$\langle I^m \rangle_n \underset{k\to\infty}{=} k^{\frac{(n-1)(2m-n-2)}{2(n+1)}} N \iiint S_o^{-m} |S_{n+1}|^{\frac{1-(4m-n^2+n)}{2(n+1)}} P_x^{(n)} (S_{n+1}, 0, 0..0, S_o)\, dS_o dS_{n+1} dx \quad (8)$$

$$N = n!^{\frac{4m-n^2+n}{2(n+1)}} (n-2)! \ldots 2! \int_{-\infty}^{\infty} ds_1 \int_{-\infty}^{\infty} ds_2 \ldots\ldots \int_{-\infty}^{\infty} ds_n$$

$$\left| \sqrt{\frac{1}{2\pi}} \int \exp i\left\{ \frac{\zeta^{n+1}}{n+1} + S_{n-1} \frac{\zeta^{n-1}}{n-1} \ldots\ldots S_1 \zeta \right\} d\zeta \right|^{2m} \quad (9)$$

The first term in (8), the wavelength dependence, was that already obtained by Berry (1977) in his original work on twinkling where the central role of catastrophes was discovered. The number N is, as shown, a definite integral depending only on m and n, namely the integral over the entire control space of the 2mth power of the standard A_n diffraction catastrophe. Finally there is the integral which identifies that particular statistic of the wavefront's randomness which is relevant for twinkling - the joint moment of S_o and S_{n+1} specified.

The fact that the powers of k, Berry's 'twinkling exponents' are different for the different orders n supplies the final simplification that means the problem is actually already solved. As $k\to\infty$, for any given moment $\langle I^m \rangle$, that particular n with the largest exponent will dominate - all other orders can be forgotten. Pictorially in Figure 4 the straight lines are the exponents as functions of m for the different values of n, the dominant one at any m being the highest there. A little algebra shows that it is given by

$$n = \text{Integer part of } \frac{1}{2} (\sqrt{16m - 7} - 1) \quad (10)$$

the dominant order therefore increasing roughly as the square root of the moment. With an important reservation this formally completes the solution to the twinkling problem, $\langle I^m \rangle$ is given by (8) with (10) substituted for n.

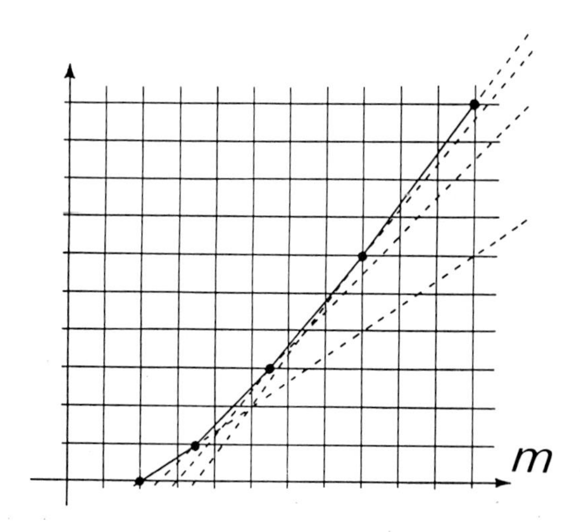

Fig. 4

The reservation concerns those 'exceptional' moments for which there is a change of n in (10), i.e. an exchange of dominance of two catastrophes. This evidently happens whenever 16m-7 is the square of an odd number, the first few (integer) values being m=2,8,11... . These moments require a rather more delicate analysis leading to a change in the numerical coefficient N (within it n is replaced by n-1) and multiplication by a factor 2 ln k/(n+1), except for m=2 where the factor is just 2 ln k. The extra logarithmic term is in fact the only k dependence in this latter case, $\langle I^2 \rangle$, which has n=2 by (10), and therefore twinkling exponent zero. Another feature of exceptional moments is the vanishing of the exponent of S_{n+1} in the integral so that the dS_{n+1} integral can be performed trivially. Thus for instance

$$\langle I^2 \rangle \underset{k\to\infty}{=} 2 \ln k \int_{-\infty}^{\infty}\int_{-\infty}^{\infty}\int_{-\infty}^{\infty} S_o^{-2} \, P_x(S_3, 0, 0, S_o) dS_o dx \qquad (11)$$

An important sequence of specializations is (i) to <u>paraxial</u> optics, where the wavenormals everywhere make only small angles with a principal direction of propagation (in this case S_o can be taken as a given constant) (ii) to a stationary random wavefront whose deviation from a plane is a <u>stationary</u> random function, (iii) to a stationary <u>Gaussian</u> random function, whose spectral components have random <u>phases</u>. Under these specializations the result (8) for $\langle I^2 \rangle$ reduces to those derived by direct (non-catastrophic) means by Shishov (1971) and Buckley (1971). The general result for $\langle I^m \rangle$ is given at the end of Hannay (1982) but contains errors ((38) and (40) should be prefaced by $(-1)^j$, (42) and (43) should have μ replaced by $2-\mu$, kindly pointed out to me by Dr. D.L. Roberts.) The corrected results are as follows. The statistics of the wavefront enter through the matrix σ inverse to $\langle S_i(x) S_j(x) \rangle$ which by stationarity is independent of x and easily obtained from the correlation function of the displacement of the wavefront. Specifically let $\sigma_{ij}^{(n)}$ be the nxn matrix with labels i and j running from 2 to n+1 defined by

$$\sum_{j=2}^{n+1} \sigma_{ij}^{(n)} \langle S_j S_k \rangle = \delta_{jk} \quad \text{for} \quad \begin{matrix} 2 \leq i \leq n+1 \\ 2 \leq k \leq n+1 \end{matrix} \tag{12}$$

Then with $\mu = (4m - n^2 + n)/2(n+1)$ and n given by (10) the non-exceptional moments are

$$\langle I_m \rangle \underset{k\to\infty}{=} k^{\frac{(n-1)(2m-n-2)}{2(n+1)}} N$$

$$S_o^{-m} \sqrt{\frac{\det \sigma^{(n)}}{(2\pi)^n}}\, \Gamma\left(1-\frac{\mu}{2}\right) \left(\frac{1}{2}\sigma^{(n)}_{n+1,n+1}\right)^{\frac{\mu-1}{2}}$$

$$\exp\left\{ \frac{\left(\sigma^{(n)}_{2,n+1}\right)^2 - \sigma^{(n)}_{2,2}\,\sigma^{(n)}_{n+1,n+1}}{2S_o^2\, \sigma^{(n)}_{n+1,n+1}} \right\} \Lambda_{2-\mu}\left\{ \frac{\sigma^{(n)}_{2,n+1}}{S_o \sqrt{\sigma^{(n)}_{n+1,n+1}}} \right\} \tag{13}$$

where Λ is defined in terms of parabolic cylinder functions,

$$\Lambda_{2-\mu}(x) = \frac{\Gamma(2-\mu)}{\Gamma\left(1-\frac{\mu}{2}\right)}\, 2^{\frac{\mu}{2}-1}\, e^{-\frac{1}{4}x^2} \left(D_{\mu-2}(x) + D_{\mu-2}(-x)\right) \tag{14}$$

Exceptional moments are obtained by the modifications stated earlier.

REFERENCES

Berry, M.V., 1977, *J.Phys.* A **10** 2061-81

Buckley, R., 1971, *Aust.J.Phys.* **24** 351-71, 373-96

Hannay, J.H., 1982 *Optica Acta* **29** 1631-49

Hannay, J.H., 1983, *J.Phys.* A **16** *L61-66*

Shishov, V.I., 1971, *Izv.Vuz.Radiofiz* **14** 85-92.

SCATTERING BY MULTISCALE SYSTEMS

E. Jakeman
(Royal Signals and Radar Establishment, Malvern)

ABSTRACT

The mathematical representation of objects with hierarchical spatial structure will be reviewed briefly and the notion of random fractal and sub-fractal scattering systems of the phase screen type will be introduced. The characteristic non-Gaussian statistical properties of waves which have been scattered by such systems in Fresnel and Fraunhofer region geometries will be described, particularly their dependence on propagation distance in the Fresnel region and illuminated area and scattering angle in the far field.

1. INTRODUCTION

Although a great deal of effort has been devoted to the analysis of the statistics of waves scattered by smoothly varying objects, which are mathematically speaking continuous and differentiable to all orders, and typically characterised by a single length scale, it is well known that the descriptors of many natural phenomena are fragmented over a wide range of length scales and are not easily described using the concepts of Euclidean geometry, based on the notion of smooth shapes, and the continuously evolving dynamics of Newton. Turbulently convecting or mixing fluids are an important class of scattering systems with this multiscale nature which have received some attention in the past, but the notion of fractal geometry recently introduced by Mandelbrot (1982) has led to new insight and a resurgence of interest in their scattering properties. Fractals are only a small sub-set of the totality of fragmented objects and are characterised by a scaling behaviour, or similarity under arbitrary magnification. This implies a simple power law relationship between the energy of each scale size, and imparts characteristic scaling relationships to the non-Gaussian statistics of waves scattered by such objects. The mathematical non-differentiability of fractal objects has important implications for the intensity patterns of scattered waves which are qualitatively different from those obtained by scattering waves from smoothly varying systems.

In this paper we first briefly review the mathematical description of two fractal scattering models based on the phase-changing screen. The statistics of waves scattered by these systems in non-Gaussian scattering regimes are discussed in following sections (3 and 4) and a development of one model to describe the extended medium problem is outlined in section 5.

2. FRACTAL PHASE SCREEN MODELS

The solution of Maxwell's equations for scattering by a random phase screen is given in a physical optics approximation by the well known Huygens-Fresnel diffraction formula (Jakeman and McWhirter, 1977)

$$\varepsilon^{+}(\underset{\sim}{R},t) = \frac{iE_o}{2\lambda R}\exp\left[i(kR - wt)\right]\int_{-\infty}^{\infty} d^2\,\underset{\sim}{r}' \times$$

$$\times\ \exp\left[ik\kappa r'^2 - ik\underset{\sim}{r}'.\underset{\sim}{r}/R + i\phi(\underset{\sim}{r}',t) - r'^2/W^2\right] \qquad (1)$$

where $$\kappa = \frac{1}{2}\left(\frac{1}{\sigma} + \frac{1}{R}\right).$$

This formula describes the free propagation to a point $\underset{\sim}{R} \equiv (\underset{\sim}{r},z)$ of a scalar wave of complex amplitude ε^{+} incident normally, with radius of curvature κ and Gaussian amplitude profile of width W on a scatterer in the z = 0 plane which introduces random phase fluctuations of magnitude ϕ. Equation (1) is restricted to situations in which small-angle scalar diffraction theory can be applied: a questionable assumption in the case of fractal models containing significant power in sub-wavelength sized inhomogeneities.

The statistical properties of ε given by equation (1) can be determined, in principle, if the statistical properties of ϕ are known. The most frequently made assumption is that ϕ is a joint Gaussian process so that a knowledge of the first order correlation function, or equivalently, the structure function of ϕ completes the model. The scaling property characteristic of a fractal model is realised when

$$D(x) = \left\langle \left(\phi(0) - \phi(x)\right)^2 \right\rangle = k^2|x|^{\nu}L^{2-\nu} \quad (0 < \nu < 2) \qquad (2)$$

For simplicity we have adopted a corrugated model; alternatively (2) can be interpreted as the structure function of a section of the scattered wavefront. Note that this formula is invariant under the transformation $x \to \ell x$, $\phi \to \ell^{\nu/2}\phi$ so that ϕ is self affine rather than self-similar (Mandelbrot, 1982). The length scale L is the topethesy (Sayles and Thomas, 1978) of a notional surface height $h = \phi/k$ and measures the length over which the rms slope of a chord between two points on the surface is unity. The parameter ν can be related to the fractal dimensions D of the surface where D measures the divergence of a length estimate of the surface as the resolution becomes finer i.e. estimated length $\propto$ (resolution)$^{d-D}$, where d is the topological dimension of the object (= 1 for the height of a surface section h(x)). One method of evaluating this relationship is to note that the divergence arises from the frequent vertical excursions of the surface. The number of these is proportional to the number of zero crossings of the surface height function which can be obtained from the telegraph wave

$$T(x) = 2\,\Theta\,[h(x)] - 1$$

The number of crossings in a distance ℓ can now be calculated for a resolution $\eta = \ell/N$:

$$N_x = \frac{1}{2}\sum_{n=1}^{N} T(x_n)[T(x_n) - T(x_n - \eta)]$$

where $x_n = n\,\eta$. Using the well known arc sine formula (Papoulis, 1965)

$$\langle N_x \rangle = \frac{2\ell}{\eta\pi}\sin^{-1}\left(\frac{1}{2}\sqrt{D(\eta)}\right) \sim \frac{\ell L^{1-\nu/2}}{\pi\eta^{1-\nu/2}} \propto \eta^{1-D}$$

$$\text{i.e.} \quad \nu = 2(2 - D) \tag{3}$$

It is clear that a phase function characterised by the structure function (2) is not differentiable so that such a model cannot lead to Geometrical optics or ray effects which are defined by normals to the scattered wavefront. A fractal model which does generate ray effects is the fractal slope or sub-fractal model in which the scattered wave slope structure function is a fractal. In this case the appropriate model for the phase structure function is (Jakeman, 1982)

$$\begin{aligned} D(x) &= k^2 m_o^2 x^2 - k^2|x|^{\nu} L^{2-\nu}/\nu(\nu - 1) \quad |x|<\xi \quad (2 <\nu< 4) \\ &= C\ . \qquad\qquad\qquad\qquad\qquad\qquad |x|>\xi \end{aligned} \tag{4}$$

Unlike model (2) we have here included an outer scale size ξ which is necessary to obtain finite results in certain scattering geometries. This leads to a maximum slope deviation

$$m_o^2 = \frac{1}{2}\left(\frac{\xi}{L}\right)^{\nu-2} \quad (5)$$

Since the slope of the scattered wavefront is defined and continuous for model (4) ray-optics effects will be predicted. However, since the local curvature remains ill- defined geometrical singularities in the scattered intensity pattern will not occur. This is an important observation since it implies that the geometrical optics or short wave limit of the scattering problem can be evaluated without recourse to diffraction smoothing. It also implies that caustics should not be observed in the intensity pattern generated by a sub-fractal scatterer, unlike the smoothly varying case (Berry, 1977; Walker, Berry and Upstill, 1983).

Although we have not included outer scale effects in the model (2) real scatterers may exhibit fractal behaviour over a finite range of length scales. Beyond the outer scale size regions of the scatterer will become uncorrelated whilst beyond the inner scale size they will appear smooth - i.e. magnification will not reveal new structure.

3. SCATTERING BY A FRACTAL PHASE SCREEN

Most results on the statistics of waves scattered by fractal phase screens relate to the dependence of the mean intensity and second normalised intensity moment on illuminated area and scattering angle in the far field and on propagation distance in the Fresnel region. A few results on spatial coherence properties exist but for brevity these will not be discussed here. We shall consider the two extreme scattering geometries separately.

3.1 Fresnel Region $kW^2/2z >> 1$

The second normalised moment of the intensity fluctuations defined by (1) in this limit can be evaluated exactly for the corrugated fractal screen (equation 2) with index unity (Jakeman and McWhirter, 1977; Berry, 1979). The phase of the scattered wave is just a Brownian walk in this case. Figure 1 shows the dependence of the contrast of the scattered intensity pattern with propagation distance for this result. Two points are worthy of note: firstly there is no "focussing" peak as found with smoothly varying models. Secondly, only one curve is found for all values of the parameters in equation (2) (ν = 1) with k, L and z appearing in the combination k^3L^2z thus reflecting the scaling property of the scattering screen. This latter characteristic can be established for all values of ν but there is a weak peak in the scintillation plot for values of $\nu > 1$ (corrugated case).

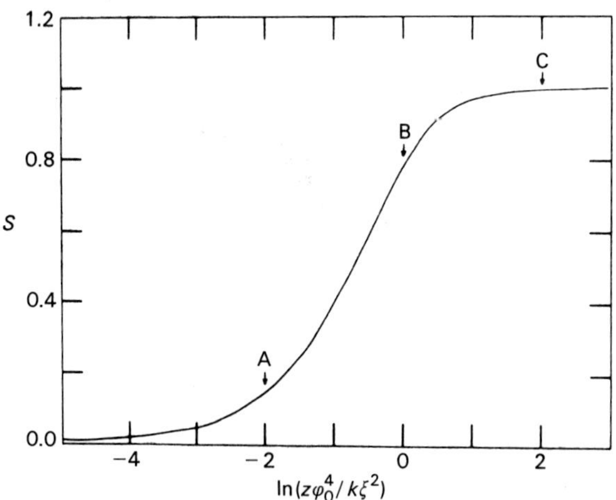

Fig. 1. Fresnel region plot of contrast ($S = \langle I^2\rangle/\langle I\rangle^2-1$) versus distance for scattering by a corrugated Brownian fractal phase screen (in the notation of this paper $2\phi_o^4/k\xi^2 = k^3L^2$) (Jakeman and McWhirter, 1977).

The largest peak is achieved in the marginal case $\nu \to 2$ shown in figure 2. A somewhat stronger peak is predicted for the isotropic fractal phase screen. Nevertheless, for the commonly used Kolmogorov model a maximum second normalised intensity moment of only about 2.4 is predicted (Furuhama, 1975). This is considerably lower than experimentally observed values for scattering systems for which this model might have been expected to be appropriate.

3.2. Fraunhofer Region $kW^2/2z \ll 1$

Relatively few calculations have been carried out for the intensity statistics in this region. It is not difficult to evaluate the mean intensity as a function of angle for the case $\nu = 1$:

$$\begin{aligned} \langle I\rangle &\propto \{k^2L^2 + 4\sin^2\theta\}^{-1} \quad \text{(corrugated)} \\ &\propto \{k^2L^2 + 4\sin^2\theta\}^{-3/2} \quad \text{(isotropic)} \end{aligned} \tag{6}$$

although other cases have to be evaluated numerically (Jakeman and McWhirter, 1977; Jordan, Hollins, and Jakeman, 1983).

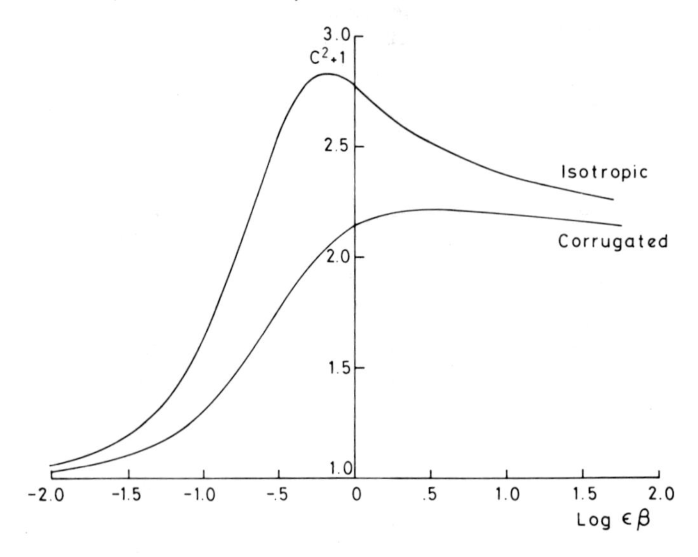

Fig. 2. Fresnel region plot of normalised second intensity moment ($C^2 + 1 = \langle I^2 \rangle / \langle I \rangle^2$) against distance ($\varepsilon\ \beta \propto (\nu-2)z$) for scattering by corrugated and isotropic fractal phase screens in the marginal case $\nu \to 2$ (Jakeman and Jefferson, 1984).

The important feature here is that unlike differentiable models, which give a distribution of intensity reflecting the slope distribution of the scattered wave front, fractal models predict a slow fall-off with increasing scattering angle which is roughly related to the diffraction lobes from small elements of the surface over which the phase changes by 2π(Jakeman and Jefferson, 1984). There is thus an easily discernible qualitative difference between the predictions of smoothly varying and fractal models (equation 2). The only other calculation relevant to the Fraunhofer geometry evaluated the contrast of the intensity pattern as a function of illuminated area in the forward scattering direction ($\Theta = 0$) for the case $\nu = 1$, Fig. 3 (Walker and Jakeman, 1982). As in the Fresnel region two points should be emphasised. Firstly, excess non-Gaussian fluctuations are not predicted, i.e. the contrast never exceeds unity. Secondly the parameters of the model now enter only in the combination k^2LW so that again a single curve describes all possible situations. This may be compared with the results predicted for smoothly varying screens which exhibit strong peaks as a function of illuminated area and different curves for different scattering strengths (Walker and Jakeman, 1982). How the contrast of the intensity varies with scattering angle when the illuminated area is small is as yet an important but unsolved problem.

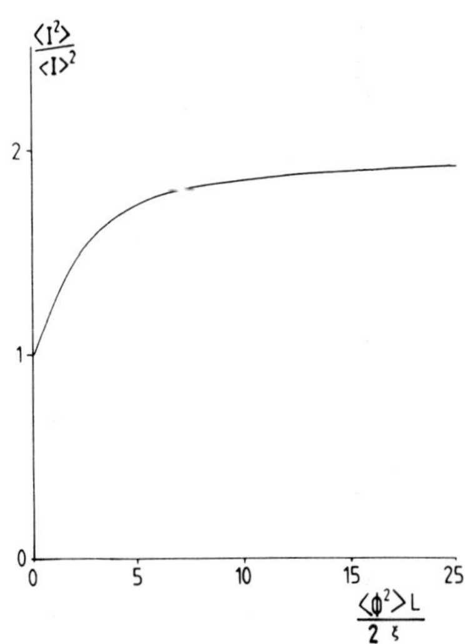

Fig. 3. Normalised second intensity moment versus illuminated area in the Fraunhofer region of a corrugated Brownian fractal phase screen (in the notation of this paper $\langle\phi^2\rangle L/2\xi = k^2LW/4$) (Walker and Jakeman, 1982).

4. SCATTERING BY A SUB-FRACTAL PHASE SCREEN

A large number of results have been obtained recently for this type of screen (equation 4) because of the relative simplicity of the short wave limit and its apparent relevance to turbulent fluid systems which are known to generate geometrical optics effects but to be characterised by a range of length scales (Jakeman, 1982; Jakeman and Jefferson, 1984; Jakeman, 1983).

4.1 Fresnel Region $kW^2/2z \gg 1$

Figs. 4 and 5 show plots of contrast versus propagation distance in the absence of outer scale effects. It is clear that (1) for $\nu \geqslant 3$ no peak is predicted (2) there is again a single curve for each value of ν reflecting the scaling property of the scatterer and (3) the curves saturate at values greater than the Gaussian speckle value of unity. This latter unusual behaviour is a result of fluctuations in the underlying ray density which even at large distances emanate from correlated regions of the scattering screen.

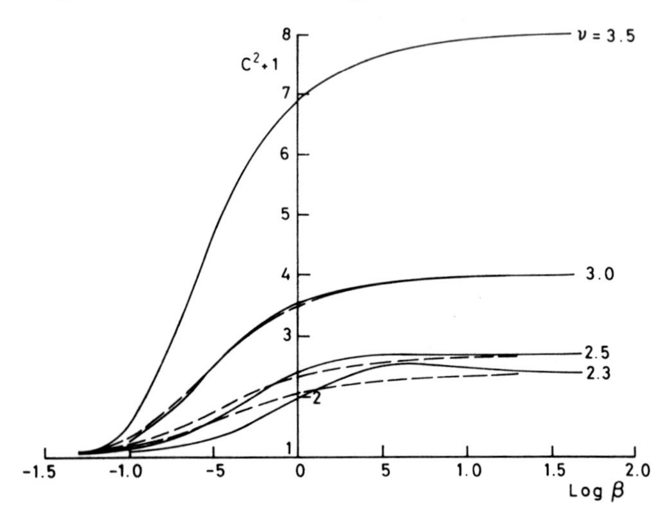

Fig. 4. Fresnel region plot of the normalised second intensity moment ($C^2+1 = \langle I^2\rangle/\langle I\rangle^2$) against distance ($\beta \propto z$) for scattering by a corrugated sub-fractal phase screen (Jakeman and Jefferson, 1984).

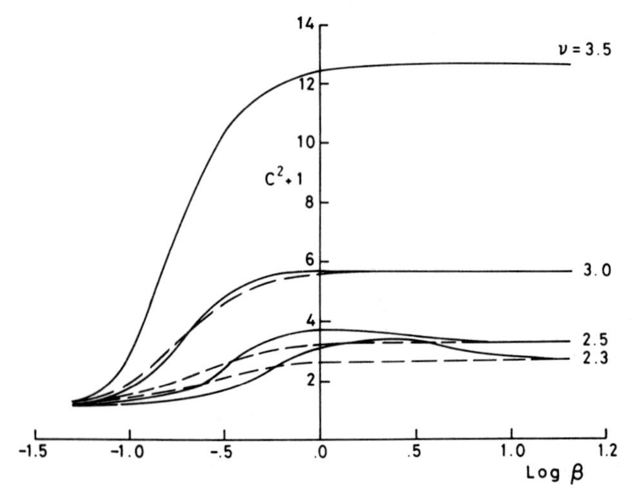

Fig. 5. Fresnel region plot of the normalised second intensity moment ($C^2+1 = \langle I^2\rangle/\langle I\rangle^2$) against distance ($\beta \propto z$) for scattering by an isotropic sub-fractal phase screen. (Jakeman and Jefferson, 1984).

The result of including a finite outer scale size is illustrated in Figure 6 which shows that convergence to Gaussian statistics is then achieved as a result of the independence of different regions of the scattered wave front. A number of interesting exact results can be obtained in the short wave limit (corrugated case) when only the fluctuations of the ray density, R need be evaluated using the definition

$$R(y,z) = \frac{1}{z}\int_{-\infty}^{\infty}\delta\left(m(x) - \frac{x-y}{z}\right) \qquad (7)$$

where m(x) is the local wave front slope. For example it may be shown that whatever the distribution of slope differences, for the structure function

$$\left\langle\left(m(0) - m(x)\right)^2\right\rangle = \left(|x|/L\right)^{\nu-2}$$
$$\langle R\rangle = 1\,,\ \text{then}\ \langle R^2\rangle = 2/(4-\nu) \qquad (8)$$

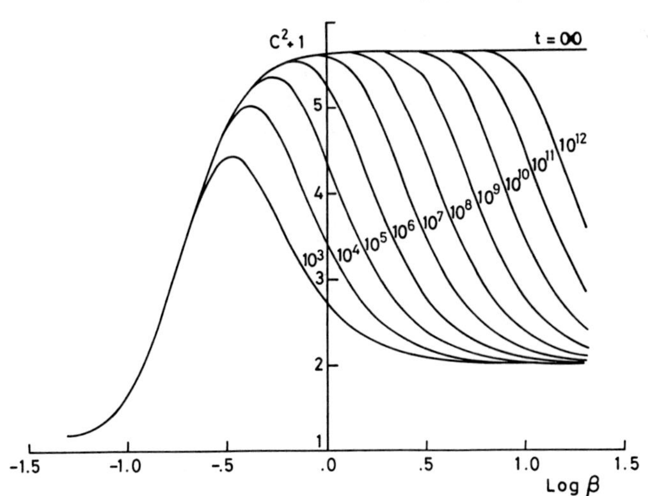

Fig. 6. Fresnel region plot of the normalised second intensity moment ($C^2+1 = \langle I^2\rangle/\langle I\rangle^2$) against distance ($\beta \propto z$) for scattering by a corrugated sub fractal phase screen showing the effect of a finite outer scale. The case $t = \infty$ corresponds to an infinite outer scale size (Jakeman and Jefferson, 1984)

When m(x) is a joint Gaussian process the entire statistical problem may be solved for the Brownian case $\nu = 3$ (Jakeman, 1982):

$$<R(y_1)R(y_2)\ \ldots > = \sum_{n} \prod_{j=1}^{N} \exp\left[-\frac{L}{z^2}|y_j - g_j^{(n)}|\right] \qquad (9)$$

where the $g_j^{(n)}$ are permutations of the y_j. This shows that R is the intensity of a circular complex Gaussian-Markov process. The third moment of R may be evaluated numerically for arbitrary ν $(2 < \nu < 4)$ and is plotted as a function of the second moment in Figure 7 for comparison with the equivalent property of the Gamma distribution. The two distributions coincide when $\nu = 3$ when

$$P(R) = e^{-R} \qquad (10)$$

and are apparently similar for a wide range of values of ν. In the saturated regime of Figure 4 the ray density fluctuations modulate a Gaussian speckle pattern producing enhanced fluctuations.

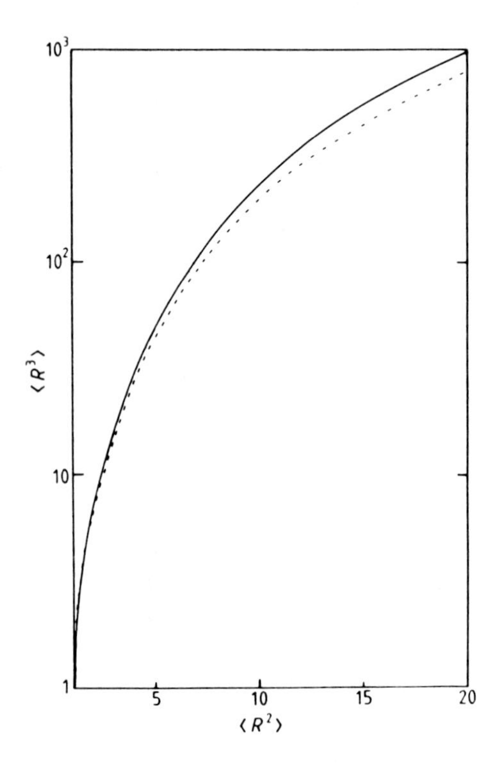

Fig. 7. Comparison of ray density moments for a corrugated surface with Brownian fractal slope (———) with those of the class of gamma distributions (-----). (Jakeman, 1982).

It has been shown that (Jakeman, 1982) the distribution of intensity fluctuations in this case is a member of the K-distribution family when $\nu = 3$

$$P(I) = K_o(2\sqrt{I}) \tag{11}$$

and it seems likely that for values of ν for which the ray density is approximately Gamma distributed, the intensity will be approximately K-distributed. If saturation is reached before outer scale effects reduce the contrast to the Gaussian value it is probable that the K-distribution model will provide an adequate description of the statistics from the peak region onwards, a behaviour which has been observed experimentally in several scattering systems (Parry, Pusey, Jakeman and McWhirter, 1977).

4.2 Fraunhofer Region $kW^2/2z << 1$

Figure 8 shows the predicted behaviour of the second normalised intensity moment with illuminated area in the far field for the model equation 4. The important feature of these results is the power law region which extends from the peak region towards the large area limit. This is a purely geometrical regime in which the incident beam is refracted through an angle limited by the rms slope m_o to generate a localised speckle pattern with wandering centre of gravity. The divergence of the scattered beam is determined by the slope structure function and a simple calculation gives the characteristic power law behaviour

$$\langle I^2\rangle/\langle I\rangle^2 \propto W^{-\nu}\exp[\sin^2\theta/2m_o^2] \tag{12}$$

The unusual area dependence here could provide an important experimental test of sub-fractal behaviour.

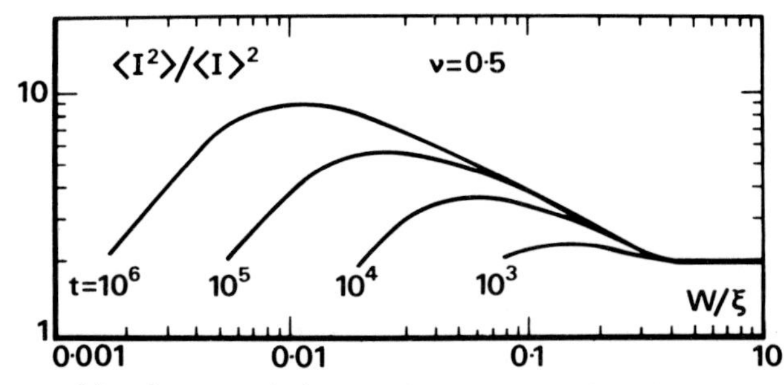

Fig. 8. Normalised second intensity moment in the Fraunhofer region of an isotropic sub-fractal diffuser as a function of illuminated area width, W, with slope structure function index ν and outer scale cut-off measured by the parameter t (Jakeman, 1983).

5. RAY PROPAGATION THROUGH A SUB-FRACTAL RANDOM MEDIUM

The sub-fractal model described above can be developed in a semi-empirical fashion to describe the propagation of rays through an extended sub-fractal medium. Consider for simplicity corrugated (one-diemsional) inhomogeneities varying in the x-direction with a plane wave incident in the z-direction. Assuming weak small angle scattering we represent the effect of the medium as a series of independent phase screens separated by an outer scale size ξ at which the direction of the rays changes by a small amount. Suppose that there is an increment of slope $m(x_o;x_r,z_r)$ of a ray incident normally at $(x = x_o,\ z=0)$ at the screen at z_r. Then the actual slope of the ray at z_n will be

$$m_n(x_o;\{x_r\},z_n) = \sum_{r=1}^{n} m(x_o;x_r,z_r) \tag{13}$$

for small angle scattering. The corresponding displacement of the ray at the z_n plane is thus

$$X(x_o,\{x_r\},z_n) = \xi \sum_{m=1}^{n} \sum_{r=1}^{m} m(x_o;x_r,z_r) \tag{14}$$

We now make the fundamental approximation that the displacement of the ray from x_o, the unperturbed ray position, is so small that $m(x_o;x_r,z_r) \simeq m(x_o;x_o z_r) = m(x_o;z_r)$. This is reasonable only if

$$\left\langle\left(m(x_o;x_r,z_r) - m(x_o;x_o,z_r)\right)^2\right\rangle = \left\langle\left(m(x_r,z_r) - m(x_o,z_r)\right)^2\right\rangle \ll 1 \tag{15}$$

We consider the significance of this approximation again later. Assuming that it is valid for some finite values of z, the ray displacement at the detection plane is

$$X(x_o,z_N) = \xi \sum_{r-1}^{N} m(x_o;z_r)(n - r + 1) \tag{16}$$

with the ray density

$$R = \int_{-\infty}^{\infty} dx_o\, \delta(X_N + x_o - y) \tag{17}$$

Here $\langle X_N \rangle = 0$ since $\langle m(x_o;z_r)\rangle = 0$ and

$$S_X(x_o - x_o') = \left\langle \Big(X(x_o,z_N) - X(x_o',z_N) \Big)^2 \right\rangle$$

$$= \xi^2 \left\langle \Big(m(x_o) - m(x_o') \Big)^2 \right\rangle \sum_{r=1}^{N} (N - r + 1)^2 \qquad (18)$$

Assuming now that m is joint Gaussian with a power law structure function i.e. fractal with a finite outer scale size ξ

$$S_X(x_o - x_o') = \frac{\xi^2}{L^\nu} \frac{N(N+1)(2N+1)}{6} |x_o - x_o'|^\nu \text{ for } |x_o - x_o'| \leqslant \xi$$

$$(0 < \nu < 2)$$

$$= \frac{\xi^{2+\nu}}{L^\nu} \frac{N(N+1)(2N+1)}{6} \text{ for } |x_o - x_o'| > \xi$$

For N >> 1 $\quad S_X(x_o - x_o') = \dfrac{z^3|x_o - x_o'|^\nu}{3L^\nu \xi}$ for $|x_o - x_o'| \leqslant \xi$

$$= \frac{z^3 \xi^{\nu-1}}{3L^\nu} \qquad \text{for } |x_o - x_o'| > \xi$$

$$(20)$$

Thus $\quad \langle X_N^2 \rangle = \dfrac{z^3 \xi^{\nu-1}}{6L^\nu} \qquad (21)$

We see that <u>in the regime where the approximations are valid the ray propagation problem is equivalent to a single screen at a distance</u> z_{eff} <u>$= (z^3/3\xi)^{\frac{1}{2}}$ from the detection plane</u> with slope structure function

$$S(x) = |x|^\nu/L^\nu \qquad |x| \leqslant \xi$$

$$0 < \nu < 2$$

$$= \xi^\nu/L^\nu \qquad |x| > \xi$$

According to (15) the fundamental approximation made in the calculations breaks down unless

$$< \left(m(X_N, z_N) - m(0,z_N) \right)^2 > \ll 1$$

i.e.
$$<X_N^2> \ll L^2$$

On the other hand it is known (Jakeman, 1982) for the screen scattering problem that the effect of a finite outer scale is to average out the ray density fluctuations when

$$<X_n^2> > \xi^2$$

where $<X_N^2>$ is given by equation (21). Thus if $\xi/L \ll 1$ the approximation (15) is valid over the whole propagation range for which ray density fluctuations exist. A more careful treatment gives the condition $m_o^2 \ll 1$ so that (15) is always always valid in the small angle approximation when outer-scale effects are taken into account in the way described above. In this situation the extended medium ray density problem reduces to a single screen problem for which many results already exist in the literature. Generalisation of this treatment to the calculation of intensity fluctuations is straightforward provided diffraction effects (as opposed to geometrical optics and interference effects) play no significant role.

6. CONCLUDING REMARKS

It is clear that multiscale models are of great importance in the description of the scattering and propagation of waves through random media. Gaussian random fractals are one special class of such models which lead to intensity patterns qualitatively different from those associated with smooth single-scale models. Considerable progress in understanding both the mathematical and physical implications of multiscale models has been made during the last decade largely as a result of the introduction of the concepts of fractal geometry (Mandelbrot, 1982), however a number of outstanding problems remain, including the question of the validity of the Kirchoff approximation in the normal treatment of scattering by fractals and also the significance of inner scale effects, which appear to play an important role in both optical and radio frequency scintillation caused by turbulent media.

REFERENCES

Berry, M.V., 1977, "Focussing and twinkling: critical exponents from catastrophes in non-Gaussian random shortwaves", *J.Phys. A* **10** 2061-2081.

Berry, M.V., 1979, "Diffractals", *J.Phys.A* **12** 781-797.

Furuhama, Y., 1975, "Covariance of irradiance fluctuations propagating through a thin turbulent slab", *Radio Science* **10** 1037-1042.

Jakeman, E., 1982, "Fresnel scattering by a corrugated random surface with fractal slope", *J.Opt.Soc.Am.* **72** 1034-1041.

Jakeman, E., 1983, "Fraunhofer scattering by a sub-fractal diffuser" *Optica Acta* **30** 1207-1212.

Jakeman, E. and Jefferson, J., 1984, *Optica Acta* in the press.

Jakeman, E. and McWhirter, J.G., 1977, "Correlation-function dependence of the scintillation behind a deep random phase screen" *J.Phys.A:Math.Gen* **10** 1599-1643.

Jordan, D.L., Hollins, R.C. and Jakeman, E., 1983, "Experimental measurements of non-Gaussian scattering by a fractal diffuser" *Appl.Phys.* **B31** 179-186.

Mandelbrot, B.B., 1982, "The fractal geometry of nature" Freeman, San Francisco.

Papoulis, A., 1965, "Probability, Random Variables and Stochastic Processes" McGraw-Hill Kogakusha Ltd, Tokyo 485.

Parry, G., Pusey, P.N., Jakeman, E. and McWhirter, J.G., 1977, "Focusing by a random phase screen" *Optics. Commun.* **22** 195-201.

Sayles, R.S. and Thomas, T.R., 1978, *Nature* **271** 431.

Walker, J.G., Berry, M.V. and Upstill, C., 1983, "Measurements of twinkling exponents of light focussed by randomly rippling water" *Optic Acta* **30** 1001-1010.

SCATTERING IN A RANDOM MEDIUM WITH A MEAN REFRACTIVE INDEX PROFILE

C. Macaskill*

(DAMTP, University of Cambridge)

*Present address: Department of Applied Mathematics, University of Sydney, Australia.

ABSTRACT

The parabolic second moment equation is used to treat the problem of determining the mean intensity of sound propagation in the ocean with an arbitrary mean refractive index profile in addition to weakly scattering random irregularities.

An approximate eikonal solution is proposed. The familiar rays of geometrical optics are obtained when there is no scattering, but in the presence of random irregularities in the medium, the ray paths are modified. In this way penetration into shadow zones and the reduction of intensity at caustics and foci are accounted for. The ray equations, with extended source initial condition, can be formulated as a boundary-value problem. Solutions to this boundary-value problem are obtained for a parabolic or a cubic channel which compare well with Monte-Carlo simulations of the stochastic wave equation. For the parabolic channel problem, an exact solution is in fact available to the moment equation, so that the simulations constitute an independent check of the moment equation formulation. For the cubic channel, it was found that for some values of the non-dimensional scattering and channel parameters, no solutions could be found for the ray equation boundary-value problem. It is not known whether this is due to deficiencies in the numerical technique, or whether, indeed, there are no solutions to the boundary-value problem, as posed, for these parameters.

INTRODUCTION

When sound propagates in the ocean, the presence of a mean sound speed profile leads to transmission along well-defined ray paths, and associated variation of intensity with depth and range that is well-understood. Some modifications must be made, however, to take into account the influence of random fluctuations in the temperature, and hence sound speed, such as are caused by the presence of internal waves.

A convenient way to approach this problem is to consider the parabolic equation for the propagation of the second moment of the sound pressure field $\langle p_1 \ p_2^* \rangle$, which is a deterministic equation that describes the effects of both scattering and channelling. This equation can be solved exactly for a parabolic sound speed profile (Beran and Whitman, 1975) but is difficult to deal with for more complicated profiles or in the case when the sound speed profile is range-dependent. A series of papers (Frankenthal et al., 1982; Beran et al., 1982; Mazar and Beran, 1983) have investigated the use of two scale expansions for this problem, but the solutions can be unwieldy. An alternative approach is to attempt an eikonal solution to the second moment equation, as was done by Macaskill and Uscinski (1981) and Macaskill, Uscinski and Freedman (1982). However, the characteristic equations obtained using this method are merely the rays of geometrical optics.

When the scattering is strong enough so that it has a dominant influence at caustics, then we expect a smooth variation of the average intensity through the caustic and significant penetration into the shadow zone. Thus there is no reason to anticipate that the ordering of terms in an eikonal approach should break down at caustics in this limit, since the basic assumption involved is that the intensity should vary slowly. It is, however, necessary that the effect of scattering should be evident in the eikonal equation since otherwise there is no mechanism by which energy can be transmitted to the shadow zones. This is a failing of the method described in Macaskill and Uscinski (1981), which is also present in the equivalent path integral solutions of Dashen (1979).

In the present paper a simple re-ordering of the equations of Macaskill and Uscinski (1981) is presented, which leads to a modified eikonal equation, with associated complex-valued ray paths. The eikonal and transport equations are then solved using standard techniques. To check these results, the same problems are considered using the direct simulation approach of Macaskill and Ewart (1984). First, good agreement is obtained with the exact solution for a parabolic sound channel, indicating that the numerical method is capable of producing useful solutions for this kind of problem. Following this, comparisons are made between the results of simulations and those of the ray method introduced here. In some special cases it appears either that there are no solutions to the ray equations or that more sophisticated numerical techniques are needed to find them.

1. THE SECOND MOMENT EQUATION AND THE EIKONAL APPROXIMATION

The second moment equation in two dimensions is

$$\frac{\partial m_2}{\partial z} = -\frac{i}{2k}\left(\frac{\partial^2}{\partial x_1^2} - \frac{\partial^2}{\partial x_2^2}\right)m_2$$

$$- ik[n_s(x_1) - n_s(x_2)]m_2$$

$$- k^2\mu^2 L_p \left[1 - \rho\left(\frac{x_1 - x_2}{L_V}\right)\right]m_2 \qquad (1)$$

where (x,y,z) are cartesian co-ordinates, z is the direction of propagation and $m = \langle p_1\, p_2^*\rangle$ with p the acoustic pressure. The wave number k is that appropriate for the mean refractive index, and $n_s(x)$ gives the deterministic variation from this mean. For example, $n_s(x) = -bx^2$ gives a parabolic channel centred on the axis. The scattering is parameterized by the mean square fluctuation μ^2 and the horizontal integral scale L_p. Finally, the transverse correlation function for the fluctuations of refractive index is given by

$$\rho\left(\frac{x_1 - x_2}{L_V}\right) = L_p^{-1}\int_{-\infty}^{\infty} f\left(\frac{x_1 - x_2}{L_V}, \zeta\right)d\zeta \qquad (2)$$

with $f(x_1 - x_2,\ z_1 - z_2)$ the correlation function of the medium, and

$$L_p = \int_{-\infty}^{\infty} f(0,\zeta)\,d\zeta\ . \qquad (3)$$

Throughout this paper, the Gaussian correlation function $\rho(x) = e^{-x^2/L_V^2}$ is used, but no major changes should be necessary to deal with other correlation functions.

The initial condition we take is

$$m_2(x_1,x_2,0) = \exp[-(x_1^2 + x_2^2)/2d^2] \tag{4}$$

which corresponds to an extended source. We look for solutions of the form

$$m_2(x_1,x_2,z) = A_0(x_1,x_2,z)e^{ik\bar{\theta}(x_1,x_2,z)} \ . \tag{5}$$

We note first that it is possible to incorporate the initial conditions directly by setting

$$A_0(x_1,x_2,0) = 1 \tag{6}$$

and
$$\bar{\theta}(x_1,x_2,0) = \frac{i}{2kd^2}(x_1^2 + x_2^2) \ . \tag{7}$$

Applying the *ansatz* (5) to equation (1) gives

$$ik\frac{\partial\bar{\theta}}{\partial z}A_0 = \frac{ik}{2}\left[\left(\frac{\partial\bar{\theta}}{\partial x_1}\right)^2 - \left(\frac{\partial\bar{\theta}}{\partial x_2}\right)^2\right]A_0$$

$$- ik[n_S(x_1) - n_S(x_2)]A_0 - k^2\mu^2L_p\left[1 - \rho\left(\frac{x_1-x_2}{L_V}\right)\right]A_0 \tag{8}$$

for terms of $O(k)$, and

$$\frac{\partial A_0}{\partial z} = \frac{\partial\bar{\theta}}{\partial x_1}\frac{\partial A_0}{\partial x_1} - \frac{\partial\bar{\theta}}{\partial x_2}\frac{\partial A_0}{\partial x_2} + \frac{1}{2}\left(\frac{\partial^2\bar{\theta}}{\partial x_1^2} - \frac{\partial^2\bar{\theta}}{\partial x_2^2}\right)A_0 \tag{9}$$

for terms of $O(1)$, with terms in $1/k$ neglected.

Equation (8) is the eikonal equation while (9) is the transport equation. These equations are similar to those given in Macaskill and Uscinski (1981), except that now the scattering term is included in the eikonal equation, i.e. we have allowed for the situation where $\mu^2 L_p \gtrsim k^{-1}$, which corresponds to strong scattering. It is now convenient to introduce the following scaled quantities

$$Z = z/kL_V^2, \quad X_1 = x_1/L_V, \quad X_2 = x_2/L_V,$$
$$\theta = k\bar{\theta} \text{ and } \Gamma = k^3\mu^2 L_p L_V^2 \tag{10}$$

so that (8) can be written

$$\frac{\partial\theta}{\partial Z} = \frac{1}{2}\left[\left(\frac{\partial\theta}{\partial X_1}\right)^2 - \left(\frac{\partial\theta}{\partial X_2}\right)^2\right] - k^2L_V^2[n_S(X_1L_V) - n_S(X_2L_V)]$$
$$+ i\Gamma[1 - \rho(X_1 - X_2)], \tag{11}$$

and (9) becomes

$$\frac{\partial A_O}{\partial Z} = \frac{\partial\theta}{\partial X_1}\frac{\partial A_O}{\partial X_1} - \frac{\partial\theta}{\partial X_2}\frac{\partial A_O}{\partial X_2} + \frac{1}{2}\left(\frac{\partial^2\theta}{\partial X_1^2} - \frac{\partial^2\theta}{\partial X_2^2}\right)A_O . \tag{12}$$

The solution as given in (11) and (12) is expected to break down near caustics in the deterministic case ($\Gamma = 0$) in the same way that geometrical optics fails. It is suggested, however, that when the scattering has a strong enough influence, the present ordering is correct, since then A_O varies slowly enough in the vertical so that the neglected higher order derivatives of A_O are not important.

2. THE RAY EQUATIONS AND THE SOLUTION FOR A_O, θ

If we differentiate the θ-equation (11) with respect to X_1 and X_2 respectively, we find

$$\left.\begin{aligned} &\frac{\partial\theta_{X_1}}{\partial Z} - \frac{\partial\theta}{\partial X_1}\frac{\partial^2\theta}{\partial X_1{}^2} + \frac{\partial\theta}{\partial X_2}\frac{\partial^2\theta}{\partial X_1\partial X_2} + k^2L_V{}^2\frac{\partial n_S}{\partial X_1}(X_1L_V) \\ &\qquad\qquad + i\Gamma\frac{\partial\rho}{\partial X_1}(X_1-X_2) = 0 \\ &\frac{\partial\theta_{X_2}}{\partial Z} - \frac{\partial\theta}{\partial X_1}\frac{\partial^2\theta}{\partial X_1\partial X_2} + \frac{\partial\theta}{\partial X_2}\frac{\partial^2\theta}{\partial X_2{}^2} - k^2L_V{}^2\frac{\partial n_S}{\partial X_2}(X_2L_V) \\ &\qquad\qquad + i\Gamma\frac{\partial\rho}{\partial X_2}(X_1-X_2) = 0 \ . \end{aligned}\right\} \tag{13}$$

If we let

$$\frac{\partial\theta}{\partial X_1} = -\frac{dX_1}{ds}\ ,\quad \frac{\partial\theta}{\partial X_2} = \frac{dX_2}{ds}\ ,\quad \frac{dZ}{ds} = 1\ , \tag{14}$$

with s a ray-parameter $\varepsilon[0,Z_0]$, then

$$\frac{d\theta_{X_1}}{ds} = -k^2L_V{}^2\frac{\partial n_S(X_1L_V)}{\partial X_1} - i\Gamma\frac{\partial\rho(X_1-X_2)}{\partial X_1} \tag{15}$$

and

$$\frac{d\theta_{X_2}}{ds} = -k^2L_V{}^2\frac{\partial n_S(X_2L_V)}{\partial X_2} - i\Gamma\frac{\partial\rho(X_1-X_2)}{\partial X_2}\ . \tag{16}$$

We can then derive the complex-valued ray equations

$$\frac{d^2X_1}{ds^2} = k^2L_V{}^2\frac{\partial n_S(X_1L_V)}{\partial X_1} + i\Gamma\frac{\partial\rho(X_1-X_2)}{\partial X_1} \tag{17}$$

and

$$\frac{d^2X_2}{ds^2} = k^2L_V{}^2 \frac{\partial n_S(X_2L_V)}{\partial X_2} - i\Gamma \frac{\partial \rho(X_1-X_2)}{\partial X_2} \tag{18}$$

We note that in the case $\Gamma = 0$, $X_1(s) = X_2(s)$ (assuming identical boundary conditions) and that (17) and (18) are then the rays of geometrical optics. When scattering is present so that $\Gamma \neq 0$, equations (17) and (18) are coupled. In the special case where the intensity only is considered, so that $X_1(Z_0) = X_2(Z_0)$ we can show that $X_1(s) = X_2{}^*(s)$ so that if $X_1 = X_R + iX_{IM}$,

$$\frac{d^2X_R}{ds^2} = k^2L_V{}^2 \operatorname{Re}\left[\frac{\partial n_S(X_1L_V)}{\partial X_1}\right] - \Gamma \operatorname{Im} G(2iX_{IM}) \tag{19}$$

and

$$\frac{d^2X_{IM}}{ds^2} = k^2L_V{}^2 \operatorname{Im}\left[\frac{\partial n_S(X_1L_V)}{\partial X_1}\right] + \Gamma \operatorname{Re} G(2iX_{IM}) , \tag{20}$$

where

$$G(X) = \frac{\partial \rho(X)}{\partial X} . \tag{21}$$

In this case we may solve for X_1 only.

In general, the ray equations must be solved numerically. We can then use these solutions to find θ and A_0. We have

$$\frac{d\theta}{ds} = \frac{dz}{ds}\frac{\partial\theta}{\partial Z} + \frac{dX_1}{ds}\frac{\partial\theta}{\partial X_1} + \frac{dX_2}{ds}\frac{\partial\theta}{\partial X_2} \tag{22}$$

which, when we substitute from (11) and (14), gives

$$\frac{d\Theta}{ds} = -\frac{1}{2}\left[\frac{dX_1}{ds}\right]^2 + \frac{1}{2}\left[\frac{dX_2}{ds}\right]^2 \tag{23}$$

$$- k^2L_V^{\,2}[n_S(X_1L_V) - n_S(X_2L_V)] + i\Gamma[1 - \rho(X_1-X_2)].$$

Integrating (23) from 0 to Z_O, noting that $\Theta(X_1,X_2,0)$ can be derived from (7), leads to

$$\Theta(Z_O) = \Theta(0) + \frac{1}{2}\left[X_2\frac{dX_2}{ds} - X_1\frac{dX_1}{ds}\right]_0^{Z_O}$$

$$+ \int_0^Z \Bigg\{- k^2L_V^{\,2}\left[n_S(X_1L_V) - \frac{1}{2}\frac{\partial n_S(X_1L_V)}{\partial X_1} - n_S(X_2L_V) + \frac{1}{2}\frac{\partial n_S(X_2L_V)}{\partial X_2}\right]$$

$$+ i\Gamma\left[1 - \rho(X_1 - X_2) + \frac{1}{2}X_1\frac{\partial\rho(X_1-X_2)}{\partial X_1} + \frac{1}{2}X_2\frac{\partial\rho(X_1-X_2)}{\partial X_2}\right]\Bigg\}\,ds, \tag{24}$$

where X_1, X_2 in the integral in (24) are implicitly functions of s and must be determined from the ray equations (17) and (18). The expression (24) is further simplified if we note that

$$\Theta(0) = \frac{1}{2}\left[X_2\frac{dX_2}{ds} - X_1\frac{dX_1}{ds}\right]_0 .$$

The transport equation (11) may be solved if we introduce the ray Jacobian J, where we define the ray parameters s, $X_1(0) = X_{10}$ and $X_2(0) = X_{20}$.

Then

$$J = \begin{vmatrix} \frac{\partial Z}{\partial s} & \frac{\partial Z}{\partial X_{10}} & \frac{\partial Z}{\partial X_{20}} \\ \frac{\partial X_1}{\partial s} & \frac{\partial X_1}{\partial X_{10}} & \frac{\partial X_1}{\partial X_{20}} \\ \frac{\partial X_2}{\partial s} & \frac{\partial X_2}{\partial X_{10}} & \frac{\partial X_2}{\partial X_{20}} \end{vmatrix}$$

$$= \frac{\partial X_1}{\partial X_{10}} \frac{\partial X_2}{\partial X_{20}} - \frac{\partial X_1}{\partial X_{20}} \frac{\partial X_2}{\partial X_{10}} . \qquad (25)$$

It can be shown readily that $A_0 J^{\frac{1}{2}}$ = constant. (See Cohen and Lewis (1967) for a similar case.) Then, since $A_0 = 1$ at the source, A_0 at a point Z may be found by obtaining the ratio $J(0)/J(Z)$.

The method for determining $m_2(X_1, X_2, Z_0)$ for an arbitrary location now becomes clear. We have a boundary condition on X_1 and X_2 at the location Z_0. To complete the system we note that from (3), (7) and (14)

$$\left.\frac{dX_1}{ds}\right|_{s=0} = - i \frac{X_1(0)}{D^2}$$

$$\left.\frac{dX_2}{ds}\right|_{s=0} = i \frac{X_2(0)}{D^2} , \qquad (26)$$

where $D = d/L_V$. We thus have a boundary-value problem to solve in equations (17) and (18). In the special case $X_1(Z_0) = X_2(Z_0) = X_R$ we can use the simpler set of equations (19) and (20), which reduces the computational labour. All the results in this paper were computed using the latter method. We note that when $n_X(x) = -\bar{b}x^2$, which describes a parabolic channel, this system can be solved exactly and the results of Beran and Whitman (1975) regained.

To obtain results from the ray equations, the boundary-value problem is solved by guessing initial values of X_R and X_{IM}, stepping out the equations to $s = Z_O$ and then comparing the computed $X_1(Z_O)$ with the required boundary value. The guess for the initial value is then updated, and the process iterated using a Newton procedure (NAG library routine CO5NBF) until the computed and required values of $X(Z_O)$ agree to better than one part in 10^{10}. A very high-accuracy Taylor expansion method is used for stepping out the equations. In this way the ray paths are known accurately and $\Theta(Z_O)$ can be determined by numerical integration. The Jacobian J is estimated by perturbing the initial values of first X_1 and then X_2 slightly and then noting the separate effects on $X_1(Z_O)$ and $X_2(Z_O)$. (For this part of the procedure the full equations (17) and (18) must be used.)

3. MONTE-CARLO SOLUTIONS

In Macaskill and Ewart (1984), a finite difference method was described for approximate Monte-Carlo solution of the stochastic wave equation

$$\frac{\partial p}{\partial Z} = -\frac{i}{2}\frac{\partial^2 p}{\partial X^2} - ik^2L_V^{\,2}n_S(X)p - ik^2L_V^{\,2}\mu Wp, \qquad (27)$$

where μW describes the fluctuating part of the refractive index and is taken to obey Gaussian statistics, with mean zero and variance μ. In that reference, results were presented only for a plane wave initial condition and with $n_S(X) = 0$. To allow comparison with the ray theory presented here we set

$$k^2L_V^{\,2}n_S(X) = -\,bX^2(1 + cX) \qquad (28)$$

which describes a cubic profile, and impose an initial condition equivalent to (4). All simulations were run with 256-point (= N) grid, with eight points per correlation length (=M) in the vertical, and with a stepping distance $\Delta Z = .033$.

4. RESULTS AND DISCUSSION

The general results discussed in the previous sections were first tested using a parabolic profile, centred on the x-axis, with $b = .217$, $c = 0$ and $Z_0 = 6.67$ and with values of the scattering parameter $\Gamma = 0.5, 2.0$. (The value of Z_0 is chosen to be near a focal point.) The initial source width is $D = 1$. For this case, if the correlation function can be approximated as

$$\rho(X) = e^{-X^2} \simeq 1 - X^2 , \qquad (29)$$

then it can be shown that the mean intensity (see Beran and Whitman (1975)) is

$$I = \frac{1}{a} e^{-X^2/a^2} , \qquad (30)$$

where

$$a^2 = \cos^2(\sqrt{2bZ} + \frac{\sin^2(\sqrt{2bZ})}{2bD^4} + \frac{\Gamma Z}{bD^2}\left[1 - \frac{\sin(2\sqrt{2bZ})}{2\sqrt{2bZ}}\right] . \qquad (31)$$

The program for the ray method was run using the approximation to the correlation function (29), and the result (31) was reproduced as expected. Fig. 1 shows the results obtained with the full form of the correlation function, using both the ray method and the simulation technique but treating a cubic channel, with $c = -0.02$. Note that, for comparison, I is approximately unity at $X = 0$ in the zero-scattering case. Sixty realisations of the simulation process were used. Prior to averaging over the realizations, the intensity $I(X_j)$ was smoothed by taking the mean of the intensity at X_j and its six nearest neighbours. This was found to reduce the noise in the estimation of the intensity without introducing any noticeable bias. Each realization took about thirty seconds on the Cambridge University IBM 3081 (with only one CPU running).

For the case $\Gamma = 2$ the agreement between the two approaches is good. For $\Gamma = 0.5$ the agreement is not so good, but the difference between the two results is probably due to the relative coarseness of the numerical mesh used in the simulations. This could be checked by increasing the resolution; only lack of computer resources precluded such an investigation.

Approximate error bars are given for $\Gamma = 0.5$. These are quite large because the medium is not statistically stationary, due to the presence of a background refractive index profile. In the cases treated in Macaskill and Ewart (1984), by contrast, averages could be taken over all the calculated points at a particular range. Therefore, for the current calculations we require roughly N/M times as many realizations to obtain the same confidence limits on the calculated moments of intensity. Because of this, moments such as $\langle I^2 \rangle$, for example, could not be adequately estimated with this number of realizations.

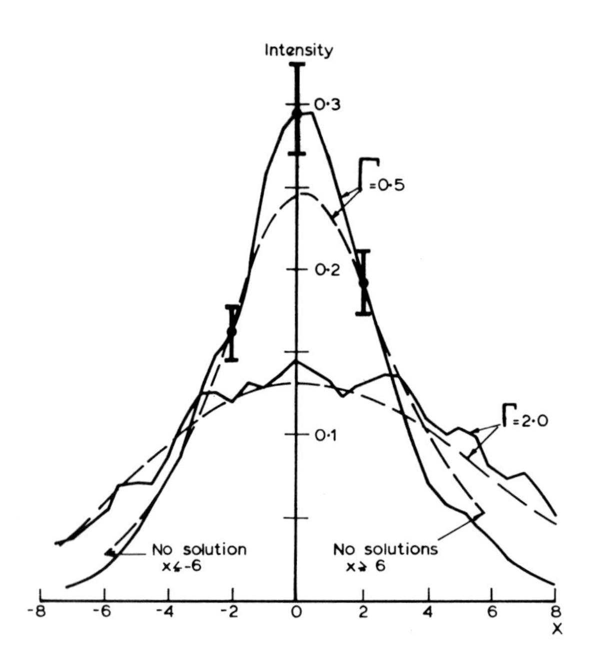

Fig. 1 Intensity plots for a cubic channel with D = 1, b = .217, c = −.02, at range Z = 6.67. The simulations are shown as (——) while the ray theory is shown as (− − −). Two values of the scattering parameter Γ are used, and approximate error bars given on the $\Gamma = 2$ plot.

A study of Fig. 1 shows that scattering does indeed degrade focussing and increases the penetration of the signal into the shadow zone. On the other hand, even for $\Gamma = 2$, there is definitely a peak in intensity at X = 0. In other words, the ray geometry is still very much evident, thus justifying the eikonal approach.

It is interesting, at this stage, to explore in more detail how the deterministic rays are changed in the presence of scatter, so as to effect the penetration into the shadow zone of the type observed in Fig. 1. For the parabolic channel case ($c = 0$), the ray equations are linear and can be solved analytically. For simplicity, we concentrate on a focal plane, given by $\sin \sqrt{2b}\, Z_0 = 0$, and expand the correlation function as in (29). It is convenient to proceed using sum and difference co-ordinates,

$$\bar{X} = \frac{X_1 + X_2}{2}, \quad \xi = X_1 - X_2 \tag{32}$$

so that we have

$$\frac{d^2\bar{X}}{ds^2} = -\,2b\bar{X} - 2i\Gamma\xi \tag{33}$$

and

$$\frac{d^2\xi}{ds^2} = -\,2b\xi\,. \tag{34}$$

We immediately obtain, since $\xi(Z_0) = 0$ for the intensity,

$$\xi(s) = C \sin \sqrt{ab}\, s\,, \tag{35}$$

where C is a constant. Use of (35) in (33) and the imposition of the initial condition

$$\frac{d\xi}{ds} = -\frac{2i\bar{X}_s}{D^2}, \quad \frac{d\bar{X}}{ds} = -\frac{i\xi_s}{2D^2} \quad \text{at } s = 0\,, \tag{36}$$

with the end-condition $\bar{X}(Z_O) = \bar{X}_O$, gives

$$\bar{X}(s) = \bar{X}_s \cos\sqrt{2b}\, s + \frac{2X_O\Gamma}{\sqrt{2b}\, D^2}\left[\frac{s\cos\sqrt{2b}\, s}{\sqrt{2b}} - \frac{\sin 2\sqrt{2b}\, s \cos\sqrt{2b}\, s}{4b}\right.$$

$$\left. + \left(\frac{\cos 2\sqrt{2b}\, s - 1}{4b}\right)\sin\sqrt{2b}\, s\right\} \tag{37}$$

and

$$\xi(x) = -\frac{i2\bar{X}_s \sin\sqrt{2b}\, s}{\sqrt{2b}\, D^2}\, . \tag{38}$$

This result may be obtained by variation of parameters, for example. The important term for our purposes is that proportional to $\Gamma s\cos\sqrt{2b}\, s$, which shows that the effect of scattering is to cause the mean ray paths to diverge as the range increases, thus reducing the 'trapping' effect of the sound channel. At the same time, this gives a mechanism for increased penetration of acoustic energy into the shadow zone, due to scattering.

Numerical results from the ray model for a cubic channel indicate a qualitatively similar change in ray path with increased scatter. In Fig. 2 the simulations and ray theory are compared for a fixed value of Γ and with several values of c. The increasing asymmetry of the intensity pattern with increasing $|c|$ is apparent and the two methods appear to agree well.

A major proviso should be made about the adequacy of the ray method at this point. Extensive exploration of the parameter regime has indicated that there are some values of the parameters for which convergence cannot be obtained in the numerical solution of the boundary-value problem. This indicates either some deficiency in the numerical method or that there is no solution to the boundary-value problem for such a choice of parameters. For example, in Fig. 2, no convergence could be obtained for $X > 6.71$ at the value of Γ shown. However, for smaller Γ, convergence was obtained. The simulations indicate, moreover, that there is no unusual behaviour of the propagating field in this region. At this stage, the precise reason for the failure to converge in some cases is not clear.

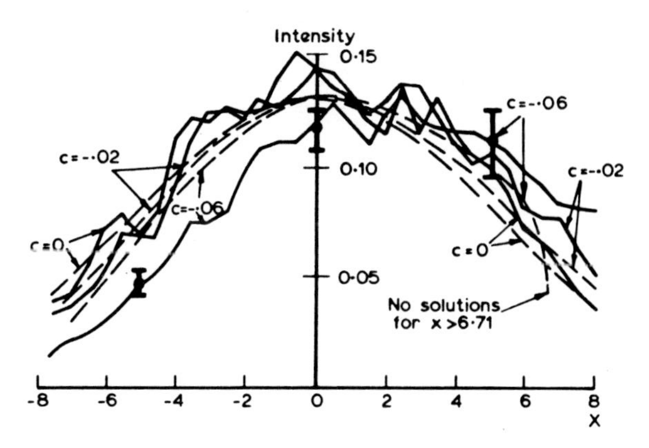

Fig. 2 Intensity plots for fixed Γ (=2.0) with c varying between -.02 and -.06, with other parameters as in Fig. 1. Simulations (——) are compared with ray theory (---). On the c = -.06 ray theory curve, no solutions could be obtained for X > 6.71.

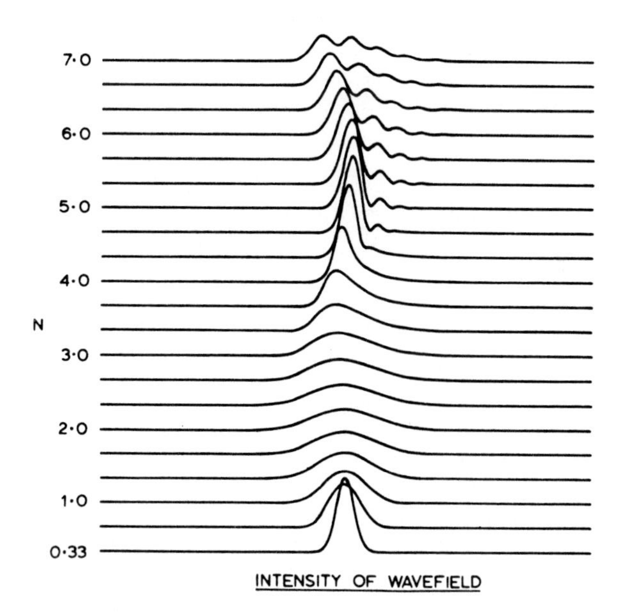

Fig. 3(a) Typical simulation output with no scattering, showing intensity at different range values Z. (Γ = 0, b = .217, c = -.06, D = 0.5.)

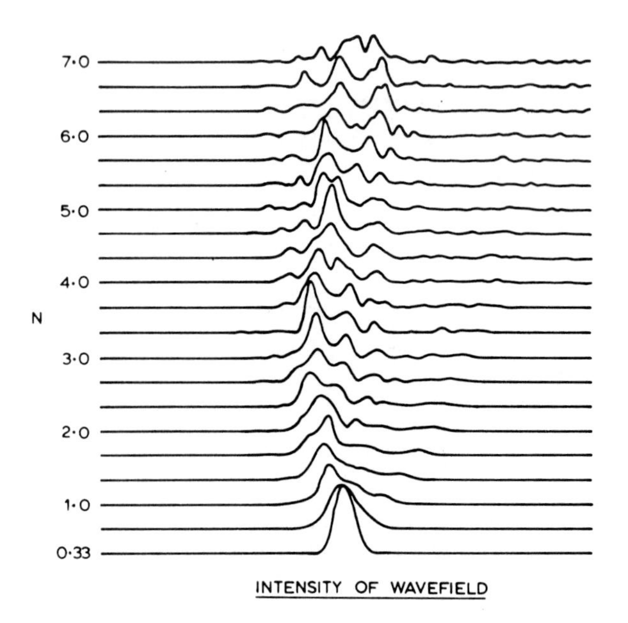

Fig 3(b) As in (a), but with Γ = .05.

To complete this section, a typical simulation output is displayed in Fig. 3. The parameters are as in Fig. 2, but with D = 0.5 in order to make the caustic more obvious. Fig. 3(a) shows the deterministic case, with $\Gamma = 0$, while in 3(b) we have $\Gamma = 0.05$. In Fig. 3(a), the caustic is clearly well-resolved at $Z_0 = 6.67$. Fig. 3(b) shows penetration of energy into the shadow zone (to the left of the picture) and a breaking-up of the regular deterministic behaviour. At the same time, it is clear, however, that the ray geometry is still important.

ACKNOWLEDGEMENT

This work has been done with the support of the Ministry of Defence (Procurement Executive).

REFERENCES

Beran, M.J. and Whitman, A.M., (1975) Scattering of a finite beam in a random medium with non-homogeneous background, *J. Math. Phys.*, **16**, 214.

Beran, M.J., Whitman, A.M. and Frankenthal, S., (1982) Scattering calculations using the characteristic rays of the coherence function, *J. Acoust. Soc. Am.*, **71**, 1124.

Cohen, J.K. and Lewis, R.M., (1967) A ray method for the asymptotic solution of the diffusion equation, *J. Inst. Maths. Applics.*, **38**, 83.

Dashen, R., (1979) Path integrals for waves in random media, *J. Maths. Phys.*, **20**, 894.

Frankenthal, S., Beran, M.J. and Whitman, A.M., (1982) Caustic corrections using coherence theory, *J. Acoust. Soc. Am.*, **71**, 348.

Macaskill, C. and Ewart, T.E., (1984) Computer simulation of two-dimensional random wave propagation, *IMA J. Applied Maths.*, **33**, 1-15.

Macaskill, C. and Uscinski, B.J., (1981) Propagation in wave-guides containing random irregularities: the second moment equation, *Proc. R. Soc. Lond.*, A377, 73.

Macaskill, C., Uscinski, B.J. and Freedman, N., (1982) Acoustic propagation in the upper sound channel, *J. Acoust. Soc. Am.*, **72**, 1544.

Mazar, R. and Beran, M.J., (1982) Intensity corrections in a random medium in the neighbourhood of a caustic, *J. Acoust. Soc. Am.*, **72**, 1269.

SOLUTION OF THE FOURTH MOMENT EQUATION. INTERPRETATION AS A SET OF PHASE SCREENS

B.J. Uscinski
(Department of Applied Mathematics and Theoretical Physics, University of Cambridge)

ABSTRACT

The variance of intensity fluctuations in a multiply scattering random medium is described by the parabolic equation for the fourth moment of the wavefield. This equation can be solved by iteration in the spatial frequency domain. Several important simplifications can be made in the case of multiple scatter and the resulting series can be summed. The solution, which then has the form of a multiple convolution of spatial frequencies, gives the fourth moment to a high degree of accuracy. The physical significance of this solution is that it represents the extended medium as a large number of equally spaced weak phase modulated screens, and exactly the same result can be obtained from such a model. This illustrates the meaning of the so-called "delta-function correlation" sometimes used in deriving the moment equations. The multiple convolution can be evaluated approximately as a single integral referred to as the fundamental solution.

Greater accuracy can be achieved by evaluating the difference between the fundamental solution and the multiple convolution. The result is a triple integral, two of which can be done analytically in many cases. The more accurate solution is presented and compared with the results of numerical simulations of the corresponding propagation experiment.

1. INTRODUCTION

In the study of wave propagation in randomly scattering media the importance of the fourth moment of the wave field has long been recognised. It is the fourth moment which yields the spectrum and the variance of the intensity fluctuations. A relatively simple equation for the propagation of the fourth moment was first derived by Shishov (1968) and subsequently by a number of other authors. Although it was possible to integrate this equation numerically for a limited range of parameters, no analytical solution was available that could allow the general behaviour of the intensity fluctuation spectrum to be studied for various types of scattering media.

Following the general approach initiated by Shishov (1971a) an analytical solution of the fourth moment equation was given by Uscinski (1982) in the form of a multiple integral, and an approximate evaluation of this integral allowed the general behaviour of the spectrum and scintillation index to be studied. In 1983 Macaskill obtained the same fundamental approximation of the solution using a method of multiscale expansion and also investigated a higher order approximation to this solution.

The aim of the present paper is to present the essentials of the multiple convolution solution, proposed by Uscinski in 1982, in such a way as to stress the relationship of the mathematics to the underlying physics. As a result the solution procedure is seen to be straightforward and the mathematics elementary. The multiple convolution solution is found to have a particularly simple interpretation. It is just that solution which would result from subdividing the medium into a set of parallel slabs transverse to the direction of propagation, and then representing each slab by a randomly modulated phase-changing screen with the appropriate transverse autocorrelation function. This clear cut correspondence provides a convincing check on the correctness of the solution procedure.

The multiple convolution solution is presented as a very accurate expression for the intensity fluctuation spectrum in conditions of multiple scatter. It provides a convenient starting point for deriving various simpler approximate representations, and it can even be evaluated directly if desired. For these reasons it is a key result in intensity fluctuation theory and occupies a central role in the present treatment.

Finally, some analytical approximations to the multiple convolution expression are presented. These are evaluated to give the intensity fluctuation spectra and scintillation indices for some typical scattering media. These results are compared with the corresponding quantities obtained by numerical simulation of wave propagation in the appropriate multiply scattering medium.

2. SERIES SOLUTION OF FOURTH MOMENT EQUATION

Let the refractive index of the random medium be

$$n(x,y,z,t) = \langle n\rangle + \mu n_1(x,y,z,t) \tag{1}$$

where x,y,z are Cartesian axes and t is time. The angle brackets denote an ensemble average and μ, the r.m.s. value of the fluctuating part of the refractive index, is assumed to be small. The fourth moment of the wave field propagating in the medium in the positive z direction

$$m^{iv} = \langle E(x_1,y_1,z,t_1)E^*(x_2,y_2,z,t_2)E(x_3,y_3,z,t_3)E^*(x_4,y_4,z,t_4)\rangle \tag{2}$$

satisfies the equation

$$\frac{\partial m^{iv}}{\partial z} = -\frac{i}{2k}[\nabla_1^2 - \nabla_2^2 + \nabla_3^2 - \nabla_4^2]m^{iv} \tag{3}$$

$$-k^2\mu^2[2\rho_o + \rho_{13} + \rho_{24} - \rho_{12} - \rho_{23} - \rho_{14} - \rho_{34}]m^{iv}$$

where

$$\rho_{ij} = \int_{-\infty}^{\infty} \langle n_1(x_i,y_i,z,t_i)n_1(x_j,y_j,z',t_j)\rangle d(z-z') \tag{4}$$

$$\nabla_i^2 = \frac{\partial^2}{\partial x_i^2} + \frac{\partial^2}{\partial y_i^2}, \quad \rho_o = \rho_{ij}(i=j)$$

and $k = 2\pi/\lambda$, where λ is the wavelength in the average medium with refractive index $\langle n\rangle$. It is assumed that the scale size of the irregularities transverse to the direction of propagation is L. For simplicity we shall consider the case when there is only one transverse coordinate x. The final result is easily extended to include the y coordinate. The following scaled variables are now introduced,

$$\xi_a = \tfrac{1}{2}(x_1-x_2-x_3+x_4)L^{-1}, \quad \xi_b = \tfrac{1}{2}(x_1+x_2-x_3-x_4)L^{-1}, \tag{5}$$

$$u = (x_1-x_2+x_3-x_4)L^{-1}, \quad X = \tfrac{1}{4}(x_1+x_2+X_3+x_4)L^{-1},$$

$$\tau_a = \tfrac{1}{2}(t_1-t_2-t_3+t_4), \quad \tau_b = \tfrac{1}{2}(t_1+t_2-t_3-t_4),$$

$$v = (t_1-t_2+t_3-t_4),\ T = \tfrac{1}{4}(t_1+t_2+t_3+t_4)$$

$$Z = z/kL^2,$$

and (3) becomes

$$\frac{\partial m^{iv}}{\partial Z} = -i\left(\frac{\partial^2 m^{iv}}{\partial\xi_a\partial\xi_b} + \frac{\partial^2 m^{iv}}{\partial U\partial X}\right) - \Gamma h(\xi_a,\xi_b,u;\ \tau_a,\tau_b,v)m^{iv} \tag{6}$$

where

$$\Gamma = k^3\mu^2\rho_o L^2 \tag{7a}$$

$$f_{ij} = \rho_{ij}\rho_o^{-1} \tag{7b}$$

and

$$h = 2(1-g), \tag{8}$$

$$2g = f(\xi_a + \frac{u}{2};\ \tau_a + \frac{v}{2}) + f(\xi_a - \frac{u}{2};\ \tau_a - \frac{v}{2})$$

$$+ f(\xi_b + \frac{u}{2};\ \tau_b + \frac{v}{2}) + f(\xi_b - \frac{u}{2};\ \tau_b - \frac{v}{2})$$

$$- f(\xi_a+\xi_b;\ \tau_a+\tau_b) - f(\xi_a-\xi_b;\ \tau_a-\tau_b).$$

Now the time variables do not appear as operators in (6) and so will be suppressed in future unless required explicitly.

The plane wave case

Consider a plane wave of unit amplitude normally incident on a half space $z>0$ of the medium described above. The resultant M^{iv} in the medium is independent of the mean transverse position X and so (6) becomes

$$\frac{\partial m}{\partial Z} = -i\frac{\partial^2 m}{\partial\xi_a\partial\xi_b} + 2\Gamma\, g(\xi_a,\xi_b)m \tag{9}$$

where

$$m^{iv} = me^{-2\Gamma Z} \tag{10}$$

and g, (8), now reduces to

$$g(\xi_a,\xi_b,\tau_a) = f(\xi_a;\tau_a) + f(\xi_b) - \tfrac{1}{2}f(\xi_a+\xi_b;\tau_a) - \tfrac{1}{2}f(\xi_a-\xi_b;\ \tau_a)$$

with the time variable given explicitly. This form of g results from the fact that in practice we are interested in average products of the intensity

$$I = EE^* \tag{11}$$

at two points separated in space and time. Such products are obtained by letting the points with subscripts 4 and 3 coincide with those having subscripts 1 and 2 respectively in the solution. From (5) this means that we are interested in solutions in which u, v, τ_b, and ξ_b are all zero.

The following Fourier transforms are now introduced

$$M(\nu_a,\nu_b,Z) = \frac{1}{(2\pi)^2}\iint m(\xi_a,\xi_b,Z)\exp\{-i(\nu_a\xi_a+\nu_b\xi_b)\}d\xi_a d\xi_b \tag{12}$$

$$G(\nu_a,\nu_b\ ) = \frac{1}{(2\pi)^2}\iint g(\xi_a,\xi_b,)\ \exp\{-i(\nu_a\xi_a+\nu_b\xi_b)\}d\xi_a d\xi_b$$

and, when used in (9), give

$$\frac{\partial M(\nu_a,\ \nu_b,Z)}{\partial Z} = i\nu_a\nu_b M + 2\Gamma\iint G(\nu_a-\nu_a';\nu_b-\nu_b')M(\nu_a',\nu_b',Z)d\nu_a' d\nu_b' \tag{13}$$

We now represent M in the form of a series

$$M = \sum_{n=0}^{\infty} M_n \tag{14}$$

where M_o is the value of M in the absence of scattering, while the other $M_n (n\neq 0)$ are contributions due to interaction with the scattering medium, and are zero at Z=0 by definition. In case of an initial plane wave

$$M_o(\nu_a,\nu_b,Z) = \delta(\nu_a)\delta(\nu_b). \tag{15}$$

When (14) is used in (13) we obtain the equation

$$\frac{\partial M_n}{\partial Z} = i\nu_a\nu_b M_n + 2\Gamma\iint G(\nu_a-\nu_a';\ \nu_b-\nu_b')M_{n-1}(\nu_a',\nu_b',Z)d\nu_a' d\nu_b' \tag{16}$$

This can now be solved for successive values of n starting with n=1. The solution for M_n is

$$M_n(\nu_a,\nu_b,Z) = \frac{(2\Gamma)^n}{(2\pi)^2} \int_0^Z \cdots \int_0^{Z_2} \int_{-\infty}^{\infty} \int M_o(\nu_{a_o};\nu_{b_o}) \tag{17}$$

$$G(\nu_{a_1}-\nu_{a_o};\nu_{b_1}-\nu_{b_o})$$

$$G(\nu_{a_2}-\nu_{a_1};\nu_{b_2}-\nu_{b_1})$$

$$\vdots$$

$$G(\nu_{a_n}-\nu_{a_{n-1}};\ \nu_{b_n}-\nu_{b_{n-1}})$$

$$\exp\{i[\nu_{a_1}\nu_{b_1}(Z_2-Z_1)+\nu_{a_2}\nu_{b_2}(Z_3-Z_2)+\ldots+\nu_{a_n}\nu_{b_n}(Z-Z_n)]\}$$

$$\exp\{-i[\xi_a(\nu_a-\nu_{a_n})+\xi_b(\nu_b-\nu_{b_n})]\}d\xi_a d\xi_b$$

$$d\nu_{a_o}\ldots d\nu_{a_n},\ d\nu_{b_o}\ldots d\nu_{b_n},\ dZ_1\ldots dZ_n.$$

The terms G in (17) are replaced by their Fourier transforms from (12) and the resulting exponentials are regrouped to give

$$M_n(\nu_a,\nu_b,Z) = \frac{(2\Gamma)^n}{(2\pi)^{2n+2}} \int_0^Z \cdots \int_0^{Z_2} \int_{-\infty}^{\infty} \int M_o(\nu_{a_o};\ \nu_{b_o}) \prod_{j=1}^{n} g(\xi_{a_j};\ \xi_{b_j}) \tag{18}$$

$$\exp\{-i(\xi_{a_j}(\nu_{a_j}-\nu_{a_{j-1}})+\nu_{b_j}[(\nu_{b_j}-\nu_{b_{j+1}})-\nu_{a_j}(Z_{j+1}-Z_j)])\}$$

$$\exp\{-i[\xi_a(\nu_a-\nu_{a_n})-\xi_{b_1}\nu_{b_o}+\xi_b\nu_b]\}$$

$$d\xi_{a_1}\ldots d\xi_{a_n}d\xi_a,\ d\xi_{b_1}\ldots d\xi_{b_n}d\xi_b$$

$$d\nu_{a_o}\ldots d\nu_{a_n},\ d\nu_{b_o}\ldots d\nu_{b_n},\ dZ_1\ldots dZ_n$$

where

$$\xi_{b_{n+1}} = \xi_b, \quad Z_{n+1} = Z \tag{19}$$

The integral with respect to $\xi_a, \nu_{a_o}, \nu_{a_n}, \nu_{b_o} \ldots \nu_{b_n} \; \xi_{b_1} \ldots \xi_{b_n}$ can now be carried out in that order, to yield

$$M_n(\nu_a, \nu_b, Z) = \frac{(2\Gamma)^n}{(2\pi)^{n+1}} \int_o^Z \ldots \int_o^{Z_2} \int_{-\infty}^{\infty} \prod_{j=1}^{n} g(\xi_{a_j}; \xi_b + Q_j) \tag{20}$$

$$\exp\{-i\xi_{a_j}(\nu_{a_j} - \nu_{a_{j-1}}) - i\nu_b \xi_b\} \, d\xi_b$$

$$d\xi_{a_1} \ldots d\xi_{a_n}, \; d\nu_{a_1} \ldots d\nu_{a_{n-1}}, \; dZ_1 \ldots dZ_n,$$

where $\nu_{a_n} = \nu_a$ and

$$Q_j = [\nu_a(Z - Z_n) + \nu_{a_{n-1}}(Z_n - Z_{n-1}) + \ldots + \nu_{a_j}(Z_{j+1} - Z_j)]. \tag{21}$$

We note that the series (14) with M_n given by (20) represents an exact solution of the fourth moment equation in its spectral form. It is possible to make some approximations that simplify this solution considerably while at the same time introducing an error that is extremely small.

3. INITIAL APPROXIMATIONS. THE MULTIPLE CONVOLUTION SOLUTION

Single scatter

The subscripts n can be usefully interpreted as the number of times the wave field contributing to a given M_n is scattered. When the total amount of scattering is small only the first term in series M is important and this is sometimes called single scatter. When the integrals with respect to ξ_a, ξ_b are carried out in the first term of the series (21) we have

$$M_1(\nu_a, \nu_b, Z) = 2\Gamma \int_o^Z G(\nu_a, \nu_b) e^{i\nu_a \nu_b (Z - Z_1)} dZ_1. \tag{22}$$

The spectrum of intensity fluctuations

As pointed out above we are mostly interested in intensity fluctuations, in which case we can set ξ_b equal to zero. Taking the inverse transform of M, (12), and setting ξ_b equal to zero we have that the spatial autocorrelation of intensity fluctuation is

$$m(\xi_a, Z) = \int M(\nu_a, Z) e^{i\nu_a \xi_a} d\nu_a \tag{23}$$

where

$$M(\nu_a, Z) = \int M(\nu_a, \nu_b, Z) d\nu_b. \tag{24}$$

In the single scatter case the spectrum of intensity fluctuations is, from (12), (22) and (24)

$$M_1(\nu_a, Z) = 2\Gamma Z \delta(\nu_a) + 4\Gamma \int_0^F F(\nu_a) \sin^2[\tfrac{1}{2}\nu_a^2 (Z - Z_1)] dZ_1 \tag{25}$$

where

$$F(\nu) = \frac{1}{2\pi} \int f(\xi) e^{-i\nu\xi} d\xi. \tag{26}$$

This is just the well known Born approximation.

Multiple scatter

A number of approximations can be made in the case when the total amount of scattering is large and terms with large n are important in the series (14).

(a) The first of these approximations concerns the integrals with respect to the Z_i. The ordering of the variables Z_i in their nested integrals is shown in Fig. 1, and Z_i is interpreted as the position where the i-th scattering of the field occurs. When n is large the Z_i can be taken to be roughly equally spaced in the interval 0 to Z and we set

$$Z_i - Z_{i-1} \simeq \frac{Z}{n} \tag{27}$$

in Q_j. This allows the Z_i integrals to be carried out and M_n becomes

$$M_n(\nu_a,\nu_b,Z) = \frac{(2\Gamma Z)^n}{n!} \frac{1}{(2\pi)^{n+1}} \iint \prod_{j=1}^{n} g(\xi_{a_j};\xi_b+Q_j) \tag{28}$$

$$e^{-i\xi_{a_j}(\nu_{a_j}-\nu_{a_{j-1}}) - i\nu_b\xi_b} d\xi_b$$

$$d\xi_{a_1} \ldots d\xi_{a_n}$$

$$d\nu_{a_1} \ldots d\nu_{a_{n-1}}$$

$$Q_j = \frac{Z}{n}[\nu_a+\nu_{a_{n-1}} + \ldots \nu_{a_j}].$$

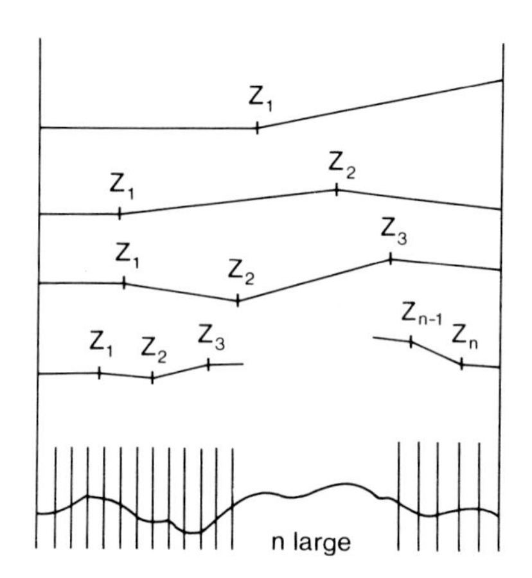

Fig. 1. Typical scattering intervals for some different orders of scattering.

This approximation is discussed in some detail by Uscinski (1982) where the error involved is estimated. It turns out to be of the order of $n^{-\frac{3}{2}}$.

(b) The second approximation relies on the fact that the coefficients $(2\Gamma Z)^n/n!$ in (28) vary much more rapidly with n than the remainder of the term for large ΓZ. It is not difficult to show that for large ΓZ

$$e^{-2\Gamma Z}\frac{(2\Gamma Z)^n}{n!} = \frac{1}{\sqrt{4\pi\Gamma Z}} \exp\{-\frac{(n-2\Gamma Z)^2}{4\Gamma Z}\} \tag{29}$$

The left hand side of (29) is just the general term in a Poisson distribution and the right hand side is its limiting form for large ΓZ. This approximation has been discussed in the context of wave scattering by Shishov (1971a) and Uscinski (1977, 1982). We now write (28) in the form

$$M_n = \frac{(2\Gamma Z)^n}{n!} \mathcal{M}(n,\Gamma,Z) \tag{30}$$

use (14) and (30) to obtain

$$M(\nu_a,\nu_b,Z)e^{-2\Gamma Z} = \frac{1}{\sqrt{4\pi\Gamma Z}} \sum_{n=0}^{\infty} e^{\frac{-(n-2\Gamma Z)^2}{4\Gamma Z}} \mathcal{M}(n,\Gamma,Z). \tag{31}$$

If the scaled variable

$$n' = n/2\Gamma Z \tag{32}$$

is introduced in the right hand side of (31) the sum effectively becomes a continuous integral and the function multiplying $\mathcal{M}$ is very sharply peaked about its mean value. For large ΓZ this multiplying function behaves like a Dirac delta function and the integral can be easily evaluated. The right-hand side of (31) then becomes $\mathcal{M}(N,\Gamma,Z)$ where N is the value of n in the neighbourhood of which the main contribution of the series arises, and is

$$N = 2\Gamma Z. \tag{33}$$

Thus

$$M(\nu_a,\nu_b,Z)e^{-2\Gamma Z} = \frac{1}{(2\pi)^{N+1}} \iint \prod_{j=1}^{N} g(\xi_{a_j};\ \xi_b+Q_j) \tag{34}$$

$$e^{-i\xi_{a_j}(\nu_{a_j}-\nu_{a_{j-1}})\ -i\nu_b\xi_b} d\xi_b$$

$$d\xi_{a_1} \dots d\xi_{a_N}$$

$$d\nu_{a_1} \dots d\nu_{a_{N-1}}$$

where

$$\nu_{a_N} = \nu_a$$

and

$$Q_j = \frac{1}{2\Gamma}[\nu_a + \nu_{a_{N-1}} \ldots + \nu_{a_j}]. \tag{35}$$

The error involved in making this approximation is discussed in Uscinski (1982) and turns out to be less than N^{-1}.

(c) Finally the multiple product in (34) can be expressed in a more convenient form (see Appendix A), to give

$$M^{iv}(\nu_a,\nu_b,Z) = \frac{1}{(2\pi)^{N+1}}\iint e^{-\sum_{j=1}^{N}\{1-g(\xi_{a_j};\xi_b+Q_j)\}+i\xi_{a_j}(\nu_{a_j}-\nu_{a_{j-1}})-i\nu_b\xi_b\}}$$

$$d\xi_b, d\xi_{a_1} \ldots d\xi_{a_N};\ d\nu_{a_1} \ldots d\nu_{a_{N-1}} \tag{36}$$

where

$$M^{iv} = M\, e^{-2\Gamma Z} \tag{37}$$

from (10) and (12).

The errors involved in making approximations (a), (b) and (c) above have been discussed in detail by Uscinski (1982) where they are estimated to be of order $(2\Gamma Z)^{-1}$. In the case of large ΓZ, large multiple scatter, this error is small and so the solution (36), while not exact, is a good approximation to the exact solution. The result (36) is therefore a very important one and will form the basis for further discussions. In what follows it will be referred to as the multiple convolution solution.

4. PHYSICAL INTERPRETATION AS A SET OF PHASE SCREENS

The multiple convolution solution (36) has a simple and clear physical interpretation. It is just the value of M^{iv} for the field of a wave that has transversed N randomly modulated phase screens, each with a mean square phase deviation $\phi_o^2 = \frac{1}{2}$ set at equal distances of Z/N from each other. The autocorrelation function of the phase screens is the same as the transverse autocorrelation function of the extended medium

(4), (76). This situation is illustrated in Fig. 2.

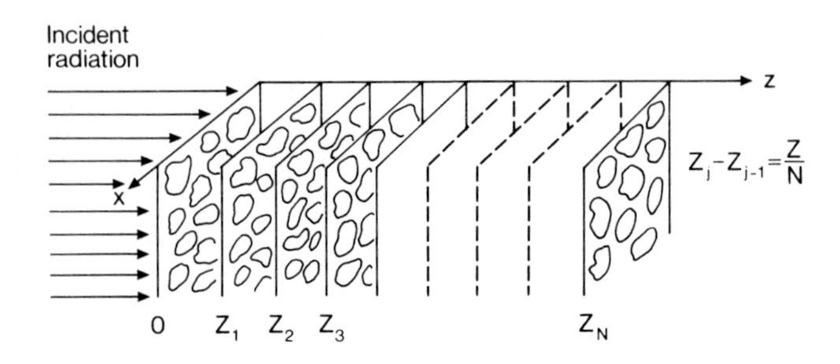

Fig. 2. Representation of multiply scattering extended medium as a set of phase modulating screens.

If an incident plane wave passes through the screen at Z = O then M^{iv} at the next screen is given by

$$M_1^{iv}(\nu_a, \nu_b, Z) = \frac{1}{(2\pi)^2} \iint e^{-[1-g(\xi_{a_1};\xi_{b_1})]-i(\nu_a\xi_{a_a}+\nu_b\xi_{b_1})} \times e^{i\nu_a\nu_b Z_1} d\xi_{a_1} d\xi_{b_1} \tag{38}$$

Now if ξ_{a_j}, ξ_{b_j}, ν_{a_j}, ν_{b_j}, refer to the j-th screen set at distance Z_j, where

$$Z_j - Z_{j-1} = Z/N, \tag{39}$$

it follows that M^{iv} for the field emerging from the N-th screen is

$$M^{iv}(\nu_a, \nu_b, Z)$$

$$= \frac{1}{(2\pi)^{2N+2}} \iint e^{-\sum_{j=1}^{N}\{[1-g(\xi_{a_j};\xi_{b_j})]-i[\xi_{a_j}(\nu_{a_j}-\nu_{a_{j-1}})+\xi_{b_j}(\nu_{b_j}-\nu_{b_{j-1}})]+i\frac{Z}{N}\nu_{a_j}\nu_{b_j}} \tag{40}$$

(cont.)

$$e^{-i[\xi_a(\nu_a-\nu_{a_N})+\xi_b(\nu_b-\nu_{b_N})]} d\nu_{a_1}\ldots d\nu_{a_N},\ d\nu_{b_1}\ldots d\nu_{b_N},$$

$$d\xi_{a_1}\ldots d\xi_{a_N}, d\xi_a d\xi_{b_1}\ldots d\xi_{b_N}, d\xi_b.$$

We can now carry out the integrals with respect to $\xi_a, d\nu_{a_N}, d\xi_{b_1}\ldots$ $d\xi_{b_N}$, $d\nu_{b_1}\ldots d\nu_{b_N}$, to obtain

$$M^{iv}(\nu_a,\nu_b,Z) = \frac{1}{(2\pi)^{N+1}} \iint e^{-\sum_{j=1}^{N}\{[1-g(\xi_{a_j};\xi_{b_j}+Q_j)]+ i\xi_{a_j}(\nu_{a_j}-\nu_{a_{j-1}})-i\xi_b\nu_b\}} \tag{41}$$

$$d\xi_b; d\xi_{a_1}\ldots d\xi_{a_N}; d\nu_{a_1}\ldots d\nu_{a_{N-1}}.$$

Here

$$\nu_{a_N} \equiv \nu_a, \tag{42}$$

$$Q_j = (\nu_a + \nu_{a_{N-1}} + \ldots + \nu_{a_j})Z/N.$$

If N the number of screens is set equal to $2\Gamma Z$ then (41) is identical with (36), the multiple convolution solution, and the total mean square phase deviation of the set of screens is equal to ΓZ, the same as that of the equivalent extended medium.

The delta correlation function

The example considered above shows that the extended medium can be represented by a set of random phase changing screens with an autocorrelation function in the plane of the screen equal to f_{ij}, (7b). The autocorrelation function in a direction normal to the screen is clearly a Dirac delta-function. This representation of the extended scattering medium has been used by some authors and the present example both justifies and clarifies this approach. While no real extended medium contains irregularities with a longitudinal autocorrelation function that is a Dirac delta function, the autocorrelations that appear in any of the moment equations for a field propagating in the extended medium have already been integrated in the longitudinal direction (see Eq. (4)). Thus the real extension of the irregularities in the longitudinal direction has already been taken into account and the integration yields the transverse correlation function f_{ij} that is delta-correlated longitudinally.

Accuracy of the screen representation

Although the multiple-convolution solution (36) of the fourth moment equation is identical with the result (41) obtained by representing the extended medium as a set of phase screens, some error was incurred in arriving at (36). This error arises from the use of approximations (a), (b) and (c) of §3 and has been calculated to be of the order of $(2\Gamma Z)^{-1}$. We thus have an estimate of the accuracy of the random screen representation of the extended medium. From the above discussion we see that the screen representation is accurate provided that the total phase fluctuation introduced by the medium, which is equal to ΓZ, is large.

5. EVALUATION OF THE MULTIPLE CONVOLUTION

In what follows we shall be chiefly interested in the spectrum of intensity fluctuations which is, from (24) and (36),

$$M^{iv}(\nu,Z) = \frac{1}{(2\pi)^N} \iint e^{-\sum_{j=1}^{N}\{[1-g(\xi_j;Q_j)]+i\xi_j(\nu_j-\nu_{j-1})\}} \quad (43)$$

$$d\xi_1 \ldots d\xi_N, \; d\nu_1 \ldots d\nu_{N-1}.$$

The subscript "a" will be omitted in what follows since all quantities with the subscript "b" have now vanished from M^{iv}. As before $\nu_N \equiv \nu$. The multiple convolution solution (43) can be regarded as an exact solution in the sense discussed above for large ΓZ. However, it cannot be easily calculated numerically and its analytical properties are not clear. There are several possible methods for evaluating the multiple convolution (43). We shall consider the method used in Appendix E of Uscinski (1982).

Fundamental form

A first estimate of the multiple convolution (43) can be obtained by considering one integral from (43) written in terms of a scaled spatial frequency $\bar{\nu}_j$ defined as

$$\bar{\nu}_j = \nu_j/2\Gamma \,, \tag{44}$$

so that

$$\bar{Q}_j = \bar{\nu} + \bar{\nu}_{N-1} + \ldots + \bar{\nu}_j . \tag{45}$$

The integral is then

$$I_j = \int_{-\infty}^{\infty} \exp\{-[1-g(\xi_j;\bar{Q}_j)] - i2\Gamma\xi_j(\bar{\nu}_j - \bar{\nu}_{j-1})\}d\xi_j . \tag{46}$$

When Γ is large the term $\exp\{-2\Gamma\xi_j(\bar{\nu}_j - \bar{\nu}_{j-1})\}$ in (46) oscillates rapidly as ξ_j changes, by contrast with the other term which varies slowly and smoothly with ξ_j. Thus it would appear that the major contribution to the integral (46) occurs for $\bar{\nu}_j \approx \bar{\nu}_{j-1}$ when the oscillations are not rapid. Thus as a first approximation we replace $\bar{\nu}_{j-1}$ by $\bar{\nu}_j$ in all the $\bar{Q}_j$. The same reasoning holds for all the terms like I_j in (43), and the $\bar{\nu}_j$ can be successively replaced in the slowly varying functions $g(\xi_j;\, \bar{Q}_j)$, so that eventually

$$\bar{Q}_j = (n-j+1)\bar{\nu}. \tag{47}$$

Written in the unscaled variables ν Eq (43) becomes

$$M_{(o)}^{iv}(\nu,Z) = \frac{1}{(2\pi)^N} \iint e^{-\sum_{j=1}^{N} \{1-g(\xi_j;\frac{N-j+1}{N}\nu Z) + i\xi_j(\nu_j-\nu_{j-1})\}} \tag{48}$$

$$d\xi_1 \ldots d\xi_N,\ d\nu_1 \ldots\ d\nu_{N-1} .$$

The subscript "(O)" will be used to indicate the first or fundamental estimate. The integrations with respect to the ν_j can now be carried out, resulting in a product of δ-functions each of the form $\delta(\xi_j - \xi_{j-1})$. The integrations with respect to all the ξ_j, with the exception of ξ_N, can then be carried out to give

$$M^{iv}_{(o)}(\nu,Z) = \frac{1}{(2\pi)}\int e^{-2\Gamma\int_0^Z[1-g(\xi;\nu Z')]dZ'} e^{-i\xi\nu}d\xi, \quad (49)$$

where ξ_N is now written simply as ξ, the summation in the exponent has been replaced by an integral and Eq. (33) has been used.

The intensity fluctuation spectrum and scintillation index

The correlation coefficient of intensity fluctuations is defined as

$$R = \frac{\langle I_1 I_2\rangle - \langle I_1\rangle\langle I_2\rangle}{\langle I_1\rangle\langle I_2\rangle} \quad (50)$$

where the subscripts can denote intensities at different points in space or time, or both. The space-time spectrum of intensity fluctuations is then

$$\Phi(\nu,\omega) = \frac{1}{(2\pi)^2}\iint R_I(\xi,\tau)e^{-i(\nu\xi+\omega\tau)}d\xi d\tau \quad (51)$$

while the normalized variance of intensity fluctuations, sometimes called the scintillation index, is

$$R(o,o) = \iint \Phi_I(\nu,\omega)d\nu\, d\omega. \quad (52)$$

The above quantities are frequently used and we shall consider them for the fundamental form of $M^{iv}_{(o)}$. It is shown in Appendix B that

$$\Phi_{(o)}(\nu,\omega) = \frac{1}{(2\pi)^2}\iint[e^{-2\Gamma\int_0^Z[1-g(\xi;\nu Z';\tau)]dZ'} - e^{-2\Gamma\int_0^Z[1-f(\nu Z')]dZ'}] \quad (53)$$

$$e^{-i(\nu\xi+\omega\tau)}d\xi\, d\tau\ .$$

From (52) and (53),

$$R_{(o)}(o,o) = \frac{1}{(2\pi)^2}\iint e^{-2\Gamma\int_0^Z[1-g(\xi;\nu Z';o)]dZ'} e^{-i\nu\xi}d\xi\, d\nu - 1. \quad (54)$$

The form of the spectrum (53) is the same as that originally obtained in Uscinski (1982) (Eq. (108)) where the second term was written simply as a Dirac delta function. The full form of the second term in (53) is given here since it turns out to be useful when investigating some analytical properties of the spectrum $\Phi(\nu,\omega)$ such as its behaviour for small distances Z.

Some properties of the fundamental solution

It is obvious that the fundamental form of the intensity fluctuation spectrum for an extended medium (53) is similar to that for the random phase screen, except for the fact that the exponent contains an integral of the transverse correlation function rather than the correlation function itself. Comparison with the expression for the screen spectrum (Eq. (21) of Buckley 1971, Eq. (6) of Shishov (1971b)) confirms this.

Much effort has been devoted to studying the intensity fluctuation spectrum of the deeply modulated phase screen, and expressions like (53) have been evaluated by various authors for different types of transverse autocorrelation functions f.

The fundamental form (53) therefore has the advantage that existing analytical and numerical techniques can be used to study its properties. The insight derived from phase screen studies can also be transferred qualitatively to the extended medium, allowing us to deduce many properties of the intensity fluctuations of the field without having to resort to elaborate computations.

The fundamental forms for $\Phi_{(o)}(\nu,\omega)$ and $R_{(o)}(o,o)$, although suffering from some degree of inaccuracy, are nevertheless important and useful expressions for the reasons given above. In what follows the accuracy of the fundamental expressions will be examined, a more exact evaluation of the multiple convolution solutions (43) will be discussed, and the results compared with scattering simulations. It turns out that for the scattering regime corresponding to the initial increase of the scintillation index, almost up to the scintillation index peak, the fundamental expressions for $\Phi_{(o)}(\nu,\omega)$ and $R_{(o)}(o,o)$ are quite accurate and further refinements in evaluating the multiple-convolution solution are not necessary.

The general behaviour of the fundamental forms for $\Phi_{(o)}(\nu,\omega)$ and $R_{(o)}(o,o)$ is illustrated in Figs. 3-4 where the spectra and scintillation indices are given for two types of media. One of the media contains irregularities with a single scale size while the other has a large range of scale sizes. The spectra are shown for a single value of Γ at different distances in the scattering medium and we notice the gradual spread of the spectrum to higher and lower spatial frequencies as distance in the medium increases. A scintillation index curve is given in Fig. 7 for the same value of Γ. It exhibits the characteristic peak whose height increases with Γ and the subsequent saturation to unit scintillation index. Examples of the fundamental form of the spectra and scintillation index curves are given in Uscinski (1982) for media with Gaussian and fourth order power-law spectra of refractive index fluctuations. Some simpler analytical forms for these quantities are also given there.

The fundamental forms of $\Phi_{(o)}(\nu,\omega)$ and $R_{(o)}(o,o)$ have further interesting properties, but these are best discussed elsewhere.

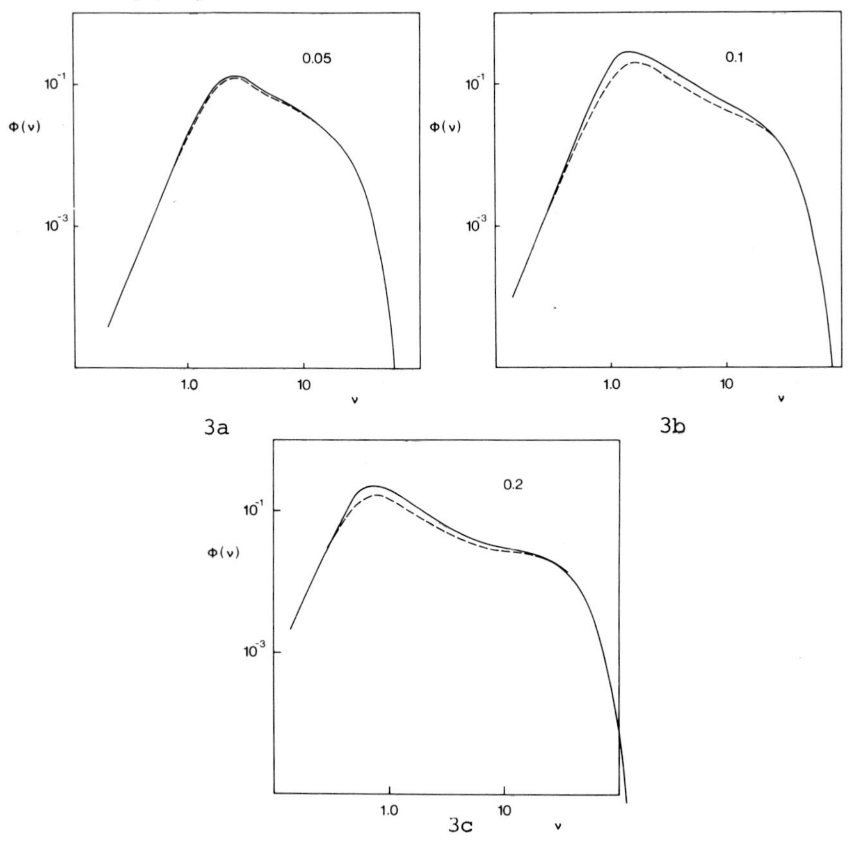

Fig. 3. Spatial frequency spectra $\Phi(\nu)$ for $\Gamma = 1000$ in a single scale medium with refractive index spectrum (65). The number by each curve gives Z scaled by Z_{fo}. The broken line gives the contribution from $\Phi_{(o)}(\nu)$ and the full line gives the sum of $\Phi_{(o)}$ and $\Phi_{(1)}$.

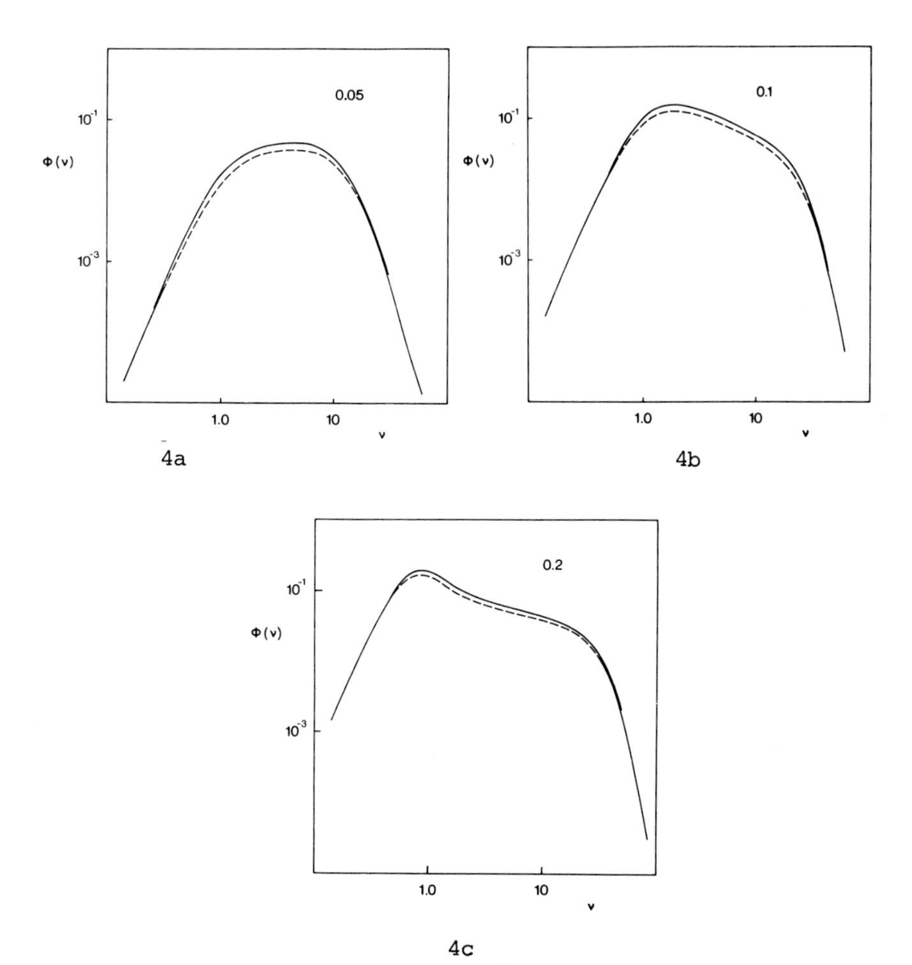

4c

Fig. 4. Spatial frequency spectra $\Phi(\nu)$ for $\Gamma = 1000$ in a medium with a range of scale sizes with refractive index spectrum (67). The number by each curve gives Z scaled by Z_{fo}. The broken line gives the contribution from $\Phi_{(o)}(\nu)$ and the full line gives the sum of $\Phi_{(o)}$ and $\Phi_{(1)}$.

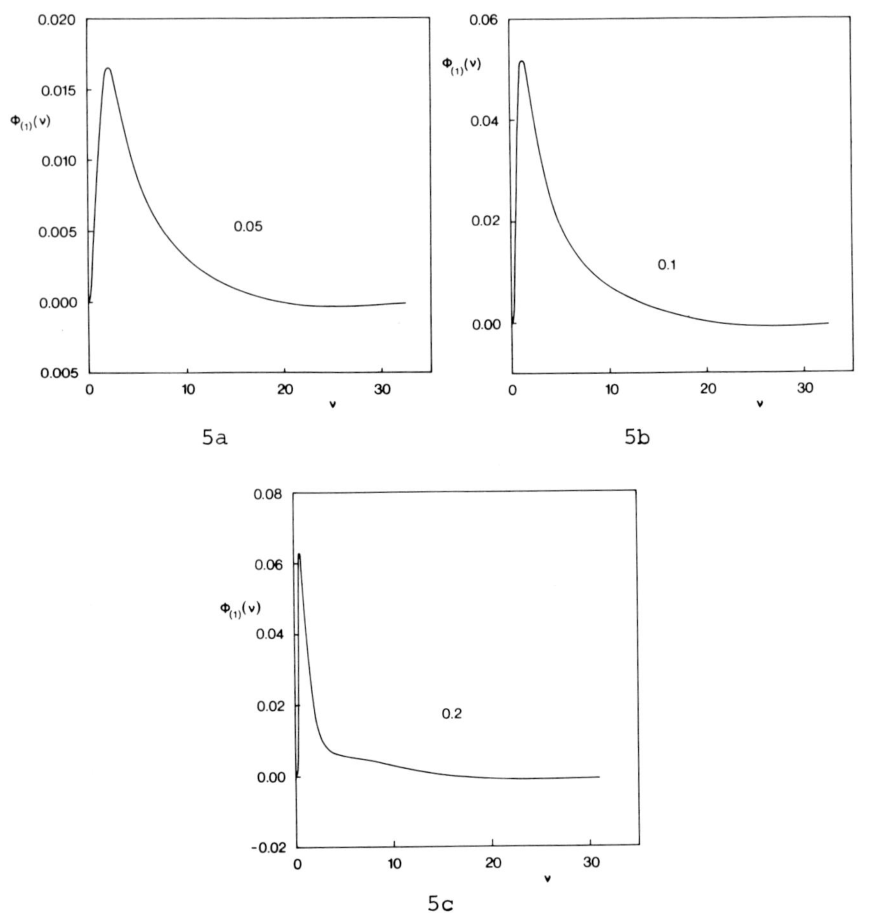

Fig. 5. First order correction to the spatial frequency spectrum $\Phi_{(1)}(\nu)$ for $\Gamma = 1000$ in a single scale medium with refractive index spectrum (65). The number by each curve gives Z scaled by Z_{fo}.

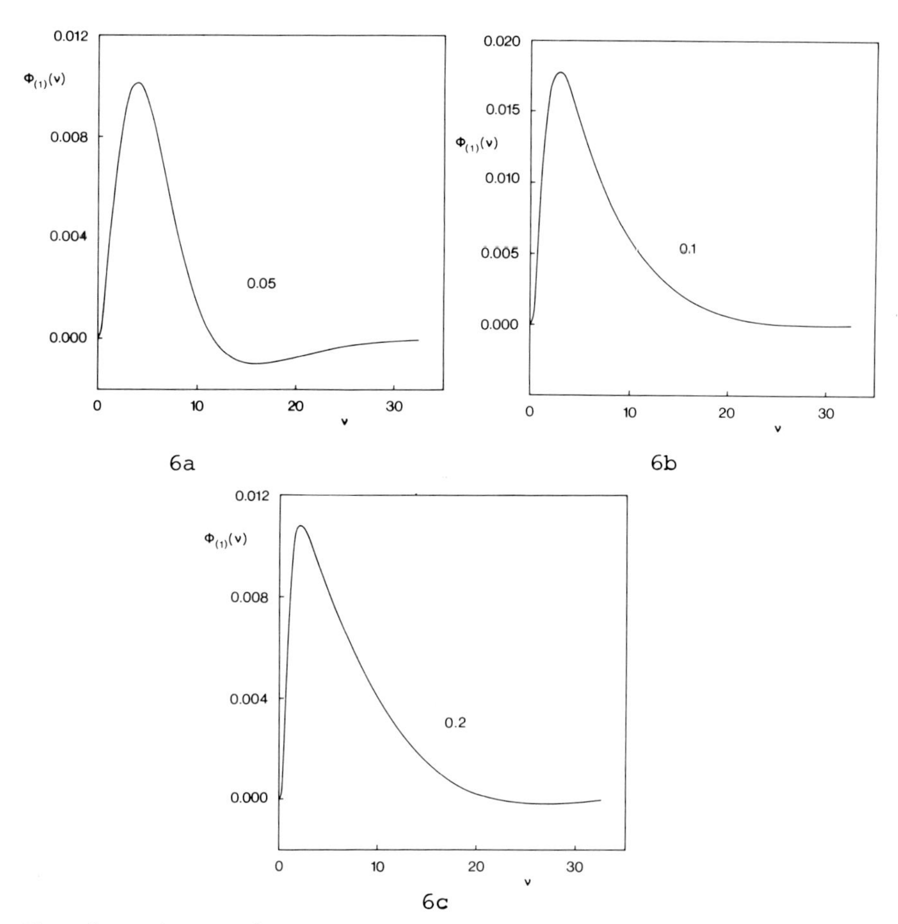

Fig. 6. First order correction to the spatial frequency spectrum $\Phi_{(1)}(\nu)$ for $\Gamma = 1000$ in a medium with a range of scale sizes with refractive index spectrum (67). The number by each curve gives Z scaled by Z_{fo}.

6. *Refined estimate of the multiple convolution*

The multiple convolution solution M^{iv}, Eq. (43), can be evaluated more accurately by obtaining an estimate of the difference between it and the fundamental form or first estimate M_o^{iv}, Eqs. (48), (49). Using Eq. (48) for M_o^{iv} we have that

$$M^{iv} - M_o^{iv} = \frac{1}{(2\pi)^N} \int..\int [e^{-D} - e^{-D_o}] e^{-i \sum_{j=1}^{N} \xi_j (\nu_j - \nu_{j-1})} \qquad (55)$$

$$d\xi_1 \ldots d\xi_N, \; d\nu_1 \ldots d\nu_{N-1}$$

where

$$D = \sum_{j=1}^{N} [1 - g(\xi_j, Q_j)], \qquad (56)$$

$$D_o = \sum_{j=1}^{N} [1 - g(\xi_j, \frac{N-j+1}{N} \nu Z)] . \qquad (57)$$

We now assume that

$$d = D_o - D \qquad (58)$$

is small, expand the terms in the square brackets in Eq. (55) in powers of d, and write

$$M^{iv} - M_{(o)}^{iv} = \frac{1}{(2\pi)^N} \int \int [e^{-D_o} [d + \frac{d^2}{2!} + \frac{d^3}{3!} + \ldots] e^{-i \sum_{j=1}^{N} \xi_j (\nu_j - \nu_{j-1})} \qquad (59)$$

$$d\xi_1 \ldots d\xi_N, \; d\nu_1 \ldots d\nu_{N-1}$$

$$= M_{(1)}^{iv} + M_{(2)}^{iv} + M_{(3)}^{iv} + .$$

When the spectrum F(q), Eq. (26), corresponding to the transverse correlation function F(ξ) is used in g of Eqs. (56), (57), and the summations converted to integrals we obtain for $M_{(1)}^{iv}$, the first term in the expansion (59) (see Appendix C),

$$M^{iv}_{(1)} = \frac{4\Gamma}{2\pi} \int_{0}^{Z} \int_{-\infty}^{\infty} \int F(q) \sin^2(q\xi/2) e^{-i\nu(\xi - qS)} \tag{60}$$

$$\left\{ e^{-2\Gamma\left(\int_0^S [1-g(\xi+q[y-S];\nu y] + \int_S^Z [1-g(\xi;\nu y)]\right) dy} - e^{-2\Gamma\int_0^Z [1-g(\xi;\nu y)]) dy} \right\} dq\, d\xi\, dS.$$

If it is assumed that d is small then d^2 and higher powers of d can be neglected. The improved estimate of M^{iv} is then, from (59),

$$M^{iv} = M^{iv}_{(o)} + M^{iv}_{(1)}. \tag{61}$$

We note that the improved estimate of $\Phi(\nu,\omega)$ is obtained by adding

$$\Phi_{(1)}(\nu,\omega) = \frac{1}{2\pi} \int M^{iv}_{(1)}(\nu, \tau) e^{-i\omega\tau} d\tau \tag{62}$$

to $\Phi_{(o)}(\nu,\omega)$, Eq. (53), and that the improved estimate of $R(o,o)$ is obtained by adding

$$R_{(1)}(o,o) = \int M^{iv}_{(1)}(\nu,o) d\nu \tag{63}$$

to $R_{(o)}(o,o)$, Eq. (54).

It is immediately clear that $M^{iv}_{(1)}$ has a much more complicated structure than $M^{iv}_{(o)}$. There is no obvious similarity with any other well known expression that could give a clue to its behaviour. Moreover, because it is a triple integral, apart from the integrals in the exponents which can probably be evaluated analytically, there can be difficulties in evaluating it numerically, especially as two of the integrals are oscillatory. For these reasons it is important to obtain some simplifications that allow $M^{iv}_{(1)}$ to be evaluated analytically. The most obvious simplifications can be made when the quantity g in the exponents of (60) possesses a Taylor series expansion with respect to ξ. This allows the ξ integral to be carried out analytically and the resulting expression is then a double integral. The integral with respect to q has an oscillatory character but its numerical evaluation in the case of a general spectrum F(g) can be carried out in the same way as that of the fundamental form (49). The remaining integral with respect to S is smooth and presents no difficulty. In the particular case when F(g) is a Gaussian spectrum the g integral can be evaluated analytically also.

It should be pointed out that the above simplifications depend upon g having a Taylor expansion, which is not always the case. It is then necessary to evaluate (60) directly. Although $M^{iv}_{(1)}$ as given by (60) has a complex structure it involves only the transverse autocorrelation of the medium and its corresponding spectrum, and does not depend on Taylor expansions of these quantities. This is important for those cases when the medium correlation function does not possess a series expansion. One example of this is the transverse autocorrelation function of internal waves in the ocean. Thus Eq. (60) for $M^{iv}_{(1)}$ can be applied to the case of acoustic propagation through ocean interval waves, whereas other estimates of $M^{iv}_{(1)}$ that rely on Taylor expansion techniques will fail in this case. Thus the form of $M^{iv}_{(1)}$ in Eq. (60) is quite general in nature, although we may wish to employ expansion techniques to evaluate it in some cases.

The evaluation of the multiple convolution solution (43) presented here is only one of several possible methods. There may well be better procedures for approximating the multiple-convolution and it is important for this question to be investigated further. In particular, a better estimate for the fundamental form $M^{iv}_{(o)}$ would be quite valuable.

7. SPECTRA AND SCINTILLATION INDICES

Some typical spatial frequency spectra are now given for different values of Γ at several distances in the scattering medium. The spectra have been calculated using the theory presented in the preceding sections, and the contribution due to the fundamental (53) and the improved estimate (62) are distinguished. The two types of scattering media mentioned in Section 5 are considered. The medium containing irregularities with a single scale size is modelled by a Gaussian autocorrelation function of refractive index inhomogeneities

$$f(\xi) = e^{-\xi^2} \tag{64}$$

with the corresponding spectrum, from Eq. (26),

$$F(\nu) = \frac{2}{\sqrt{\pi}} e^{-\nu^2/4}. \tag{65}$$

The medium with a range of scale sizes is modelled by the autocorrelation function

$$f(\xi) = (1 + |\xi|)e^{-|\xi|}, \tag{66}$$

which has the corresponding spectrum

$$F(\nu) = \frac{2}{\pi(1 + \nu^2)^2}. \tag{67}$$

We note that (67) behaves like ν^{-4} for $\nu > 1$ and is very similar to the Kolmogorov spectrum of a turbulent medium

$$F(\nu) = 0.033\, C_n^2(1 + \nu^2)^{-\frac{11}{6}} \exp\{-\nu^2/\nu_m^2\} \tag{68}$$

Here C_n is a constant and ν_m is the high spatial frequency cut-off. This spectrum behaves like $\nu^{-\frac{11}{3}}$ for large ν. Thus (66), (67) give a good representation of a turbulent type medium containing a range of scale sizes, while at the same time possessing a very simple form that appreciably simplifies analytical evaluations.

The spatial frequency spectra $\Phi(\nu)$, (51), for a medium with the Gaussian autocorrelation function (64) are shown in Fig. 3 at different distances in the scattering medium. These distances are given as fractions of the distance at which the scintillation index is a maximum, Z_{fo}. The spectra appear as the sum of the fundamental form $\Phi_{(o)}(\nu)$, (53), and the first order correction $\Phi_{(1)}(\nu)$, (62), i.e.

$$\Phi(\nu) = \Phi_{(o)}(\nu) + \Phi_{(1)}(\nu) \tag{69}$$

with the contribution from the fundamental being given by the broken line and the sum (69) by the full line. Similar results are presented in Fig. 4 for the medium with autocorrelation function (66) possessing a range of scale sizes.

It is clear from Figs. 3 and 4 that the fundamental $\Phi_{(o)}(\nu)$ gives quite an accurate representation of the spectrum. The maximum discrepancy with the improved estimate occurs for the case when the scintillation index is a maximum, i.e. $Z/Z_{fo} = 1$, and is at most of the order of 18% at the peak of the spectrum and much less elsewhere. Thus, if high accuracy is not required, the fundamental form $\Phi_{(o)}(\nu)$ provides a simple and quick method for investigating how the different autocorrelation functions of refractive index irregularities affect the form of the intensity fluctuation spectra.

The first order correction $\Phi_{(1)}(\nu)$ to the fundamental spectrum is shown in greater detail in Figs. 5 and 6 for the single scale and multiscale media respectively. These curves were obtained from (60) by expanding the functions g in small powers of ξ and carrying out the resulting ξ integral analytically. The subsequent integrations with respect to g were easily evaluated numerically since the integrands are smooth non-oscillatory functions.

The scintillation index for a single scale medium is given in Fig. 7 as a function of Z for a value of $\Gamma = 1000$.

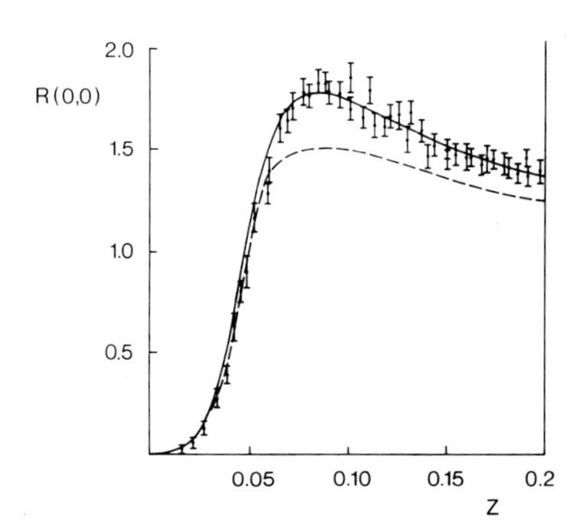

Fig. 7. The scintillation index R(o,o) as a function of Z for $\Gamma = 1000$ in a single scale medium with refractive index spectrum (65). The broken curve gives the fundamental approximation and the full curve includes the first order correction. The points and error bars are the results of numerical simulations of the corresponding scattering experiment (Macaskill and Ewart, 1984).

The scintillation index R(o,o), (52), is the integral of the spectrum $\Phi(\nu)$ and in the case Γ = 1000 the maximum discrepancy between the fundamental and the improved estimate is of the order of 14% near the scintillation index peak. The scintillation index obtained by computer simulations of propagation in a randomly irregular medium with the Gaussian autocorrelation function (64) (Macaskill and Ewart 1984) are also shown in Fig. 7.

Finally, it would be desirable to evaluate the second order correction, corresponding to the d^2 term in (59), and show whether, and under what conditions, it can be neglected, or whether it may be necessary to take it into account.

ACKNOWLEDGEMENTS

This work was carried out with the support of the Ministry of Defence, Procurement Executive.

APPENDIX A

The multiple product

$$P = \prod_{j=1}^{N} g(\xi_{a_j}; \xi_b + Q_j) \tag{A1}$$

can be expressed in a different form. First g is expanded in powers of ξ_a to obtain

$$g(\xi_a, \xi_b + Q) = 1 + g_r(\xi_a, \xi_b + Q) \tag{A2}$$

where

$$g_r(\xi_a; \xi_b + Q) = \sum_{k=1}^{\infty} \frac{\partial^k g}{\partial \xi_a^k}(o, \xi_b + Q) \frac{\xi^k}{k!}$$

and we note that $g(o, \xi_b + Q) = 1$. Taking the logarithm of both sides of (A1) we have

$$\ell nP = \sum_{j=1}^{N} \ell n(1 + g_r(\xi_{a_j}, \xi_b + Q_j)).$$

Now g is a decreasing function of ξ, and since P is the product of a very large number of terms it falls off rapidly as ξ increases from zero. Since P is of interest only for small values of ξ, and consequently for small values of g, we may write

$$\ell nP = \sum_{j=1}^{N} g_r(\xi_{a_j}, \xi_b + Q_j)$$

which gives, from (A2)

$$P = \exp\{-\sum_{j=1}^{N} [1-g(\xi_{a_j}, \xi_b + +_j)]\} .$$

APPENDIX B

In the case of the fundamental approximation (49)

$$<I_1 I_2> = \frac{1}{2\pi} \iint e^{-2\Gamma\int_0^Z [1-g(\xi', \nu Z'; \tau)dZ'] - i(\xi'-\xi)\nu} d\xi' d\nu, \qquad \text{(B1)}$$

where the time separation τ has now been written explicitly in g (Eq. (11)). Then introducing

$$s = \xi' - \xi \qquad \text{(B2)}$$

and remembering that $<I_1> <I_2>$ is obtained by letting ξ, τ tend to infinity in $<I_1 I_2>$ we have that

$$R_{(o)}(\xi,\tau) = \frac{1}{2\pi}\iint [e^{-2\Gamma\int_0^Z [1-g(s+\xi, \nu Z';\tau)]dZ'} - e^{-2\Gamma\int_0^Z [1-f(\nu Z')]dZ'}] \qquad \text{(B3)}$$

$$e^{-i\nu s} ds\, d\nu .$$

When (B3) is used in the Fourier transform (51) we obtain the space-time spectrum (53).

APPENDIX C

The first term in the expansion (59) is considered, and the spectrum F(q) corresponding to the transverse correlation function f(q) is used in g to give

$$M_{(1)}^{iv} = \frac{2}{(2\pi)^N} \iint e^{-D_o} \sum_{k=1}^{N} F(q) \sin^2(\xi_k q/2)[e^{iqQ_k} - e^{iq\nu Z(1-\frac{k-1}{N})}]$$

(cont.)

$$e^{-i\sum_{j=1}^{N}\xi_j(\nu_j-\nu_{j-1})} d\nu_1..d\nu_{N-1}, d\xi_1..d\xi_{N-1},$$

$$= M_{(a)} + M_{(b)} \ . \qquad (C1)$$

The integrals with respect to ν_j, ξ_j can be done in $M_{(b)}$ and the summations converted to integrals to give

$$M_{(b)} = \frac{4\Gamma}{2\pi}\int_0^Z\int_{-\infty}^{\infty} e^{-2\Gamma\int_0^Z[1-g(\xi,\ \nu(Z-Z')]dZ'} F(q)\sin^2(\xi q/2) \qquad (C2)$$

$$e^{i\nu(q[Z-Z'']-\xi)} dZ'' \, dq \, d\xi \ .$$

The first term in (C1) is evaluated in a similar manner. First Q_k, Eq. (42), is written out fully and the ν_j integrals carried to give the product of deltafunctions

$$\delta(\xi_1-\xi_2)\delta(\xi_2-\xi_3)\ldots\delta(\xi_{k-1}-\xi_k)\delta(\xi_k-\xi_{k+1}-Zq/N) \qquad (C3)$$

$$\delta(\xi_{k+1}-\xi_{k+2}-Zq/N)\ldots\delta(\xi_{N-1}-\xi_N-Zq/N) \ .$$

The ξ_j integrals are then done, with the exception of that with respect to ξ_N, leading to the following set of substitutions

$$\xi_j = \xi_N + (N-k)qZ/N, \quad (1 \leqslant j \leqslant k) \qquad (C4)$$

$$\xi_j = \xi_N + (N-j)qZ/N. \quad (k < j \leqslant N)$$

When ξ_j, ξ_k in $M_{(a)}$ are replaced in accordance with (C4) and the summations converted into integrals we obtain

$$M_{(a)} = \frac{4\Gamma}{2\pi}\int_0^Z\int_{-\infty}^{\infty}\int e^{-2\Gamma\int_0^{Z'}[1-g(\xi+(Z-Z'')q;(Z-Z')\nu)]dZ'} \qquad (C5)$$

$$e^{-2\Gamma\int_{Z'}^{Z}[1-g(\xi+(Z-Z')q;(Z-Z')\nu]dZ'}$$

(Cont.)

$$F(q)\sin^2([\xi + (Z-Z'')q]q/\ 2)e^{-i\nu\xi}dZ''\ dq\ d\xi$$

since exp $\{i\ Zq\nu/N\} \approx 1$ for N large.

On setting

$$S = Z-Z'' \tag{C6}$$

$$y = Z-Z' \tag{C6}$$

in (C2), (C5) and combining them we obtain Eq. (60).

REFERENCES

Buckley, R., 1971, "Diffraction by a random phase screen with very large R.M.S. phase deviation", *Aust. J. Phys.*, **24**, 351.

Macaskill, C., 1983, "An improved solution to the fourth moment equation for intenstiy fluctuations", *Proc. Roy. Soc. Lond.* **A386**, 461.

Macaskill, C., and Ewart, T.E., 1984, "Computer simulation of two-dimensional random wave propagation", *I.M.A. Journal of Appl. Maths.*, **33**, 1.

Shishov, V.I., 1968, "Theory of wave propagation in random media", *IZv. Vyssh. Ucheb, Zaved. Radiofizika,* **11**, 866.

Shishov, V.I., 1971a, "Strong intensity fluctuations of a plane wave propagating in a random refractive medium", *Zk. Eksp. Teor. Fiz.* **61**, 1399.

Shishov, V.I., 1971b, "Diffraction of waves by a strongly refracting random phase screen", *Izv. Vyssh. Ucheb. Zaved. Radiofiz,* **14**, 85.

Uscinski, B.J., 1977, "The Elements of Wave Propagation in Random Media", McGraw-Hill, New York.

Uscinski, B.J., 1982, "Intensity fluctuations in a multiple scattering medium, solution of the fourth moment equation", *Proc. Roy. Soc. Lond.* **A380**, 137.

A FOKKER-PLANCK EQUATION FOR THE PROBABILITY DISTRIBUTION OF THE FIELD SCATTERED BY A RANDOM PHASE SCREEN

R. Buckley
(Institute of Sound and Vibration Research, Southampton University)

ABSTRACT

I consider a one-dimensional random phase changing screen with the exponential type of autocorrelation function. No limitation is placed on the depth of modulation. The statistical properties of a wave field diffracted by the screen are considered. It is shown that the joint probability distribution of the field components co-phased and in phase quadrature can be determined from the solution of an equation of the Fokker-Planck type, in which the time like variable becomes the transverse distance across the screen, and the diffusive variable is the phase change itself. In an accompanying paper, Roberts (1985) considers the numerical solution of the equation.

1. INTRODUCTION

The propagation of waves through a medium containing random irregularities of refractive index has, on many occasions, been successfully modelled for some three decades by a thin "screen" which modulates the phase of the wave in a manner with prescribed statistical properties of any desired kind. The history and applications of this method are described by Uscinski and Macaskill (1983) and by Uscinski (1984) in this volume.

In this paper I consider a special type of screen for which the problem of the complete probability distribution of the diffracted field can be tackled without approximation.

Consider a wave of unit amplitude, and wavelength $\lambda=2\pi/k$ propagating parallel to the z-axis from negative z. Its complex amplitude, with the time dependent factor suppressed, is

e^{ikz}. On reaching the plane z=0, its phase is modified in a one-dimensional x dependent fashion, so that its complex amplitude becomes,

$$\exp[i\phi(x)], \tag{1}$$

where $\phi(x)$ is assumed to be real.

We assume that the wavelength is considerably smaller than the spatial scale of $\phi(x)$, so that the forward scattering, or paraxial approximation can be used. Roberts (1984) discusses this point in the present context more fully. The Fresnel-Kirchoff formula enables the field amplitude at positive z to be expressed in the form

$$\exp(ikz)\,E(x,z), \tag{2}$$

where

$$E(x,z) = \left(\frac{k}{2\pi z}\right)^{\frac{1}{2}} \int_{-\infty}^{\infty} dx' \exp[i\{k(x-x')^2/2z + \phi(x') - \pi/4\}] \tag{3}$$

The most illuminating way to derive this expression is to exploit the concept of the angular spectrum of plane waves generated by the screen. This was introduced by Booker and Clemmow (1950), and Ratcliffe (1956), and is described by Buckley (1975).

If certain statistical properties are ascribed to the function $\phi(x)$, what are the corresponding properties of the amplitude E? If ϕ remains (statistically) much less than one radian, $\exp(i\phi)$ can be approximated by $1 + i\phi$ and the relation between ϕ and E in (3) becomes linear. In this case the problem is relatively simple. But if, as is often the case in practice, ϕ is not small, the problem is very much more complicated. Nearly all approaches have concentrated on the low order moments of E, in particular, those of the <u>intensity</u> $|E|^2$, rather than the full probability distribution of E.

In section 2, a formal path integral expression is derived for the latter, and it is shown that provided the stochastic process $\phi(x)$ is of the Ornstein-Uhlenbeck form (Gaussian with autocorrelation function $\exp(-|x|/\ell)$), the path integral can be calculated from the solution of a linear partial differential equation of the Fokker-Planck form. The formalism is summarised in equations (34)-(37). In a certain sence, as $\ell \to \infty$, one recovers the "Brownian fractal" screen discussed by Berry (1979).

Some of the properties of this equation are discussed in section 3. However it has not proved possible to solve the equation analytically. In an accompanying paper in this volume, Roberts (1985) discusses the equation further, and presents results of a numerical solution of it.

2. THE DEVELOPMENT OF THE FOKKER-PLANCK EQUATION

2.1 Preliminaries

Since $\phi(x)$ will be assumed to be a stationary stochastic process, all transverse positions x are equivalent for the single point statistics of the scattered field. For simplicity therefore I take x to be zero in (3). Let ℓ be a length scale, later to be identified with the correlation length of the screen. Scale the transverse x coordinate in units of ℓ, but the propagation (z) coordinate in units of $k\ell^2$. In the paraxial approximation it is implicit that $k\ell^2 >> \ell$. From (3) the real (in phase) component of the field is given by

$$E_R = \int_{-\infty}^{\infty} F(X, \phi(X))\,dX, \tag{4}$$

where

$$F(X,\phi) = (2\pi z)^{-\frac{1}{2}} \cos(X^2/2z + \phi - \pi/4) \tag{5}$$

The imaginary (phase-quadrature) component E_I is given by the analogous formula with sine replacing cosine.

E_R is expressed in (4) as a non-linear functional of $\phi(x)$. The following development is, as will be seen, not dependent on the particular form for $F(X,\phi)$ in (5). The problem to be faced is: what is the joint probability distribution of E_R and E_I when $\phi(x)$ is taken to be a stochastic process with definite properties? For simplicity in the development, I shall consider the distribution of E_R alone. The elementary extensions to both E_R and E_I will be indicated at the end of the section.

The field E_R will be expressed as a <u>sum</u> rather than <u>integral</u> of random variables. For this reason we replace the infinite range of integration by a finite one.

$$E_R = \int_{-a}^{X} F(X,\phi(X))\,dX \tag{6}$$

The formalism now represents a finite screen of width $x + a$. Eventually we shall allow a and $x \to \infty$. The upper limit x will turn out to be the time like independent variable in the final development.

2.2 The Formal Probability Distribution and Characteristic Function of E_R

Representing (6) as a Riemann sum, we have

$$E_R = \underset{\Delta \to 0}{\mathrm{Lt}} \sum_{j=-M}^{N} \Delta F_j(\phi_j), \tag{7}$$

where

$$X_j = j\Delta, \quad \phi_j = \phi(X_j),$$

$$x = N\Delta, \quad a = M\Delta,$$

and

$$F_j(\phi_j) = F(X_j, \phi_j), \tag{8}$$

We shall omit the "Lt" symbol henceforth. E_R is now represented as a sum of N+M+1 functions of the values ϕ_j of $\phi(x)$ at the X_j. We regard these ϕ_j as dependent random variables with joint probability density $P([\phi])$. The distribution of E_R follows formally simply by integrating P over the sample space of $[\phi]$, restricting the sum in (2.4) to be equal to E_R

$$P(E_R, x) = \int_{-\infty}^{\infty} [d\phi] \; P([\phi]) \; \delta\{E_R - \Delta \sum_{-M}^{N} F_j(\phi_j)\} \tag{9}$$

where

$$[d\phi] = \prod_{-M}^{N} d\phi_j$$

and δ is the Dirac delta function.

The latter being awkward to deal with, it is natural to introduce the <u>characteristic function</u> of E_R

$$C(q,x) = \int_{-\infty}^{\infty} dE_R \; P(E_R, x) e^{iq \, E_R} \tag{10}$$

the inverse Fourier transform

$$P(E_R, x) = \frac{1}{2\pi} \int_{-\infty}^{\infty} dq C(q,x) e^{-iqE_R}, \tag{11}$$

yielding P. In (9)-(11), the dependence of P and C on the upper limit of integration x in (6) is indicated explicitly because of its fundamental character which will appear later.

If (9) is inserted into (10) and the δ function integral carried out, there results,

$$C(q,x) = \int_{-\infty}^{\infty} [d\phi]\ P([\phi])\ \exp[i\Delta q \sum_{-M}^{N} F_j(\phi_j)], \qquad (12)$$

an integral over N+M+1 variables. In the limit $\Delta \to o$, $N,M \to \infty$ (8) it becomes a <u>functional</u> or <u>path</u> integral.

We assume henceforth that $\phi(X)$ is a Gaussian stochastic process. Recall that the usual time variable in such a process is represented here by the transverse distance X across the diffracting screen. The joint distribution of the N+M+1 values ϕ_j is explicitly,

$$P([\phi]) = [(2\pi\Phi^2)^{N+M+1} ||\rho||]^{-\frac{1}{2}} \times$$

$$\times \exp[-\frac{1}{2} \sum_{i,j=-M}^{N} (\rho^{-1})_{ij}\phi_i\phi_j/\Phi^2] \qquad (13)$$

Here Φ is the RMS value of $\phi(x)$ used temporarily in place of the usual symbol ϕ_o to avoid confusion with the dummy variable ϕ_o in (12), (13).

ρ is the correlation matrix of ϕ,

$$\rho_{ij} = \langle\phi_i\phi_j\rangle/\Phi^2 = \rho(|X_i - X_j|) \qquad (14)$$

ρ^{-1} and $||\rho||$ are respectively the inverse and determinant of ρ. Every statistic of $\phi(x)$ is expressible in terms of Φ and $\rho(X)$ the <u>autocorrelation function</u>. The latter is restricted only in that it has a real non-negative Fourier transform (the power spectrum of $\phi(X)$).

2.3 The Ornstein-Uhlenbeck Process

Progress with (12), (13) is possible if the inverse matrix ρ^{-1} has a simple enough form. One such standard process (and probably the only one) for which this is the case is that of Ornstein and Uhlenbeck (1930), which has played a central role in the development of stochastic theory. In different variables it is equivalent to the Wiener process of Brownian motion. It is specified by the autocorrelation function

$$\rho(x) = e^{-|x|} \tag{15}$$

(In dimensional terms $\rho(x) = e^{-|x|/\ell}$ where ℓ is the characteristic length used to scale the transverse dimension). Any realisation $\phi(X)$ of the screen phase subject to (15) contains structure down to arbitrarily small scales, and is thus of fractal nature, below the scale ℓ. This point is amplified in Robert's paper and others in this volume.

Letting

$$\rho(\Delta) = e^{-\Delta} = \lambda, \tag{16}$$

we see that the matrix ρ (of order N+M+1) in (13) is

$$\rho = \begin{pmatrix} 1 & \lambda & \lambda^2 & \cdots & \lambda^{N+M} \\ \lambda & 1 & \lambda & & \\ \lambda^2 & \lambda & 1 & & \\ \vdots & & & \ddots & \\ \lambda^{N+M} & & & & 1 \end{pmatrix} \tag{17}$$

The inverse matrix is tri-diagonal,

$$\rho^{-1} = \frac{1}{1-\lambda^2} \begin{pmatrix} 1 & -\lambda & & & 0 \\ -\lambda & 1+\lambda^2 & -\lambda & & \\ & -\lambda & \ddots & \ddots & \\ & & \ddots & 1+\lambda^2 & -\lambda \\ 0 & & & -\lambda & 1 \end{pmatrix} \tag{18}$$

This simple form enables progress to be made with (12).

Further, the determinant is

$$||\rho|| = (1-\lambda^2)^{N+M} . \tag{19}$$

It can be seen that, in view of (18), the Gaussian exponent in (13) can be expressed as a sum of squares,

$$\sum_{-M}^{N} (\rho^{-1})_{ij}\, \phi_i \phi_j = \phi_{-M}^2 + \frac{1}{1-\lambda^2} \sum_{-M}^{N-1} (\phi_{j+1} - \lambda\phi_j)^2 \tag{20}$$

Substituting (20) into (13) and the latter into (12) we get,

$$C(q,x) = \int_{-\infty}^{\infty} [d\phi]\,(2\pi\Phi^2)^{-\frac{1}{2}} \exp(-\phi_{-M}^2/2\Phi^2) \; x$$

$$x \prod_{-M}^{N-1} K_j(\phi_{j+1}, \phi_j), \tag{21}$$

where

$$K_j(\phi', \phi) = [\,2\pi\Phi^2\;(1-\lambda^2)]^{-\frac{1}{2}} \; x$$

$$x \; \exp\left\{-\frac{(\phi' - \lambda\phi)^2}{2(1-\lambda^2)\Phi^2} + i\Delta q\; F_j(\phi)\right\} \tag{22}$$

Anticipating the limit $\Delta \to o$, a term $\exp\,[\,i\Delta q\; F_N(\phi_N)]$ has been omitted from (21).

The product form in (21) is reminiscent of inhomogeous Markov chain theory with K_j playing the role of transition probability.

2.4 Reduction to a Differential Equation

There are N+M+1 "ϕ" variables in (21), but N+M K's. It is necessary to separate that ϕ_j corresponding to x, namely ϕ_N, and I write

$$C(x) = \int_{-\infty}^{\infty} d\phi_N \; C(\phi_N, x), \tag{23}$$

where

$$C(\phi_N, x) = \int_{-\infty}^{\infty} (2\pi\Phi^2)^{-\frac{1}{2}} \exp(-\phi_{-M}^2/2\Phi^2) \prod_{-M}^{N-1} K_j(\phi_{j+1}, \phi_j)\, d\phi_j \tag{24}$$

We now show that the function $C(\phi,x)$ obeys a linear partial differential equation of Fokker-Planck type. The demonstration of this has its inspiration in Feynmans' (1965) path integral development of quantum mechanics (see also Schulman 1981). To summarise the latter very briefly, Feynman considers a particle in a potential well. Whatever path in position/velocity space it follows between points A and B, the classical action S can be defined as the time integral of the Lagrangian. Feynman conjectures that the following quantity

$$\psi = \langle \exp(i\, S/\hbar) \rangle, \tag{25}$$

is the quantum mechanical wave function. $\hbar$ is Planck's constant and <> indicates the unweighted average over all possible paths. The latter concept leads to "path integrals". Feynman shows by a method similar to that described below, that ψ so defined satisfies the Schrodinger wave equation. The classical limit of quantum mechanics follows satisfactorily from this. When $\hbar$ is "small", the average (25) is dominated by those paths for which S is close to extremal. On the other hand, Hamilton's principle for classical mechanics states that the (classical) path is that for which S <u>is</u> extremal. It appears that the path integral is central to the modern development of quantum field theory.

Returning to our development of (23), (24), suppose the screen is incremented in length by Δ. A new quantity ϕ_{N+1} appears, associated with the point $x+\Delta$. ϕ_N, at x, is now itself integrated over. The function C at $x + \Delta$ and ϕ_{N+1} is, from (24),

$$C(\phi_{N+1}, x+\Delta) = \int_{-\infty}^{\infty} (2\pi\Phi^2)^{-\frac{1}{2}} \exp(-\phi_{-M}^2/2\Phi^2) \prod_{-M}^{N} K_j(\phi_{j+1}, \phi_j)\, d\phi_j \tag{26}$$

By comparison with (24) we have therefore the (single) integral equation,

$$C(\phi, x+\Delta) = \int_{-\infty}^{\infty} d\phi_N\, K_N(\phi,\phi_N) C(\phi_N, x), \tag{27}$$

which is reminiscent of the Chapman-Kolmogrov equation of Markov theory.

We are now in a position to take the limit $\Delta\to o$ in (27). We expand each side in powers of Δ. To leading order, as $\Delta\to o$, $\lambda\to 1$, K (22) tends to $\delta(\phi'-\phi)$ and we have an identity. In order to recover the next order, make the change of variable

$$\phi_N = \phi/\lambda + \tau\Phi\frac{\sqrt{1-\lambda^2}}{\lambda} \tag{28}$$

in (27). It becomes, without approximation,

$$C(\phi,x+\delta) = \frac{1}{\lambda}\int_{-\infty}^{\infty} d\tau(2\pi)^{-\frac{1}{2}} \exp(-\frac{1}{2}\tau^2)\times$$

$$\times \exp\left[i\delta qF(x, \frac{\phi}{\lambda} + \tau\Phi\frac{\sqrt{1-\lambda^2}}{\lambda}\right] C(\frac{\phi}{\lambda} + \tau\Phi\frac{\sqrt{1-\lambda^2}}{\lambda}, x) \tag{29}$$

Recally that $\lambda = e^{-\Delta}$ and so $1-\lambda^2 = 2\Delta + O(\Delta^2)$, we expand the exponent, and function C in powers of Δ, and perform the Gaussian integrals over τ. On equating to zero the first power of Δ (the terms independent of Δ having cancelled) we get eventually,

$$\frac{\partial C}{\partial x} = \frac{\partial}{\partial\phi}(\phi C + \Phi^2\frac{\partial C}{\partial\phi}) + i\, qF(x,\phi)C \tag{30}$$

This is the key result of the paper. It enables the distribution of E_R (eqn 4) to be derived if $\phi(X)$ is the Ornstein-Uhlenbeck process, whatever form the function F (5) takes (provided of course (4) exists).

The boundary conditions for (30) are now discussed. If we evaluate (21) for the case of a "closed" screen with $x = -a$, only one dummy variable remains, and

$$C(q, -a) = \int_{-\infty}^{\infty} d\phi_{-M}(2\pi\phi^{-2})^{-\frac{1}{2}} \exp(-\phi_{-M}^2/2\phi^2) = 1 \tag{31}$$

This is correct, since when $x = -a$, $E_R = 0$ with probability 1, $P(E_R) = \delta(E_R)$, and $C(q) = 1$ (10). By comparison with (23), we see that the boundary condition for (30) is

$$C(\phi, -a) = (2\pi\Phi^2)^{-\frac{1}{2}} \exp(-\phi^2/2\Phi^2) \tag{32}$$

The subsidiary characteristic function C thus takes Gaussian form in this limit.

Having derived the equation with a finite screen, we can now allow a and $x \to \infty$ to deduce results for the infinite one.

The joint probability density for E_R and E_I is treated similarly. The full scheme is now quoted. The complex field amplitude is given by

$$E_R + iE_I = \int_{-\infty}^{\infty} [F(x,\phi(x)) + i\, G(x,\phi(x))\}dx \qquad (33)$$

with $F(x,\phi) + iG(x,\phi) = (2\pi z)^{-\frac{1}{2}} \exp[i\{x^2/2z + \phi(x) - \pi/4\}]$

The joint distribution of E_R, E_I and associated characteristic function are,

$$P(E_R,E_I) = \frac{1}{(2\pi)^2} \int dq_R dq_I\ C(q_R,q_I) \exp[-i(q_R E_R + q_I E_I)\} \qquad (34)$$

$C(q_R,q_I)$ is given by

$$C = \int_{-\infty}^{\infty} d\phi\ \mathcal{C}(\phi,\infty), \qquad (35)$$

where $\mathcal{C}(\phi,x)$ satisfies the Fokker-Planck equation,

$$\frac{\partial \mathcal{C}}{\partial x} = \frac{\partial}{\partial \phi}\left(\phi \mathcal{C} + \phi_o^2 \frac{\partial \mathcal{C}}{\partial \phi}\right) + i(q_R F(x,\phi) + q_I G(x,\phi))\mathcal{C}, \qquad (36)$$

with boundary condition

$$\mathcal{C}(\phi,-\infty) = (2\pi\phi_o^2)^{-\frac{1}{2}} \exp(-\phi^2/2\phi_o^2) \qquad (37)$$

ϕ_o is now used for the RMS value of $\phi(x)$ to conform to the usual notation. Note that the propagation distance z, as well as the parameters q_R, q_I appear parametrically in this formalism. The distance x across the screen plays the role of time in conventional stochastic process theory.

2.4 The Brownian fractal screen

The Ornstein-Uhlenbeck screen possesses no inner length scale; its autocorrelation function is not differentiable at zero spacing. Having as it does an outer scale ℓ, realisations of it are not fractal. A truly fractal screen is recovered if the limit $\ell \to \infty$ is taken suitably, and there then results the Brownian fractal screen (Berry 1979, section 2) with the fractal dimension 1.5. This case can be recovered from the present formalism. Not surprisingly, our fundamental equation becomes one of pure diffusion rather than of Fokker-Planck type. We proceed as follows. All dimensional factors are retained in the development. The mean square phase difference at separation x for our screen is

$$<(\phi(x+X) - \phi(x))>^2 = 2\phi_o^2(1-e^{-|X|/\ell}) \quad (38)$$

$$\simeq \frac{2\phi_o^2|X|}{\ell} \quad \text{for } |X|<<\ell \quad (39)$$

If we now allow ℓ and ϕ_o^2 to diverge in such a way that their ratio remains constant, we recover the Brownian fractal. Specifically, to conform with Berry's (1979) definition, introduce the scale L ("topothesy") via

$$\frac{2\phi_o^2}{\ell k^2} = L \quad (40)$$

where k is the original wave-number.

Turning to the Fokker-Planck equation (36), restore the dimensionality of x, z ($x \to x/\ell$, $z \to z/k\ell^2$), substitute for ℓ from (40), divide by $2\phi_o^2$ and allow $\phi_o \to\infty$. There results the equation of diffusion type,

$$\frac{1}{k^2L}\frac{\partial C}{\partial x} = \frac{1}{2}\frac{\partial^2 C}{\partial\phi^2} + \frac{1}{(2\pi k^3L^2z)^{\frac{1}{2}}}\{q_R\cos(\frac{kx^2}{z}+\phi-\pi/4)$$

$$+q_I\sin(\frac{kx^2}{z}+\phi-\pi/4)\}\,C \quad (41)$$

from which the "drift" term $\partial(\phi C)/\partial\phi$ has disappeared. Note however that it is apparently necessary still to retain ϕ_o in the boundary condition (37), taking the limit $\phi_o\to\infty$ only after this has been applied. It is hoped to consider these matters further in future work.

3. DISCUSSION

Equation (36) appears to be intractable analytically. With a good deal of effort it is possible to calculate low order moments of E_R, E_I and compare with known (and much more simply computed) results. In a companion paper in this volume, Roberts (1985) discusses these matters, as a preamble to a numerical attack on the equation. Such moments follow by expanding the solution of the equation in powers of q_R, q_I. The coefficients lead via (34) to these moments in the usual way.

Two alternative ways of expressing the equation are described, which are each of use for the expansion in q_R, q_I.

(a) Integral equation

Regarding the final term in (36) as a source for the conventional Fokker-Planck operator we can use Green's function methods to derive

$$\mathcal{C}(\phi,x) = \mathcal{C}(\phi,-\infty) + i\int_{-\infty}^{\infty} d\phi' \int_{-\infty}^{x} dy\, G(\phi,\phi',x-y)\,\times$$

$$\times[q_R F(y,\phi') + q_I G(y,\phi')]\,\mathcal{C}(\phi',y), \tag{42}$$

the boundary condition (37) now being built into the equation. G is the conventional Fokker-Planck Green's function,

$$G(\phi,\phi',x) = \{2\pi\phi_o^2(1-e^{-2|x|})\}^{-\frac{1}{2}} \exp\left\{-\frac{(\phi-\phi' e^{-|x|})^2}{2(1-e^{-2|x|})\phi_o^2}\right\} \tag{43}$$

The characteristic function C(35) follows from (42) in the form

$$C = 1 + i\int_{-\infty}^{\infty} d\phi' \int_{-\infty}^{\infty} dy\,(q_R F(y,\phi') + q_I G(y,\phi'))\,\mathcal{C}(\phi',y), \tag{44}$$

but is of use only perturbatively, as $\mathcal{C}$ appears still inside the integral.

(b) Differential difference equation

This follows, as Roberts (1985) has pointed out, from a Fourier transform with respect to ϕ, and assumes its form because the functions F and G are periodic in this variable. Let,

$$\mathcal{D}(\rho,x) = \int_{-\infty}^{\infty} d\phi\, e^{-i\rho\phi}\mathcal{C}(\phi,x), \tag{45}$$

and introduce the polar coordinates $\mathcal{Q}$ and α corresponding to q_R, q_I,

$$q_R + iq_I = \mathcal{Q}e^{i\alpha}, \tag{46}$$

Then (36) transforms to,

$$\frac{\partial \mathcal{D}}{\partial x} + \rho \frac{\partial \mathcal{D}}{\partial \rho} + \phi_o^2 \rho^2 \mathcal{D} = \frac{i\phi}{2\sqrt{2\pi z}} \left\{ \exp\left[i\left(\frac{x^2}{2z} - \frac{\pi}{4} - \alpha\right)\right] \mathcal{D}_- \right.$$

$$\left. + \exp\left[-i\left(\frac{x^2}{2z} - \frac{\pi}{4} - \alpha\right)\right] \mathcal{D}_+ \right. \qquad (47)$$

where

$$\mathcal{D}_\pm = \mathcal{D}(\rho \pm 1, x) \qquad (48)$$

The boundary condition (37) becomes,

$$\mathcal{D}(\rho, -\infty) = \exp(-\tfrac{1}{2}\phi_o^2 \rho^2), \qquad (49)$$

and the desired characteristic function C (35) is simply

$$C = \mathcal{D}(0, \infty), \qquad (50)$$

It is possible to reduce (47) to an <u>ordinary</u> differential difference equation if the characteristic variable ρe^{-x} is used in place of ρ. Roberts (1985) exploits this fact for his perturbation expansion of the solution. However, the variable x then appears in both arguments of the functions on the right hand side of (47). If ρe^{-x} replaces $\underline{\underline{x}}$ as variable, the resulting ordinary differential-difference equation does not suffer from this disadvantage. Nevertheless it does not appear to be of a type susceptible to standard methods of solution.

Finally we note that if (47) is formally integrated along the characteristics ρe^{-x} = constant, there results the "integral-difference" equation,

$$\mathcal{D}(\rho,x) = \mathcal{D}(\rho, -\infty) + \frac{iQ}{2\sqrt{2\pi z}} \int_{-\infty}^{x} dy \exp\left[-\tfrac{1}{2}\phi_o^2 \rho^2 (1-e^{-2(x-y)})\right] \times$$

$$\times \left[\exp\left[i\left(\frac{y2}{2z} - \frac{\pi}{4} - \alpha\right)\right] \mathcal{D}(\rho e^{-(x-y)} - 1, y) \right.$$

$$\left. + \exp\left[-i\left(\frac{y^2}{2z} - \frac{\pi}{4} - \alpha\right)\right] \mathcal{D}(\rho e^{-(x-y)} + 1, y) \right] \qquad (51)$$

This equation results also, as it must do, by Fourier transforming the integral equation (42). The final result corresponding to (44) is,

$$C = \mathcal{D}(o,\infty)$$

$$= 1 + \frac{iQ}{2\sqrt{2\pi z}} \int_{-\infty}^{\infty} dy[\exp[i(\frac{y^2}{2z} - \frac{\pi}{4} - \alpha)]\mathcal{D}(-1,y)$$

$$+ \exp[-i(\frac{y^2}{2z} - \frac{\pi}{4} - \alpha)]\mathcal{D}(+1,y)] \tag{52}$$

It should be possible to extend this formalism to the consideration of two point statistics (joint distributions of E at two values of x in (3)). This would enable spatial correlation functions and power spectra to be calculated. Correspondingly statistical properties at 2 different frequencies should be capable of consideration.

The method seems however to be tied inextricably to a one dimensional Ornstein-Uhlenbeck screen. It will not, unfortunately, extend to any other type of screen, or to the two dimensional case, in which corresponding phase ACF would have the form

$$\rho(x,y) = \exp[-(x^2+y^2)^{\frac{1}{2}}] \tag{53}$$

The results of a numerical attack on the equations are described by Robters (1985) in the accompanying paper.

Finally there remains the possibility of tackling the original path integral with the W.K.B.J. approximation (Schulman 1981, chapter 13). The full exponent in (12) with (13) is expanded to 2nd order in $[\phi - \phi_s]$ about its complex stationary points $[\phi_s]$, and the resulting Gaussian integrals performed. For the exponential screen, this will result, in the continuous limit $\Delta \to o$, in the solution of a certain ordinary differential equation. It is not yet clear what physical circumstance this approximation describes. There remains however the possibility of applying this method to more general screens than the exponential, when it is expected that a certain integral equation and its associate Fredholm determinant (Kac, 1959) will arise. It is hoped to consider these matters in a future publication.

ACKNOWLEDGEMENTS

The bulk of this work was performed at the Physics Department, University of Adelaide, South Australia, in 1969 when I was supported by a Queen Elizabeth II Fellowship. I am grateful to the Institute of Mathematics and its Applications for financing my attendance at this conference, and to Dr. B.J. Uscinski for inviting me to it.

REFERENCES

Berry, M.V., 1979, *J.Phys A:Math.Gen.* **12**, 781.

Booker, M.G., and Clemmow, A., 1950, *Proc. Inst. Rad. Engrs,* **97**, 11.

Buckley, R., 1975, *J.Atmos.Terr.Phys.* **37**, 1431.

Feynman, R.P., and Hibbs, A.R., 1965, Quantum Mechanics and Path Integrals, New York, McGraw Hill.

Kac, M., 1959, Probability and related topics in physical sciences, New York, Interscience.

Ratcliffe, J.A., 1956, *Rep. Prog. Phys.* **19**, 188.

Roberts, D.L., 1985, (This vol).

Schulman, L.S., 1981, Techniques and Applications of Path Integrals, John Wiley & Sons

Uhlenbeck, G.E., and Ornstein, L.S., 1930, *Phys. Rev.* **36**, 823.

Uscinski, B.J., and Macaskill, C., 1983, *J.Atmos.Terr.Phys.* **45**, 595.

Uscinski, B.J., 1985, (This vol).

THE PROBABILITY DISTRIBUTION OF INTENSITY FLUCTUATIONS DUE TO A RANDOM PHASE SCREEN WITH AN EXPONENTIAL AUTOCORRELATION FUNCTION

D.L. Roberts
(Department of Applied Mathematics and Theoretical Physics, University of Cambridge; Present address: Meteorological Office, Bracknell)

ABSTRACT

The statistics of the intensity fluctuations due to an exponential random phase screen are investigated via a new approach devised by Buckley (reported in this volume). His partial differential equation for the auxiliary characteristic function of the complex field is analysed using a series expansion. A method of obtaining the moments of intensity by solving Buckley's equation numerically is described, and some results are presented. Together with data from Monte Carlo simulation experiments, they indicate that the intensity distribution is at least a close approximation, and perhaps identical, to a Rice distribution.

1. INTRODUCTION

In a companion paper in this volume, Buckley has introduced a novel approach to the problem of finding the statistics of the wave field diffracted by a certain random phase screen. The present paper begins the exploration of some of the ramifications of Buckley's idea. In particular, it presents the results of a numerical investigation. Using these, a form for the probability distribution of intensity fluctuations is suggested.

The situation to be studied will now be outlined. Let (x,y,z) be the usual Cartesian coordinates. Consider a monochromatic plane wave, of wavelength $\lambda = 2\pi/k$, propagating in the z direction. Without much loss of generality we shall take its amplitude to be unity. At $z=0$ the wave encounters a thin plane layer, or screen, of infinite extent in the x and y directions. The screen randomly modulates the wave's phase but does not (directly) alter its amplitude. The phase change, ϕ, will be supposed independent of y; the problem is thus two-dimensional. The wave field just beyond the phase screen is then $\exp(i\phi(x))$. For $z>0$ the wave propagates once more in a uniform medium; nevertheless, fluctuations in the wave amplitude will develop, as explained, for example, in Uscinski (1977).

This phase screen model, and variants of it, have been used in many studies of random wave propagation in diverse physical contexts. Some of the other articles in this volume give an idea of its scope. The emphasis here, though, will be on theoretical development rather than practical applications.

The problem is to determine the statistical properties of the random wave field in $z > 0$, given some model for the stochastic process $\phi(x)$. In common with almost all previous investigations, we will suppose that $\phi(x)$ is a real, stationary, normal process. (See, e.g., Papoulis 1965.) It can therefore be completely described by specifying its r.m.s. value ϕ_o and autocorrelation function (a.c.f.) $\rho(x)$. This paper will be concerned with the case of an exponential a.c.f.:

$$\rho(x) = e^{-\frac{|x|}{r_o}} . \tag{1}$$

The screen thus defined will be referred to as the exponential screen.

The reason for confining the discussion to this particular case is that it appears that (1) may be the only autocorrelation function for which Buckley's approach is fruitful; see his article. This restriction is of course unfortunate and frustrating. However, the exponential screen is a worthwhile object of study in any case. Its main characteristics will now be discussed.

The parameter r_o which appears in (1) is a measure of the largest separation at which variations in ϕ are significantly correlated: the outer scale size. This can also be seen by referring to the spectrum of ϕ, which is the Fourier transform of $\phi_o^2\rho$:

$$\Gamma(f) = \frac{2r_o\phi_o^2}{1 + 4\pi^2 r_o^2 f^2} . \tag{2}$$

Observe also that at high spatial frequency f the spectrum (2) falls off rather slowly, as f^{-2}. This means that realisations of the screen contain structure on arbitrarily fine scales; unlike the case (for instance) of a Gaussian spectrum, where screen realisations are very smooth if viewed on a sufficiently small scale. One implication of this is that ray theory is completely inapplicable, for the scattered wavefront just beyond the screen is so rough - not even once differentiable - that it does not have well-defined normals. Thus there is no possibility of obtaining the strong intensity fluctuations, due to random focusing events associated with the convergence and intersection of rays, which are well known (Mercier 1962; see also the articles by Berry and by Hannay in this volume) in the case of a deep phase screen with a Gaussian spectrum.

The self-similar behaviour implied by the spectrum's power-law decay, combined with the roughness caused by the slowness of that decay, are characteristics of a fractal function. The only reason why $\phi(x)$ is not quite a true fractal is that it has a finite outer scale size: the self-similarity does not persist out to arbitrarily large scales. Berry (1979) has considered diffraction from fractal phase screens with unmodified power-law spectra and thus no outer scale. He paid special attention to the case of a f^{-2} spectrum - the Brownian fractal - which clearly is very closely related to the screen being examined here. Much information about the fascinating properties of fractals can be found in the definitive book by Mandelbrot (1977).

We return now to the random wave field produced by the phase screen. There is no need to repeat Buckley's analysis here. However, it is worth noting that his treatment (and hence the present paper) relies on making the paraxial approximation, i.e. assuming that nearly all the scattered wave energy travels in directions inclined at small angles to the z axis. This is customary in random phase screen theory, but it is not obvious that it can be justified for a screen so rough that ray theory is inapplicable. Berry (1979) discussed this question and derived a condition for the approximation to be valid in the case of a fractal screen. Following his approach, one can find an analogous condition in the present case. If $\phi_o^2 \gg 1$ ($\phi_o^2 = 10$ is sufficient) this turns out to be $kr_o \gg \phi_o^2$.

Having made the paraxial approximation, it is usual to use the wave component that travels parallel to the z axis as the phase reference. This just amounts to writing the actual complex field as $E(x,z)\exp(ikz)$ and working with the quantity E, which is often referred to as "the field". This will be understood from now on.

Among the quantities describing the field, the statistical properties of the intensity $I = |E|^2$ are of particular interest. Perhaps the most important of these are the one-point probability distribution of intensity and the transverse spectrum of intensity fluctuations; we shall concentrate on the former. Note that by conservation of energy, the unit amplitude of the incident wave and the assumption that ϕ is stationary we have $\langle I \rangle = 1$. (The angle brackets denote an average over the ensemble of realisations of $\phi(x)$.) The normalised variance of the intensity, often known as the scintillation index, is

$$\sigma_I^2 = \frac{\langle I^2 \rangle - \langle I \rangle^2}{\langle I \rangle^2} = \langle I^2 \rangle - 1. \tag{3}$$

Buckley's method offers a way of investigating the joint probability density function of the real and imaginary parts of the field, $p(E_R, E_I)$, and through this the intensity distribution. Let us now see how this relates to previous studies of the exponential screen and related types of screen.

Jakeman & McWhirter (1977) considered (inter alia) the screen with a truncated linear a.c.f.,

$$\rho(x) = \begin{cases} 1 - \dfrac{|x|}{r_o}, & |x| \leqslant r_o \\ 0, & |x| \geqslant r_o \end{cases} . \tag{4}$$

They were concerned with deep phase screens, defined as those with $\phi_o >> 1$. However, for the particular statements to be quoted here it is actually sufficient that $\phi_o^2 >> 1$. They explained that with a deep screen the behaviour of its a.c.f. near the origin is the dominant influence on both σ_I^2 and the intensity spectrum. Now (4) and (1) behave similarly in the neighbourhood of x=0. They therefore asserted that their approximate results for σ_I^2 and the intensity spectrum, although derived for the deep truncated linear screen, would also hold for the deep exponential screen.

Uscinski, Booker & Marians (1981) considered the exponential screen itself. Assuming $\phi_o^2 >> 1$, they found an approximation to the intensity spectrum which tallies with the expression given in Jakeman & McWhirter (1977) (to within a printing error in the latter). A more complete treatment was presented by Uscinski & Macaskill (1983), who confirmed that the error incurred by using Jakeman & McWhirter's approximation for the scintillation index was negligible. Macaskill & Uscinski (1983) obtained numerical results which agreed with this.

Booker & MajidiAhi (1981) studied a screen with an exponential a.c.f. modified by incorporating an inner scale r_I, so that screen realisations are smooth on scales much smaller than r_I. Again the main emphasis was on the intensity spectrum. However, the introduction of r_I significantly changes the properties of the scattered field - a point stressed by Macaskill & Uscinski (1983) - so the work of Booker & Majidi Ahi is not strictly relevant in the present context.

The paper of Berry (1979), which has already been mentioned, included results for the scintillation index and the intensity spectrum in the case of the Brownian fractal screen. These are compatible with the results of Jakeman & McWhirter (1977). In addition, Berry drew attention to the fact that the probability distribution of intensity was unknown. All that could be said was that the scattered field did not obey Gaussian statistics, except (possibly) far from the screen.

Thus, rather than superseding existing work on the exponential screen, Buckley's approach allows the investigation of previously intractable aspects of the problem: the functional forms of the distributions of the field and the intensity. It is hoped that the present paper will assist in this process.

The plan of the rest of the paper is as follows. In the next section, Buckley's equation for the auxiliary characteristic function of the field is discussed, and a series expansion technique is applied to it. The results provide some useful clues to the nature of the solution. Section 3 describes and explains a method of solving the equation, or at least extracting desirable information from it, by computational means. Section 4 presents results gained with this approach. They are compared with data obtained from numerical simulations carried out by the author. Finally, section 5 summarises the conclusions, points to some areas where improvements would be useful, and speculates on future developments.

2. THEORY

We begin by introducing more notation. The joint characteristic function of the real and imaginary parts of the field is (see, e.g. Papoulis 1965):

$$C(q_R,q_I) = \int_{-\infty}^{\infty}\int_{-\infty}^{\infty} p(E_R,E_I).\ \exp\{i(q_R E_R+q_I E_I)\}dE_R dE_I. \tag{5}$$

Note that C is not an analytic function of the complex variable $q = q_R + iq_I$. It will often prove convenient to discuss the dependence of C on Q and α, the modulus and argument of q. Thus (abusing notation slightly)

$$C(Q,\alpha) = \int_{-\infty}^{\infty}\int_{-\infty}^{\infty} p(E_R,E_I) \,.\, \exp\{iQ(E_R\cos\alpha + E_I\sin\alpha)\}\, dE_R dE_I . \qquad (6)$$

Buckley invented a function $\mathcal{C}(x,\phi;Q,\alpha)$ which we shall call the auxiliary characteristic function. The second argument, though closely related to the phase change due to the screen (as in section 1), is here a deterministic variable. The first argument corresponds to a length of screen rather than a coordinate of the point at which the field is observed: see Buckley's article for details.

One can think of $\mathcal{C}$ as depending on two real variables, x and ϕ, and a complex parameter q. The latter provides the link with C through the relation

$$C(Q,\alpha) = \int_{\infty}^{\infty} \mathcal{C}(\infty,\phi;Q,\alpha)\, d\phi . \qquad (7)$$

Buckley, in this volume, has shown that (under the circumstances described in section 1) $\mathcal{C}$ satisfies a certain partial differential equation of Fokker-Planck type, namely

$$\frac{\partial \mathcal{C}}{\partial x} = \phi_o^2 \frac{\partial^2 \mathcal{C}}{\partial \phi^2} + \phi \frac{\partial \mathcal{C}}{\partial \phi} + \mathcal{C} + \frac{iQ\mathcal{C}}{\sqrt{2\pi Z}} \cos\left(\frac{x^2}{2Z} + \phi - \frac{\pi}{4} - \alpha\right); \quad (x,\phi) \in \mathbb{R}^2 . \qquad (8)$$

The initial condition for this equation is (suppressing the q dependence)

$$\mathcal{C}(-\infty,\phi) = G(\phi) \equiv \frac{1}{\phi_o\sqrt{2\pi}} \exp\left(-\frac{\phi^2}{2\phi_o^2}\right) . \qquad (9)$$

In (8), the parameter Z is the distance beyond the phase screen scaled by the Fresnel length corresponding to the outer scale size,

$$Z = \frac{z}{kr_o^2} . \qquad (10)$$

Let $a = \sqrt{2Z}$. A simple manipulation ($x' = \frac{x}{a}$, followed by replacing x' by x) enables (8) to be rewritten as

$$\frac{\partial C}{\partial x} = a\phi_o^2 \frac{\partial^2 C}{\partial \phi^2} + a\frac{\partial(\phi C)}{\partial \phi} + \frac{iC}{\sqrt{\pi}} .Q\cos(x^2 + \phi - \frac{\pi}{4} - \alpha). \qquad (11)$$

The initial condition is still (9), and (7) is also unchanged.

Let us pause to contemplate this equation. It is of second order, linear, and homogeneous (though with an inhomogeneous initial condition). The evolution of the solution from $x = -\infty$ to $x = \infty$ may be described in qualitative terms as follows. Initially, the first two terms on the right hand side of (11) balance one another. When x is finite but large in magnitude and negative, the last term on the right hand side tries to drive the solution away from the initial condition; however the self-cancellation of this term, caused by the rapid oscillation of the cosine factor, means that change is gradual at first. Near x=0, though, the driving term can produce a substantial effect (this is "stationary phase" in operation), and it is in this region that the important changes in C occur. As $|x|$ increases again, self-cancellation of the driving term once more comes into play, and the solution "spirals in" towards a limit function $C(\infty,\phi)$.

Now $C(\infty,\phi)$ must satisfy the equation

$$\phi_o^2 \frac{d^2}{d\phi^2} (C(\infty,\phi)) + \frac{d}{d\phi} (\phi C(\infty,\phi)) = 0. \qquad (12)$$

Two linearly independent solutions of (12) are $G(\phi)$ and

$$H(\phi) \equiv G(\phi) \int_0^{\phi} \exp(\frac{\phi'^2}{2\phi_o^2}) \, d\phi'. \qquad (13)$$

But $\int_{-\infty}^{\infty} H(\phi)d\phi$ does not exist, for one can easily show that

$$H(\phi) \sim \frac{\phi_o}{\sqrt{2\pi}} . \frac{1}{\phi} \text{ as } \phi \to \infty. \qquad (14)$$

Therefore $H(\phi)$ should not occur, and $C(\infty,\phi)$ should be proportional to $G(\phi)$.

(A remark concerning the numerical solution of the Buckley equation, described in the next section, is in order here. It is conceivable that the numerical solution might approach a limit function which is some linear combination of $G(\phi)$ and

$H(\phi)$. This might arise because of the inevitable errors due to using a finite grid spacing Δ_ϕ and range $-L_\phi \lesssim \phi \lesssim L_\phi$. Should this happen, the fact that $H(\phi)$ is an odd function means that it will make no contribution to the estimation of the final integral, to within round-off error.)

Then $C(\infty,\phi;Q,\alpha) = G(\phi) . C(Q,\alpha)$. (15)

We now consider how we may extract information from (11). Ideally, one would solve the equation for all (Q,α); and hence find $p(E_R,E_I)$ by inverting the double Fourier transform. This would be an ambitious program. Fortunately, most of the interesting information in $C(Q,\alpha)$ is concentrated in one region of the q-plane: the neighbourhood of the origin. For, as is well known (see, e.g. Papoulis 1965), the derivatives of C at $q = 0$ specify the moments of the distribution. In particular, the moments of intensity are given by

$$\langle I^n \rangle = (-\nabla_q^2)^n C|_{q=0}, \tag{16}$$

where
$$\nabla_q^2 \equiv \frac{\partial^2}{\partial q_R^2} + \frac{\partial^2}{\partial q_I^2} . \tag{17}$$

Let us therefore examine the case $Q \ll 1$. The natural approach is to try a perturbation expansion in powers of Q. If ξ and η are the real and imaginary parts of C, then the structure of (11) is such that

$$\xi = \xi_o + \xi_1 Q^2 + \xi_2 Q^4 + \ldots , \tag{18}$$

$$\eta = \eta_1 Q + \eta_2 Q^3 + \ldots . \tag{19}$$

Here $\xi_j = \xi_j(x,\phi;\alpha)$, $\eta_j = \eta_j(x,\phi;\alpha)$.

At zeroth order we have $\xi_o = G(\phi)$. (20)

At first order, writing $\beta \equiv x^2 - \frac{\pi}{4} - \alpha$, we have

$$\frac{\partial \eta_1}{\partial x} = a\phi_o^2 \frac{\partial^2 \eta_1}{\partial \phi^2} + a\frac{\partial(\phi\eta_1)}{\partial \phi} + \pi^{-\frac{1}{2}}\cos(\beta+\phi)G(\phi), \tag{21}$$

$$\text{with } \eta_1(-\infty,\phi;\alpha) = 0. \tag{22}$$

One can now proceed by integrating (21) over all ϕ, assuming (with reasonable confidence) that the Gaussian factor which seems to be incorporated in C will ensure that all the integrals converge, and that the contributions from the first two terms on the right hand side will vanish. This produces

$$\frac{d}{dx}\int_{-\infty}^{\infty} \eta_1(x,\phi)\, d\phi = \pi^{-\frac{1}{2}}\exp(-\tfrac{1}{2}\phi_o^2)\cdot\cos\beta, \tag{23}$$

$$\int_{-\infty}^{\infty} \eta_1(x,\phi)d\phi = \frac{1}{2}\exp\left(-\frac{\phi_o^2}{2}\right).\{(1 + C_1(x) + S_1(x))\cos\alpha + (S_1(x) - C_1(x))\sin\alpha\}. \tag{24}$$

The functions C_1 and S_1 are the Fresnel integrals defined by

$$C_1(x) \equiv \sqrt{\frac{2}{\pi}}\int_o^x \cos(t^2)\, dt\ , \quad S_1(x) \equiv \sqrt{\frac{2}{\pi}}\int_o^x \sin(t^2)\, dt. \tag{25}$$

(See Abramowitz & Stegun 1965, §7.3, P.300.)

$$\text{Hence} \int_{-\infty}^{\infty} \eta_1(\infty,\phi)\, d\phi = \exp\left(-\frac{1}{2}\phi_o^2\right)\cos\alpha. \tag{26}$$

This approach would be fine if we were interested only in η_1, but there is no way to continue the expansion. So we must backtrack and solve (21) completely. This can be done by Fourier transforming with respect to ϕ and applying the method of characteristics. Details are given in the Appendix, where it is shown that, writing $\beta_s \equiv s^2 - \frac{\pi}{4} - \alpha$,

$$\eta_1(x,\phi) = \pi^{-\frac{1}{2}}e^{-\frac{1}{2}\phi_o^2}\, G(\phi)\int_{s=-\infty}^{x} \exp\{\tfrac{\phi_o^2}{2}e^{2a(s-x)}\}\cos\{\beta_s + \phi e^{a(s-x)}\}ds. \tag{27}$$

This is consistent with (24). Moreover, letting $x \to \infty$ we obtain

$$\eta_1(\infty,\phi) = \exp\left(-\frac{1}{2}\phi_o^2\right) G(\phi)\cos\alpha, \tag{28}$$

as expected.

Proceeding to the next order, we find that ξ_1 satisfies

$$\frac{\partial\xi_1}{\partial x} = a\phi_o^2\frac{\partial^2\xi_1}{\partial\phi^2} + a\frac{\partial(\phi\xi_1)}{\partial\phi} - \pi^{-\frac{1}{2}}\cos(\beta+\phi)\eta_1, \tag{29}$$

$$\text{with } \xi_1(-\infty,\phi;\alpha) = 0. \tag{30}$$

This equation can be approached in the same ways as the η_1 equation. The solution is given in the Appendix. The analogue of (24) is (writing $\beta_j \equiv s_j^2 - \frac{\pi}{4} - \alpha$)

$$\int_{-\infty}^{\infty} \xi_1(x,\phi)\, d\phi = -\frac{1}{2\pi}\exp(-\phi_o^2).\{X_1 + X_2\}, \tag{31}$$

where

$$X_1(x;\alpha) = \int_{s_2=-\infty}^{x}\int_{s_1=-\infty}^{s_2} \exp[-\phi_o^2 e^{a(s_1-s_2)}]\cos(\beta_1+\beta_2)\, ds_1 ds_2, \tag{32}$$

$$X_2(x) = \int_{s_2=-\infty}^{x}\int_{s_1=-\infty}^{s_2} \exp[\phi_o^2 e^{a(s_1-s_2)}]\cos(\beta_1-\beta_2)\, ds_1 ds_2. \tag{33}$$

Evidently, ϕ_o does not need to be much larger than 1 for X_2 to be the dominating contribution. In fact,

$$\frac{1}{2\pi}\exp(-\phi_o^2)\; X_1(\infty;\alpha) = \frac{\exp(-\phi_o^2)}{2\sqrt{2\pi}}\int_0^{\infty}\exp[-\phi_o^2 e^{-a\sigma}]\cos(\frac{\sigma^2}{2} - 2\alpha - \frac{\pi}{4})\, d\sigma \tag{34}$$

whereas

$$\frac{1}{2\pi}\exp(-\phi_o^2)X_2(\infty) = \frac{1}{4}. \tag{35}$$

Thus as $Q \to 0$,

$$C(Q,\alpha) \sim 1 + i\exp(-\tfrac{1}{2}\phi_o^2)Q\cos\alpha$$
$$- Q^2\{\frac{1}{4} + \frac{e^{-\phi_o^2}}{2\sqrt{2\pi}}\int_0^{\infty}\exp(-\phi_o^2 e^{-a\sigma})\cos(\frac{\sigma^2}{2} - 2\alpha - \frac{\pi}{4})\, d\sigma\} \tag{36}$$

This result implies that $\langle E_R\rangle = \exp(-\frac{1}{2}\phi_o^2)$, $\langle E_I\rangle = 0$, and $\langle I\rangle = 1$. (Note that it is the X_2 term alone which contributes to $\langle I\rangle$.) These relations are well known and can be obtained with far less effort by other means, for instance by the method of moment equations (see, e.g., Uscinski 1977). However, the expansion process is not a waste of time, because it provides information about the behaviour of $\mathcal{C}$ and C which may be useful when solving (8) numerically. Two examples of this will now be given.

The first is the observation that the coefficients of odd and even powers of Q in the expansion of C depend on trig. functions of odd and even multiples of α, respectively. (Thus ξ_2 may only depend on 1, cos 2α, sin 2α, cos 4α and sin 4α.) This can be checked easily by considering the structure of (21), (29) and the higher order equations, and using induction. There is a contrast here with the behaviour of any function analytic in a neighbourhood of q=0, for which the coefficient of Q^n may depend only on $\cos(n\alpha)$ and $\sin(n\alpha)$.

Secondly, consider (31) when $\phi_o^2 >> 1$ and $a\phi_o^2$ is not small compared with 1. We have already noted that X_1 is exponentially small. Substituting $\sigma = s_2 - s_1$, $S = \frac{1}{2}(s_1 + a_2)$,

$$X_2(x) = \int_{S=-\infty}^{x} \int_{\sigma=0}^{\infty} \exp(\phi_o^2 e^{-a\sigma}).\cos(2\sigma S)\, d\sigma dS. \tag{37}$$

The dominant contribution comes from the neighbourhood of σ=0. Hence

$$\int_{-\infty}^{\infty} \xi_1(x,\phi)d\phi \sim -\frac{1}{2\pi}\int_{S=-\infty}^{x} \mathrm{Re}\Big\{\int_{\sigma=0}^{\infty} \exp[(-a\phi_o^2+2iS)\sigma]\, d\sigma\Big\}\, dS \tag{38}$$

$$\sim -\frac{1}{4\pi}\Big\{\frac{\pi}{2} + \tan^{-1}\Big(\frac{2x}{a\phi_o^2}\Big)\Big\}. \tag{39}$$

This has the correct limits as $x \to +\infty$, $x \to -\infty$. It gives some quantitative feeling for how $\mathcal{C}$ evolves, and incidentally supports the qualitative description put forward earlier. The practical use of (39) will be illustrated in the next section.

If one tries to extend the expansion to find $\langle I^2 \rangle$, the algebra becomes quite lengthy. We therefore turn to consider a computational approach.

3. NUMERICAL METHODS

This section describes how the first few moments of intensity were estimated by numerically solving a variant of equation (11). The process is quite involved, and it will soon become apparent that there are several sources of numerical error. Moreover, some of these errors are not easily quantified. Under these circumstances, the best way to assess the accuracy of the results is to take advantage of the fact that $\langle I\rangle$ and $\langle I^2\rangle$ are known from theory. Thus the general approach adopted during development was to improve the method and refine the finite difference grid until good agreement with theory was obtained. One hopes that this means that the estimates of the higher moments are also not too far away from the true values, although the relative errors are likely to be larger.

The procedure used can be divided into two main parts. The first consists of estimating C_R, the real part of C, at several points of the q-plane. (For present purposes C_I is unimportant, since it is C_R alone that determines the moments of intensity, as can be seen from (16), (18) and (19).) The second consists of combining the estimates of C_R to form estimates of the intensity moments.

In section 2 it emerged that $\mathcal{C}(x,\phi)$ seems to contain a factor $G(\phi)$. See, for instance, equations (9), (15), (27) and (A.7). It is thus natural to consider the function F defined by

$$\mathcal{C}(x,\phi;Q,\alpha) = G(\phi)F(x,\phi;Q,\alpha). \tag{40}$$

From (11), F must satisfy

$$\frac{\partial F}{\partial x} = a\phi_o^2\frac{\partial^2 F}{\partial\phi^2} - a\phi\frac{\partial F}{\partial\phi} + \frac{iF}{\sqrt{\pi}}Q\cos(x^2+\phi-\frac{\pi}{4}-\alpha), \tag{41}$$

$$F(-\infty,\phi) = 1. \tag{42}$$

In numerical work it is better to proceed via (41) than to solve (11) directly, because with (41) the finite difference scheme does not have to cope with the variation due to $G(\phi)$. Evidence for this is provided by the case $Q=0$, where F correctly remains identically equal to 1 as x increases, whereas $\mathcal{C}$ deviates from $G(\phi)$.

To find an approximation to F, equation (41) was solved on the rectangular domain $x \in [-L_x, L_x]$, $\phi \in [-L_\phi, L_\phi]$, by "marching" forward from $x = -L_x$. The finite difference lattice consisted of lines of constant x and ϕ at intervals Δ_x and Δ_ϕ apart. Though Δ_ϕ was uniform, it is not sensible to use a uniform Δ_x because of the x^2 dependence of the forcing term. Instead, it was decided to take

$$\Delta_x = \begin{cases} k_1 & : \quad |x| \leq x_o \\ k_1 \frac{x_o}{|x|} & : \quad |x| \geq x_o, \end{cases} \tag{43}$$

where k_1 and x_o are constants. (Thus the last row of grid points will not be at exactly $x=L_x$. For brevity, this detail will be glossed over in the following account.) Although the effect on the solution of choosing various values for k_1, x_o and Δ_ϕ was closely monitored, it is unnecessary to rely totally on trial and error. For instance, (39) (which is valid for the parameter ranges investigated) indicates that it is necessary to take $k_1 << \frac{1}{2} a\phi_o^2$, and in practice this proved to be a useful criterion. Also, one must choose $\Delta_\phi << \min.(\phi_o, \pi)$ in order to resolve the cosine forcing and accurately estimate the final integral over ϕ (see later).

To help describe the "marching" process, we shall let

$$W(x,\phi) = \frac{Q}{\sqrt{\pi}} \cos(x^2 + \phi - \frac{\pi}{4} - \alpha), \tag{44}$$

and call $F(x_k, \phi_j) = F_{k,j}$ the numerical approximation to F. (Here k is not the wavenumber of section 1.) Write

$$\overline{x}_k = x_k + \frac{1}{2} \Delta_x(x_k). \tag{45}$$

To advance from row k to row k+1, the following 3-stage algorithm was used.

(i) Replace $F_{k,j}$ by $F_{k,j} \exp\{\frac{1}{2} i\Delta_x W(\overline{x}_k, \phi_j)\}$ for all j.

(ii) Writing the result of (i) as $F_{k,j}$, put

$$\frac{F_{k+1,j} - F_{k,j}}{\Delta_x} = \frac{a\phi_o^2}{2\Delta_\phi^2} \{F_{k+1,j+1} - 2F_{k+1,j} + F_{k+1,j-1} + F_{k,j+1} - 2F_{k,j} + F_{k,j-1}\}$$
$$- \frac{a\phi_j}{4\Delta_\phi} \{F_{k,j+1} - F_{k,j-1} + F_{k+1,j+1} - F_{k+1,j-1}\}. \tag{46}$$

(iii) Replace $F_{k+1,j}$ by $F_{k+1,j} \exp\{\frac{1}{2} i\Delta_x W(\overline{x}_k, \phi_j)\}$ for all j.

Stages (i) and (iii) handle the change due to the forcing term, stage (ii) the terms involving derivatives w.r.t. ϕ. This arrangement is a compromise between accuracy and computational efficiency. Stage (ii) is a simple extension of the Crank-Nicolson scheme for the diffusion equation (see, e.g. Richtmyer & Morton 1967). The overall truncation error is $O(\Delta_x) + O(\Delta_\phi^2)$.

We now come to the problems with boundary conditions caused by solving on a finite domain instead of the whole (x,ϕ) plane. These were dealt with in a rather ad hoc manner. First, at the edges $\phi = \pm L_\phi$ the condition $\frac{\partial F}{\partial \phi} = 0$ was specified. The hope here is that if one chooses L_ϕ large enough - in particular $L_\phi >> \phi_o$ - the resulting errors will not be important.

The situation at $x=L_x$ is more complicated. Although (15) shows that $F(x,\phi;Q,\alpha) \to C(Q,\alpha)$ as $x \to \infty$ irrespective of the value of ϕ, $F(L_x,\phi)$ will not be independent of ϕ. So a reasonable approach is to use the numbers $F(L_x,\phi_j)$ to compute an approximation to

$$\int_{-\infty}^{\infty} G(\phi)F(L_x,\phi)\, d\phi,$$

and use this as an estimate of $C(Q,\alpha)$. Unfortunately this produced poor results, partly because of the slow convergence of F as $x\to\infty$ (which can be inferred from (39)), partly because (43) makes it hard to choose a very large L_x. The course therefore adopted was to estimate

$$\int_{-\infty}^{\infty} G(\phi)F_R(x,\phi)\, d\phi \quad \text{(where } F_R \text{ is the real part of } F\text{)},$$

at 3 values of x such as (L_x-2), (L_x-1) and L_x; and thence predict

$$\int_{-\infty}^{\infty} G(\phi)F_R(\infty,\phi)\, d\phi = C_R(Q,\alpha) \tag{47}$$

by a simple extrapolation process. Consistency was checked by extrapolating from data at more than one triad of x-values. (Notice that it would be more difficult to extrapolate the imaginary part, as (24) shows.)

At $x = -L_x$ tha natural initial condition F=1 was used. However, it was found that the numerical solution behaved as though starting from an initial condition $F_R = 1+\varepsilon(Q,\alpha)$ at $x=-\infty$. This was compensated for by estimating ε using backwards extrapolation, and dividing the existing estimate of C_R by $(1+\varepsilon)$. This works because (41) is linear. Incidentally, one must not extrapolate back from too near the start line $x=-L_x$ because of the initial adjustment of the solution.

The second part of the method - finding estimates of the intensity moments from values of C_R - will now be described. From the work in section 2 it follows that

$$C_R \sim 1 + Q^2\psi_1(\alpha) + Q^4\psi_2(\alpha) + Q^6\psi_3(\alpha) + Q^8\psi_4(\alpha) + \ldots \tag{48}$$

as $Q \to 0$, where

$$\psi_1(\alpha) = C_{1,0} + C_{1,1}\cos 2\alpha + C_{1,2}\sin 2\alpha, \tag{49}$$

$$\psi_2(\alpha) = C_{2,0} + C_{2,1}\cos 2\alpha + C_{2,2}\sin 2\alpha + C_{2,3}\cos 4\alpha + C_{2,4}\sin 4\alpha, \tag{50}$$

etc., the $C_{j,k}$ being constants. By (16), $\langle I^n \rangle$ is determined by $C_{n,0}$ alone.

Now the results presented in the next section that were obtained with this method are for the case $\phi_0=10$, when an important simplification occurs. Because ϕ_0 is then reasonably large compared with π, $p(E_R,E_I)$ should be virtually independent of θ, the phase of E. But it is easily shown from (6) that if p is independent of θ, C is independent of α. Experiments confirmed that C_R is indeed effectively independent of α when $\phi_0=10$, and (36) is also consistent with this. So we may take

$$C_R \sim 1 + \psi_1 Q^2 + \psi_2 Q^4 + \psi_3 Q^6 + \psi_4 Q^8 + \ldots, \tag{51}$$

where the ψ_n are now constants.

Digressing a little, we can show that (51) is actually a convergent series for all Q. For, by (16),

$$\langle I^n \rangle = (-\nabla_q^2)^n \psi_n Q^{2n} = (-1)^n \psi_n (2^n n!)^2. \tag{52}$$

But there are good reasons for believing that, in the case of the exponential screen, $\langle I^n \rangle \lesssim n!$, as will become clear in section 4. Thus

$$|\psi_n| \lesssim \frac{1}{4^n n!}, \tag{53}$$

and (51) converges by comparison with the exponential series. In the limiting case $\langle I^n \rangle = n!$, attained when $\phi_o \to \infty$, we have $C_R = \exp(-\frac{1}{4}Q^2)$.

We return to the main argument. The remaining idea is very simple: estimates of C_R at several values of Q are used to eliminate terms in (51) systematically, producing tableaux of approximations to the coefficients ψ_n. The process is similar to Romberg integration (see, e.g., Fox & Mayers 1968), but with the difference that several tableaux (one for each ψ_n to be estimated) are generated, instead of one. The best estimate of ψ_n then yields an approximation to the corresponding intensity moment, through (52). Note that because $|\psi_n|$ decreases rapidly as n increases (see (53)), one can readily ensure that the terms neglected in (51) are small compared with those retained.

In principle, it is easy to extend this procedure to the general case, where C_R depends on α. By a judicious selection of values of α, using (49), (50) etc., one can minimise the number of times (41) must be solved to estimate the moments up to a given order. Even so, the computer time required would be significantly greater than when ϕ_o is large.

4. RESULTS

This section presents some results obtained using the method described in section 3, for the case ϕ_o=10. For comparison, two other sets of results, for $\phi_o = 1$ and $\sqrt{10}$, will also be given. These were obtained by Monte Carlo simulation, so a few remarks concerning that method are now in order.

The idea of investigating the random phase screen problem by a direct computer simulation is very natural. One creates a number of realisations of the screen using pseudo-random number generation, computes the resulting field in each case, and estimates the statistical properties of interest by averaging over the set of realisations. Several accounts of such experiments have been published, for example by Whale (1973,1974), Buckley (1975), Fujii et al (1976), Fujii & Asakura (1977), Uozumi et al (1977), Fujii (1979) and Ohtsubo (1982). The problems studied differ in varying ways from that considered here, but share a family resemblance. In view of this, and also because the main emphasis of this article is on the Buckley equation approach, a detailed description of the simulation method used in the current investigation will not be given here. It is reasonably close to that of Buckley (1975), with some refinements.

The application of simulation techniques to the exponential screen appears to be new, and is not entirely straightforward because of the screen's quasi-fractal nature, as will now be explained. Clearly, the computer can only represent a finite band of spatial frequencies in the x-direction. For a realistic simulation of the intensity, this band must span all frequencies which significantly contribute to the intensity spectrum. The graphs of intensity spectra given by Macaskill & Uscinski (1983) show that this is a demanding requirement in the case of the exponential screen (compare their results for screens with Gaussian and modified exponential a.c.fs.), mainly because of the slow decline (as f^{-2}) of the intensity spectrum at high spatial frequency f. For $\phi_o=10$ one would need to represent a frequency band over 4 decades wide, and the position worsens as ϕ_o is increased. This is why the only simulation results presented here are for moderate values of ϕ_o. There is another aspect to the same difficulty. The inevitable introduction of a numerical inner (spatial) scale, even though it is arranged to be small compared with r_o, creates a certain amount of doubt about the qualitative correctness of the simulation. Recall that in section 1 we noted that Macaskill & Uscinski (1983) emphasized that the presence of an inner scale in the screen could significantly affect the properties of the scattered field. However, we shall soon see that the numerical resolution employed here ($\Delta_x=r_o/64$ with $\phi_o=1$, $\Delta_x=r_o/512$ with $\phi_o=\sqrt{10}$) seems to have been sufficient to yield reasonably good results.

Another new feature of the present simulation work is that, whereas the previous studies cited did not include results for intensity moments higher than the second, here the chief aim was to estimate the third and fourth moments. To achieve this within reasonable error bounds required larger samples than former workers have needed. For the record, the results to be shown were obtained from 100 screen realisations with 8,192 (ϕ_o=1) and 16,384 ($\phi_o = \sqrt{10}$) points in the x-direction.

We now come to the results themselves, beginning with the mean intensity. As noted in section 1, this is known to be unity at all ranges. The simulation algorithm is such that this result is obtained almost exactly, with a very small error due to round-off. The method of section 3 does not produce such an exact result; remember that the accuracy with which $\langle I \rangle$ and $\langle I^2 \rangle$ can be estimated is our best means of gauging the errors involved. With the final version of the method the worst case obtained was $\langle I \rangle = 0.9981$, which leaves room for improvement but is acceptable.

Turning now to the scintillation index, the three sets of results are shown together in Figure 1, plotted against the scaled range

$$\mu = 4\phi_o^4 Z. \tag{54}$$

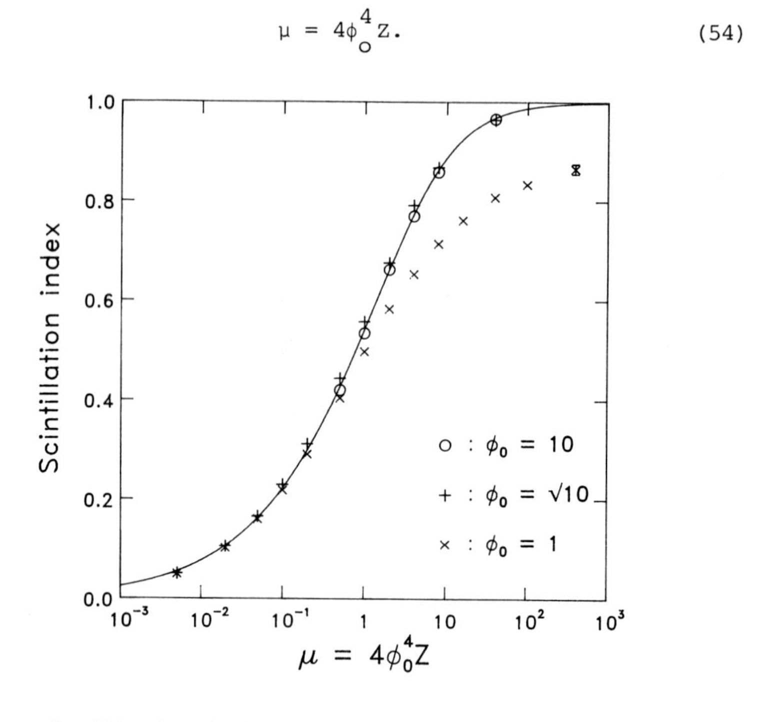

Fig. 1. Scintillation index versus scaled range. Circles: results obtained by numerical solution of Buckley's equation. Crosses (both types): results from Monte Carlo simulations. Note error bar at top right. Curve: Jakeman & McWhirter's analytical approximation.

The solid line is the analytical approximation for $\phi_o^2 >> 1$, due to Jakeman & McWhirter (1977) (see section 1). The formula for this is

$$\sigma_I^2 = 1 - 2\{[\tfrac{1}{2} - C_1(\tfrac{1}{2}\sqrt{\mu})]^2 + [\tfrac{1}{2} - S_1(\tfrac{1}{2}\sqrt{\mu})]^2\}, \tag{55}$$

where C_1, S_1 were defined in (25). Standard errors of the simulation estimates were obtained, and the largest of these is shown as an error bar on the corresponding result, at $\mu=400$. The standard errors gradually shrink to about one-tenth of this size as μ is decreased to 0.1.

We now comment on the three cases in turn. The $\phi_o=1$ results appear to be new, and are consistent with the general result (Jakeman & McWhirter (1977); derived for a Gaussian a.c.f. by Mercier (1962))

$$\sigma_I^2 \to 1 - \exp(-2\phi_o^2) \text{ as } Z \to \infty. \tag{56}$$

The $\phi_o = \sqrt{10}$ results seem a little high, even allowing for the standard errors; one would expect these points to lie between the two other sets. It is possible that the numerical inner scale is responsible for this. Still, the discrepancy between these results and theory is not great.

The $\phi_o=10$ results, produced by the method of section 3, show good agreement with theory: the worst deviation from (55) is by 0.0086 (at $\mu=1$). This encourages us to believe that the errors in the estimates of the higher moments will be tolerable. Results for smaller values of μ are not shown because of uncertainty about the consistency of the extrapolation process at these values.

We turn next to the third and fourth moments of intensity, which are displayed in Figure 2, plotted against $\langle I^2 \rangle$. Before the implications of these results are discussed, it is appropriate to mention that the standard errors of the simulation estimates are fairly small. The worst case is $\langle I^4 \rangle$ for $\phi_o=1$, for which the standard errors rise from about 0.001 at small $\langle I^2 \rangle$ through about 0.1 at around $\langle I^2 \rangle = 1.6$ to about 0.5 at around $\langle I^2 \rangle = 1.85$. Of course, the uncertainties in the estimates of $\langle I^2 \rangle$ are the same as those described for Figure 1.

The most striking feature of Figure 2 is that for both the third and fourth moments the three sets of results lie on virtually identical curves. This is doubly significant: for not only are the results of two independent methods in mutual agreement, they were also obtained using three different values of ϕ_o. The latter point strongly suggests that the probability distribution of intensity depends on only two parameters, since specifying its mean and second moment appears to fix the higher moments.

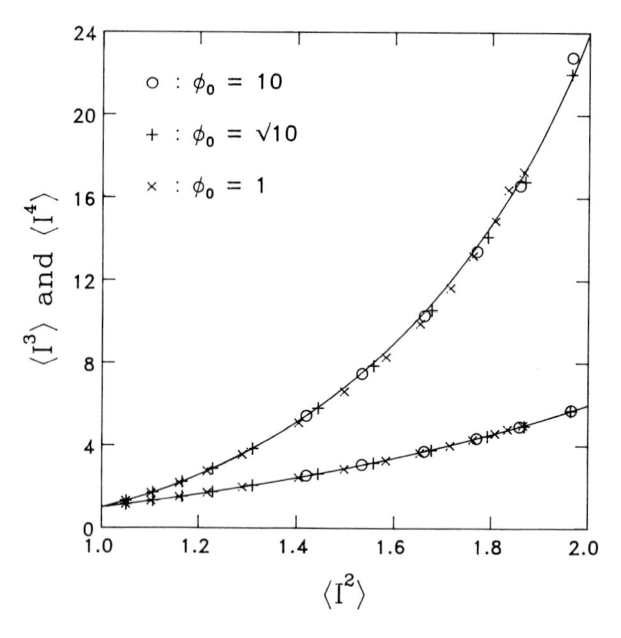

Fig. 2. Third and fourth intensity moments versus the second moment. Symbol code as in Figure 1. Curves: third and fourth moments of the Rice distribution with mean unity.

To appreciate that this property is noteworthy, we recall that in general at least three parameters are necessary to describe the intensity distribution. For instance, a diagram analogous to Figure 2 for the screen with a Gaussian a.c.f. shows a separate curve for each value of ϕ_o. Of course, it is possible that the same phenomenon might be seen in the present case if highly accurate methods or results over a wider range of values of ϕ_o were available; that would imply a relatively weak dependence on a third parameter. However, that would still be an interesting property.

Figure 2 thus poses a question: does any well-known 2-parameter probability distribution have moment curves resembling the data? It must be strongly emphasized that "curve-fitting" in the usual sense is not involved here; there are no free parameters for one to adjust. In fact, a given 2-parameter distribution leads to a unique curve on the diagram for each moment higher than the second, since $\langle I \rangle$ is fixed at unity. Thus one can make an unequivocal comparison between a suggested distribution and the data.

Various distributions have been tested. Much the closest agreement with the data was obtained with the moment curves shown in Figure 2. These belong to the distribution derived by Rice (1944) for the intensity that results when Gaussian noise is superimposed on a sinusoidal signal. We shall refer to this as the Rice distribution, although in the literature this terminology is quite often used for the corresponding distribution of the amplitude (the square root of intensity). The probability density function of the Rice distribution, $P_R(I)$, is given (for general $\langle I \rangle$) by

$$\langle I \rangle P_R(I) = \frac{1}{1-y} \exp\left(-\frac{\frac{I}{\langle I \rangle} + y}{1-y}\right) I_o\left(\frac{2}{1-y}\sqrt{y\frac{I}{\langle I \rangle}}\right). \tag{57}$$

Here I_o is the modified Bessel function of the first kind of order zero, and the parameter y is related to the second moment by

$$y = \sqrt{2 - \frac{\langle I^2 \rangle}{\langle I \rangle^2}} = \sqrt{1 - \sigma_I^2}. \tag{58}$$

The normalised moments of (57) are

$$\frac{\langle I^n \rangle}{\langle I \rangle^n} = n! \ (1-y)^n L_n\left(-\frac{y}{1-y}\right) \tag{59}$$

$$= \sum_{m=0}^{n} \frac{(n!)^2}{(m!)^2 (n-m)!} y^m (1-y)^{n-m}. \tag{60}$$

In (59), L_n denotes the Laguerre polynomial of degree n.

So Figure 2 strongly suggests that the intensity has a Rice distribution (the extension from $\langle I \rangle = 1$ to arbitrary values of $\langle I \rangle$ is immediate). Now this distribution already occupies a prominent position in random phase screen theory, thanks to the classic paper by Mercier (1962). Amongst other things, he proved that at great distances (i.e. in the limit as $z \to \infty$) from a screen with a Gaussian (or similar) a.c.f., the intensity has a Rice distribution. Although Mercier's proof is inapplicable to the exponential screen, Jakeman & McWhirter (1977) have since presented an illuminating discussion which shows that the result is actually true for an extremely large class of screens which includes the present case. In contrast, the proposal here is that in the particular case of the exponential screen the intensity distribution may be Rice at all ranges. However, the two propositions are closely connected, as the following reasoning makes clear.

Let us assume that the distribution of intensity due to an exponential screen does indeed depend only on the parameters $\langle I \rangle$ and $\langle I^2 \rangle$. Pick any (positive) values for ϕ_o, Z and $\langle I \rangle$, and consider the resulting value of $\langle I^2 \rangle$. Observe that the scintillation index due to an exponential screen never exceeds unity. Thus, by (56), there exists ϕ_1 such that if the screen had r.m.s. phase deviation ϕ_1 instead of ϕ_o, the value of $\langle I^2 \rangle$ currently attained at Z would be the limiting value at infinite range. So the intensity distribution for (ϕ_o, Z) is the same as that for (ϕ_1, ∞). But the latter is a Rice distribution, by Mercier's theorem. Hence the distribution must be Rice at all ranges.

On the other hand, if the intensity distribution depends weakly on a third parameter, then the same argument suggests that it should be approximated well by a Rice distribution. Therefore the results shown in Figure 2, interpreted either way, are consistent with Mercier's theorem.

Finally, it is worth mentioning a well-known special case of the Rice distribution. When $\sigma_I^2 = 1$, (57) reduces to the exponential distribution, which has normalised moments

$$\frac{\langle I^n \rangle}{\langle I \rangle^n} = n! \tag{61}$$

Thus it corresponds to the right-hand ends of the curves in Figure 2.

5. CONCLUSIONS

The principal conclusion reached in this paper is that the intensity distribution due to an exponential random phase screen is, either exactly or to a good approximation, a Rice distribution; for any values of ϕ_o and Z. This has not yet been proved analytically, but strong numerical evidence for it has been presented. The result appears to be original, although it was already known in the particular case of infinite Z, from Mercier's theorem on the limiting distribution far from a stationary screen (Mercier 1962, Jakeman & McWhirter 1977).

Another conclusion is that the approach developed by Buckley (this volume) has considerable practical value. Without it, any deductions made from the results of the Monte Carlo simulations alone would have rested on less solid ground, because of a residual uncertainty about the adequacy of the simulation technique in the case of the exponential screen, discussed in the previous section. On the other hand, the numerical method explained in section 3 is also not entirely immune from criticism. Thus the close agreement between the results of these completely independent methods, visible in Figure 2, is a vital part of the evidence.

Clearly, there are several areas where further work is desirable. There is much scope for analytical investigation of the Buckley equation. The numerical work described here could certainly be improved upon, by refining the methods and extending the range of values of ϕ_o covered. Perhaps most useful would be a clearer physical explanation of why the Rice distribution arises in this context.

We end with two conjectures. First, it seems probable that the higher-order statistical properties of the intensity (e.g. the joint distribution of the intensities $I(x_1,Z)$ and $I(x_2,Z)$) are those appropriate to the original Rice (1944) problem of Gaussian noise superimposed on a sinusoidal signal. This idea could be tested with more Monte Carlo simulations. Secondly, there must be a good chance that the intensity distribution due to a Brownian fractal screen is also Rice, in view of the close connection between that situation and the problem considered here.

6. ACKNOWLEDGEMENTS

I am very grateful to my research supervisor, Dr. B.J. Uscinski, for suggesting this problem and for his continuing advice and encouragement. It is also a pleasure to thank Dr. R. Buckley, for communicating his results prior to publication; and Dr. C. Macaskill, for many stimulating discussions during the course of this work. I am grateful for the support provided by a Science and Engineering Research Council studentship and by the Ministry of Defence (Procurement Executive).

7. REFERENCES

Abramowitz, M. and Stegun, I.A. (eds) 1965, "Handbook of Mathematical Functions", Dover, New York.

Berry, M.V., 1979, "Diffractals", *J.Phys. A: Math.Gen.* **12**, 781-797.

Booker, H.G. and MajidiAhi, G., 1981, "Theory of refractive scattering in scintillation phenomena", *J.atmos. terr.Phys.* **43**, 1199-1214.

Buckley, R., 1975, "Diffraction by a random phase-changing screen: A numerical experiment", *J.atmos. terr.Phys.* **37**, 1431-1446.

Buckley, R., 1985, "A Fokker-Planck equation for the probability distribution of the field scattered by a random phase screen". In this volume.

Fox, L. and Mayers, D.F., 1968, "Computing Methods for Scientists and Engineers", Oxford University Press, Oxford.

Fujii, H., Uozumi, J. and Asakura, T. 1976, "Computer simulation study of image speckle patterns with relation to object surface profile", *J.Opt.Soc.Am.* **66**, 1222 - 1236.

Fujii, H. and Asakura, T., 1977, "Effect of the point spread function on the average contrast of image speckle patterns", *Opt. Commun.* **21**, 80-84.

Fujii, H., 1979, "Non-Gaussian speckle with correlated weak scatterers: a computer simulation" *J.Opt.Soc.Am.* **69**, 1573-1579.

Jakeman, E. and McWhirter, J.G., 1977, "Correlation function dependence of the scintillation behind a deep random phase screen", *J.Phys.A: Math.Gen.* **10**, 1599-1643.

Macaskill, C. and Uscinski, B.J., 1983, "Intensity fluctuations due to a deeply modulated phase screen - II. Results". *J.atmos. terr. Phys.* **45**, 607-615.

Mandelbrot, B.B., 1977, "Fractals", Freeman, San Francisco.

Mercier, R.P., 1962, "Diffraction by a screen causing large random phase fluctuations", *Proc.Camb.Phil.Soc.* **58**, 382-400.

Ohtsubo, J., 1982, "Non-Gaussian speckle: a computer simulation", *Appl. Opt.* **21**, 4167-4175.

Papoulis, A., 1965, "Probability, Random Variables, and Stochastic Processes", McGraw-Hill Kogakusha, Tokyo.

Rice, S.O., 1944, "Mathematical analysis of random noise", *Bell System Tech J.* **23**, 282-333.

Richtmyer, R.D. and Morton, K.W., 1967, "Difference methods for initial-value problems" (2nd ed.) Wiley (Interscience), New York.

Uozumi, J., Fujii, H., and Asakura, T., 1977, "Further computer simulation study of image speckle patterns with relation to object surface profile", *J.Opt.Soc.Am.* **67**, 808-815.

Uscinski, B.J., 1977, "The Elements of Wave Propagation in Random Media", McGraw-Hill, New York.

Uscinski, B.J., Booker, H.G. and Marians, M., 1981, "Intensity fluctuations due to a deep phase screen with a power-law spectrum", *Proc.R.Soc.Lond.A.* **374**, 503-530.

Uscinski, B.J. and Macaskill, C., 1983, "Intensity fluctuations due to a deeply modulated phase screen - I. Theory", *J.atmos. terr.Phys.* **45**, 595-605.

Whale, H.A., 1973, "Diffraction of a plane wave by a random phase screen", *J.atmos.terr.Phys.* **35**, 263-274.

Whale, H.A., 1974, "Near-field statistics for the field diffracted by a thin one-dimensional random phase screen", *J.atmos.terr.Phys.* **36**, 1045-1057.

APPENDIX

To solve equation (21), Fourier transform with respect to ϕ, viz.

$$\hat{\eta}_1(x,\nu) \equiv \int_{-\infty}^{\infty} \eta_1(x,\phi)e^{-i\nu\phi}\, d\phi. \tag{A.1}$$

Then $\hat{\eta}_1$ satisfies

$$\frac{\partial\hat{\eta}_1}{\partial x} + a\phi_o^2\nu^2\hat{\eta}_1 + a\nu\frac{\partial\hat{\eta}_1}{\partial\nu} = \pi^{-\frac{1}{2}}\exp\left[-\frac{1}{2}(1+\nu^2)\phi_o^2\right]\cosh(\phi_o^2\nu+i\beta) \quad (A.2)$$

subject to the initial condition

$$\hat{\eta}_1(-\infty,\nu) = 0. \quad (A.3)$$

Equation (A.2) can be solved by the method of characteristics. In terms of a parameter s, the equations of the characteristics are

$$x(s) = s\ ,\ \nu(s) = \nu(o)e^{as}\ . \quad (A.4)$$

Along a characteristic,

$$\frac{d}{ds}\{\hat{\eta}_1\exp(\tfrac{1}{2}\phi_o^2\nu^2(o)e^{2as})\} = \pi^{-\frac{1}{2}}\exp(-\tfrac{1}{2}\phi_o^2)\cosh\{\phi_o^2\nu(o)e^{as}+i\beta_s\} \quad (A.5)$$

Since $\nu(o)$ is determined by $\nu(o) = \nu e^{-ax}$,

$$\hat{\eta}_1(x,\nu) = \frac{1}{\sqrt{\pi}}\exp\left\{-\frac{\phi_o^2(\nu^2+1)}{2}\right\}\int_{s=-\infty}^{x}\cosh\{\phi_o^2\nu e^{a(s-x)}+i\beta_s\}\,ds. \quad (A.6)$$

Inverting the transform now leads to (27).

Solving equation (29) in the same way, one finds that the characteristic curves are again given by (A.4).

Hence

$$\xi_1(x,\phi) = -\frac{1}{2\pi}\exp(-\phi_o^2)G(\phi)\int_{s_2=-\infty}^{x}\int_{s_1=-\infty}^{s_2}\{\exp[\phi_o^2\{\tfrac{1}{2}W_1^2 - e^{a(s_1-s_2)}\}]$$

$$.\cos[\beta_1+\beta_2+\phi W_1] + \exp[\phi_o^2\{\tfrac{1}{2}W_2^2 + e^{a(s_1-s_2)}\}]$$

$$.\cos[\beta_2-\beta_1+\phi W_2]\}\,ds_1 ds_2. \quad (A.7)$$

Here $W_1 \equiv e^{-ax}(e^{as_2}+e^{as_1})$, $W_2 \equiv e^{-ax}(e^{as_2}-e^{as_1})$. (A.8)

A MODEL FOR NON-RAYLEIGH SCATTERING STATISTICS

C.J. Oliver

(Royal Signals and Radar Establishment, Malvern, Worcs.)

ABSTRACT

A surface model is outlined based on a many-dimensional random-walk process. When electromagnetic waves are scattered by this surface, the detected intensity is shown to be K-distributed. In addition, the effects of correlations within the scattering surface are introduced and the dependence of the observed intensity on resolution or illuminated length derived. The findings of this model are compared with measured high-resolution radar images of land clutter. The general features of the model are shown to represent the observed behaviour adequately. The directions of future study of surface scattering models is indicated.

1. INTRODUCTION

When electromagnetic waves are scattered from a large number of randomly distributed targets the mutual interference between individual returns gives rise to a total field which has a Gaussian probability distribution, as one would expect from the central limit theorem. Envelope-detection of this field leads to a Rayleigh-distributed envelope and a negative-exponential intensity (the square of the envelope). This process is the cause of the well-known speckle in coherent imaging. However, if we allow the number of scatterers to fluctuate or introduce variations in the scatterer cross-section then these excess fluctuations will be reflected in the field and intensity distributions. Various models have been proposed which introduce this type of fluctuation (Jakeman and Pusey, 1973; Jakeman, 1974; Oliver, 1982; Pusey, 1977; Jakeman and Pusey, 1978; Jakeman, 1980a; Jakeman and Pusey, 1976; Oliver, 1983; Oliver, 1984). One particularly valuable model is that in which the local cross-section of the scattering surface is assumed to be Γ-distributed (Jakeman and Pusey, 1978; Jakeman, 1980a; Jakeman and Pusey, 1976; Oliver, 1984). On detection, the scattered field is then found to have K-distributed intensity fluctuations. This approach has proved very successful in representing a diverse selection of scattering experiments at both optical and microwave frequencies (Jakeman and Pusey, 1976; Jakeman and Pusey, 1975; Parry, Pusey, Jakeman and McWhirter, 1977; Jakeman, Pike and Pusey, 1976). The factorisation into Γ-distributed cross-section fluctuations with Gaussian speckle superimposed has also been demonstrated (Ward, 1981; Ward, 1982) for radar scattering from the sea surface.

The major drawback of previous models has been the inbuilt assumption that the scattering in neighbouring pixels is totally independent. In earlier publications we introduced correlations into the surface cross-section; first in terms of a single thermal-noise process (Oliver, 1983) and, more recently, in terms of the interference between an arbitrary number of such independent noise processes (Oliver, 1984). These methods offer the advantage of introducing correlations within the scattering surface ab initio. This enables prediction of the dependence of the detected intensity on the operating conditions, such as the resolution or illumination length. While the independent-scatterer approach was forced to derive a separate effective value of the order parameter ν for the (assumed) K-distributed intensity under every operating condition, this new model performed the analytical integration of the scattering of a finite beam size by a correlated surface.

In the present paper we shall outline the derivation of the statistical and coherence properties of the detected scattered radiation in section 2, followed, in section 3, by a discussion of the implications of this theory for experimental interpretation. In section 4 we shall introduce some high-resolution radar imagery obtained from the RSRE manned aircraft Synthetic-Aperture Radar (SAR). The extent to which these real data follow the predictions of the simple model is discussed and directions in which future thinking should take place are identified.

2. THE SURFACE MODEL

Fundamental to the understanding of the properties of the scattered radiation is the derivation of a plausible model to represent the surface. Following Oliver (1984) we choose to define a parameter, a, which represents the local scattering amplitude of the surface. This is essentially a phenomenological approach since it makes no attempt to describe the real surface, contenting itself with a representation that models the observed properties of that surface as far as possible. Thus the scattering amplitude is defined at a lattice of points in the surface which may have arbitrarily small separation. Each elemental scatterer is assumed to have a random phase factor associated with it depending on random range variations of order of the radiation wavelength. In typical radar applications, particularly with grazing-incidence illumination, the resolution cell, or illumination length, is at least two order of magnitude greater than this wavelength. Within the resolution cell, therefore, there are a large number of randomly-phased scatterers leading to a randomly-phased resultant.

Let us now derive an expression for the observed intensity when radiation is scattered by the surface. Since the scattering amplitude at position j is given by

$$\underset{\sim}{a}_j \equiv a_j e^{i\phi_j} \tag{1}$$

then the scattered and received field is given by

$$\varepsilon_j \equiv E_O a_j e^{i\phi_j} \tag{2}$$

where E_O contains all the imaging and illumination factors. Summing the contributions of all the elemental scatterers within the beam the total field is given by

$$\varepsilon = \sum_j E_O a_j e^{i\phi_j}. \tag{3}$$

The intensity of this received signal is equal to the envelope squared of this field, i.e.

$$I \equiv |\varepsilon|^2 = |E_O|^2 \sum_{j,k} a_j a_k^* e^{i(\phi_j - \phi_k)}. \tag{4}$$

On averaging over the ensemble of all possible scatterer configurations within the surface we obtain for the average intensity

$$\langle I \rangle = |E_0|^2 \sum_j \langle |a_j|^2 \rangle = |E_0|^2 \sum_j \langle \sigma_j \rangle \tag{5}$$

since all the phase cross terms $\phi_j - \phi_k$ cancel out. The same effect is obtained for all higher moments of the detected intensity so that the properties of the intensity can be expressed in terms of incoherent summation of the results of scattering from the individual elements, characterised by an elemental cross-section parameter, σ_j. It is the behaviour of this cross-section parameter which therefore determines the properties of the detected radiation.

Let us next postulate a model for the fluctuations in this cross-section. At the jth element we shall assume that the total scattering amplitude is made up of the sum of a large number of contributions in N dimensions. The total scattering amplitude is then made up of the sum of the contributions from each of the N dimensions. Thus

$$\underset{\sim}{a} \equiv \sum_{j=1}^{N} \underset{\sim}{a}_j . \tag{6}$$

These contributions will be assumed to be derived from the interference between two narrowband thermal noise processes at different frequencies. This allows us to introduce a single frequency fluctuation into the surface; more complicated forms may be introduced by extension of the method. For simplicity we assume that the noise processes have identical statistical and spatial coherence properties. Hence, the individual contributions to the scattering amplitude take the form

$$\underset{\sim}{a}_j = \alpha_j^{(1)} e^{i\omega_1 dj} + \alpha_j^{(2)} e^{i\omega_2 dj} \tag{7}$$

where ω_1 and ω_2 are the mean frequencies of the two thermal noise processes and d is the sample spacing. The cross-section contribution is now given by

$$\sigma_j \equiv |a_j|^2 = \sigma_j^{(1)} + \sigma_j^{(2)} + \alpha_j^{(1)}\alpha_j^{(2)*}\, e^{i\omega dj} + \alpha_j^{(2)}\alpha_j^{(1)*}\, e^{-i\omega dj} \tag{8}$$

where

$$\omega = \omega_1 - \omega_2. \tag{9}$$

In his analysis of non-Rayleigh scattering statistics, Jakeman (1980 b) demonstrated that derivation of the properties of a 2ν-dimensional random walk gave precisely equivalent results to those predicted for a population growth model based on birth, death and migration. In each case the probability distribution of the detected intensity was a K distribution characterised by the single parameter ν. Since each contribution (a) is made up of two interfering thermal noise variables $\alpha^{(1)}$ and $\alpha^{(2)}$, the variable N in Eq. (6) is equivalent to the order parameter, ν, in the previous analysis (Jakeman, 1975). For the multi-dimensional random walk analysis to be complete we invoke the principle of analytic continuation to extend the analysis to cover non integer and fractional values of ν. The model proposed in the present paper leads to identical statistical properties to the previous work but, in addition, included the effects of spatial correlations, previously ignored. As demonstrated by Jakeman (1980 a), the normalised moments of the detected intensity are given by

$$I^{(n)} = n!\,\frac{\Gamma(n+\nu)}{\nu^n\Gamma(\nu)}\,. \tag{10}$$

Let us next examine the effect of the surface correlations on the detected intensity. For the single frequency difference the normalised surface auto-correlation function can be shown to be given by Oliver (1984)

$$\frac{\langle\sigma_j\sigma_k\rangle}{\langle\sigma_j\rangle^2} = 1 + \frac{1}{2\nu}\,|g_{jk}|^2\,(1 + \cos\omega d(j-k)) \tag{11}$$

where $|g_{jk}|^2$ is the autocorrelation function of the narrowband noise processes $\alpha^{(1)}$ and $\alpha^{(2)}$. If we assume that these contain only a single exponential term so that

$$|g_{jk}|^2 \equiv e^{-\Gamma d|j-k|} \tag{12}$$

where Γ is the spatial linewidth, then the normalised autocorrelation function of the detected intensity can be shown to be given by

$$\frac{\langle I_j I_{j+r}\rangle}{\langle I_j\rangle^2} = 1 + \exp\left[-\frac{r^2 d^2}{\alpha}(\alpha^2+\beta^2)\right] + \frac{1}{\nu}\exp\left[-\alpha r^2 d^2\right]$$

$$(f_1+f_2+f_3+f_4+f_5+f_6+f_7+f_8) \tag{13}$$

for one-dimensional Gaussian illumination where $1/\sqrt{\alpha}$ is the 1/e width of the illumination and β is the wavefront curvature parameter (Oliver (1984)). The first two terms describe the Gaussian component of the process resulting from the interference between large numbers of random scatterers filling the illumination beam. Similar Gaussian behaviour was derived for two point receivers viewing radiation scattered from a continuous rough surface undergoing transverse motion through an illuminating laser beam (Jakeman, 1975). The third term in Eq. (13) represents the effect of fluctuations in the number of scatterers within the beam, or of fluctuations in their cross section (Oliver, 1984). This term has a profile identical to the beam shape and corresponds to the transit of individual scatterers through the beam. Only this non-Gaussian contribution is affected by the properties of the surface itself; the Gaussian component, which corresponds to classical speckle, is unaffected. The coefficients f represent the interaction of the surface and illumination properties. The form of these coefficients is given by the general summation

$$\sum_{t=0} \exp\left[-(At^2+Bt)\right] \equiv \frac{1}{2}\left(\frac{\pi}{A}\right)^{\frac{1}{2}} f(z) \tag{14}$$

where

$$f(z) \equiv \exp[z^2]\ \mathrm{erfc}\ (z) \tag{15}$$

and

$$z \equiv \frac{B}{2\sqrt{A}} \tag{16}$$

The appropriate values for z for the different coefficients were derived previously (Oliver (1984)) and are listed in the Appendix. If we study the interaction of the illumination length and the surface cross-section fluctuations we observe that when the illuminating or resolution length is small compared with the surface fluctuations then the value of f tends to unity. As the illumination length increases the value reduces asympotically to zero leaving only the classical speckle contribution. Thus, the intensity autocorrelation function carries information about those surface fluctuations that are on a scale that is large compared with the illumination length or resolution.

Similar analysis may be employed to derive the normalised intensity moments for such a process. For convenience the results were verified using the algebraic processor REDUCE running on a DARKSTAR microcomputer. The normalised moments can be shown to be given by

$$I^{(2)} \equiv \frac{<I^2>}{<I>^2} = 2\left[1 + \frac{1}{2\nu}\left\{f_9 + \frac{f_{10} + f_{11}}{2}\right\}\right] \quad (17)$$

$$I^{(3)} \equiv \frac{<I^3>}{<I>^3} = 6\left[1 + \frac{3}{2\nu}\left\{f_9 + \frac{f_{10} + f_{11}}{2}\right\} + \frac{1}{2\nu^2}\left\{\frac{f_{12}}{2}(1 + f_{13}) + \frac{3}{4}(f_{14}f_{15} + f_{16}f_{17} + f_{18}f_{19} + f_{20}f_{21})\right\}\right] \quad (18)$$

and

$$I^{(4)} \equiv \frac{<I^4>}{<I>^4} = 24\left[1 + \frac{1}{2\nu}\left\{6f_9 + 3(f_{10} + f_{11})\right\} + \frac{1}{4\nu^2}\left\{4f_{12}(1+f_{13}) + 6(f_{14}f_{15} + f_{16}f_{15} + f_{16}f_{17} + f_{18}f_{19} + f_{20}f_{21}) + 3f_9^2 + 3f_9(f_{10} + f_{11}) + \frac{3}{4}(f_{10} + f_{11})^2\right\} + \frac{3}{4\nu^3}(h_1 + 4h_2 + 2h_3 + h_4)\right] \quad (19)$$

As above, the coefficients f, defined as in Eq. (15) with the appropriate z value taken from the Appendix, describe the effects of the interaction between the surface and illumination properties. The terms h_1, h_2, h_3 and h_4 represent different fourfold summations (Oliver, 1984) which can only be evaluated numerically.

In order to make contact with previous work let us consider the limiting conditions of these distributions. First, when ν is infinite the contribution of the non-Gaussian components reduces to zero corresponding to the large N limit of the number fluctuations, and yields the classical speckle result. If, on the other hand, ν takes the value unity which corresponds to a thermal noise process (a two-dimensional random walk) then the result corresponds to Gaussian-Gaussian scattering (Bertollotti, Crossignani and di Porto, 1970; Jakeman, Pike, Parry and Saleh, 1976). For arbitrary distributions we may deduce effective values of ν from the observed value of the second moment by assuming that the scale of the surface fluctuations is longer than the illumination length so that the values of f are all unity. Hence from Eq. (17)

$$\nu_{eff} \equiv \frac{2}{I^{(2)} - 2} \qquad (20)$$

This approach has been successfully adopted in the description of sea clutter by K-distributed noise (Ward, 1981; Ward 1982). It essentially treats all scatterers as independent so that it is unable to make any prediction of the dependence of clutter statistics on the illuminated area, for example, which is the purpose of the present analysis. In general, the more spikey the distribution of the intensity the smaller the value of the parameter ν.

3. PREDICTED PROPERTIES FOR THE OBSERVED INTENSITY

The dependence of the intensity moments and autocorrelation function on the operating conditions may be evaluated from Eqs. (13) and (17) to (19) as previously demonstrated (Oliver (1984)). In this section we summarise some of the appropriate previous results in order to set the scene for the subsequent application of these techniques to real radar images. Obviously many of the assumptions made in the derivation will not be appropriate for this imagery which is obtained from the RSRE manned aircraft Synthetic-Aperture Radar. However the general features would be still expected to apply.

In Figure 1 we show a typical surface cross-section autocorrelation function, as defined in Eq. (11), for a surface having a correlation length $(1/\Gamma)$ which is half of its period $(2\pi/\omega)$ and a number-fluctuation parameter $\nu = 0.2$. When this surface is illuminated by a Gaussian profile beam of 1/e width $1/\sqrt{\alpha}$, then the interaction of the scale of the illumination with the surface fluctuations leads to the result of figure 2. Results are shown for four ratios of illuminated length to surface scale: a) $\gamma \equiv \Gamma/\sqrt{\alpha} = 0.025$, $\Omega \equiv \omega/2\pi\sqrt{\alpha} = 0.05$, b) $\gamma = 0.125$, $\Omega = 0.25$, c) $\gamma = 0.25$, $\Omega = 0.5$ and d) $\gamma = 1.0$, $\Omega = 2.0$. It is apparent that the progressive increase in the illumination length gradually averages out the effects of the fluctuations in the surface cross-section. Thus the observed intensity autocorrelation function contains information about those fluctuations in the surface which occur on length scales longer than the illumination. Smaller-scale fluctuations are averaged out so that the observed autocorrelation function consists only of the Gaussian contributions, namely the first two terms of Eq. (13).

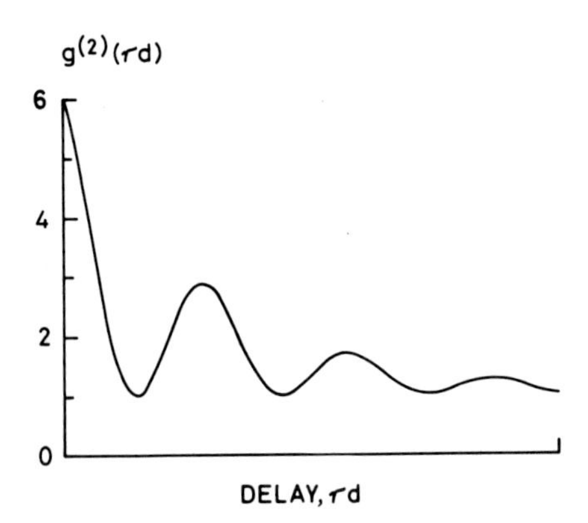

Fig. 1

The autocorrelation function for the surface model for $\nu = 0.2$. The correlation length $(1/\Gamma)$ is half the period $(2\pi/\omega)$.

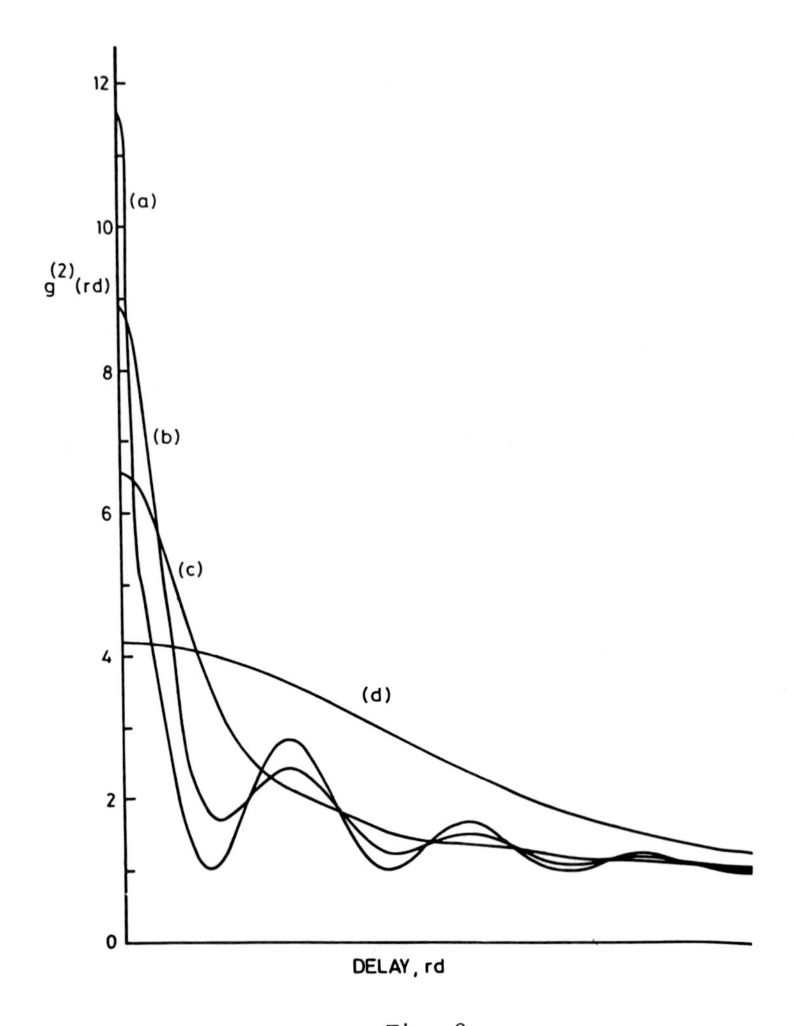

Fig. 2

The autocorrelation function of the intensity when the surface of figure 1 is illuminated by a one-dimensional Gaussian profile beam, a) $\gamma = 0.025$, $\Omega = 0.05$, b) $\gamma = 0.125$, $\Omega = 0.25$, c) $\gamma = 0.25$, $\Omega = 0.5$ and d) $\gamma = 1.0$ $\Omega = 2.0$ (Oliver, 1984).

As one would expect, similar effects may be observed in the dependence of the normalised intensity moments on the scale of the illumination. In figure 3 we show such results based on the model in which the parameters take the values $\nu = 0.2$, a) $\Omega = 0$ and b) $\Omega = 2\gamma$. As the parameter γ is increased the value of the moments decreases to the limiting value which is characteristic of classical speckle. As the illuminated length decreases the moments increase towards their limiting values when $f = 1$.

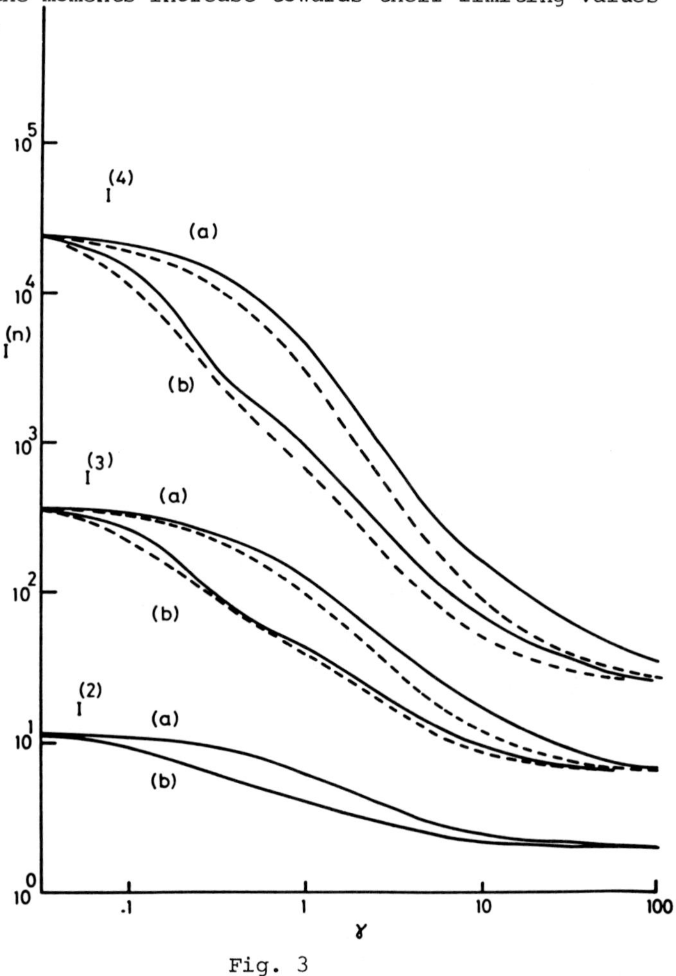

Fig. 3

The dependence of the normalised intensity moments on the illumination length ($1/\sqrt{\alpha}$) for $\nu = 0.2$; a) $\Omega = 0$ and b) $\Omega = 2\gamma$. The full lines denote the present theory, the dashed line is derived by assuming that the distribution is K-distributed and fitting ν to the value of $I^{(2)}$ (Oliver, 1984).

The fact that the surface has correlations within it is apparent from the observed saturation in the increase in these moments. Once the intensities in successive samples are completely correlated no further increase in the moments is expected. The detail of the structure of the dependence of the moments on the illuminated area is an important tool in investigating the properties of real surfaces as opposed to the simple model adopted here. The dashed lines in figure 3 are derived using a common form of analysis in which an effective value of ν is deduced, as shown in Eq. (20), for each illumination length, and the higher moments derived for this value by substitution into Eqs. (18) and (19). This approach was first used with optical measurements (Bedard, Chang and Mandel, 1967) and obviously ignores the spatial averaging effects since it assumes that an integrated K distribution is also K distributed. The errors introduced by making this assumption are not excessive; more significant is the fact that the method is unable to predict the dependence of the second moment on the illuminated length which is fundamental to the present theory.

4. APPLICATION TO HIGH-RESOLUTION RADAR IMAGERY

Having outlined a simple model for scattering by a rough surface, let us now seek to apply this approach to the understanding of real, high-resolution, radar imagery. Inherent in the previous analysis is the assumption that the scattering surface may be represented as arising from a stationary ergodic process. In the real world this assumption cannot be valid. However, one might expect that sea-surface fluctuations, which do indeed arise from the interference of competing noise-like processes, may perhaps be adequately represented in this fashion provided that one only studies the region over which conditions are consistent. In order to apply the same analysis to ground clutter one is unfortunately making even less justifiable assumptions about the nature of the terrain. It is still worthwhile, however, at least to consider how such a model might represent the observed properties without embarking on a detailed justification of the model in that context. The periodic component of the model, introduced in order to describe the sea, obviously no longer applies to the general land-clutter environment, though it might be appropriate for describing the layout of streets within a town. Thus we shall only make comparison with the model for the situation when $\Omega = 0$.

A typical image of a region of open country obtained from the RSRE manned aircraft Synthetic-Aperture Radar system is used for the comparison. The probability distribution of the modulus of the received field for ten lines selected throughout the image is shown in figure 4. The scale is changed at the large-modulus end so that the contribution of individual pixels is clearly visible. From this result, combined with many others, it is apparent that a modulus value of more than about 240 occurs only very rarely (1 in 1000) and corresponds to isolated targets. If we assume that our scene is indeed made up of a clutter background together with these dominant targets, then the statistics of the clutter may be deduced by ignoring these strong scatterers.

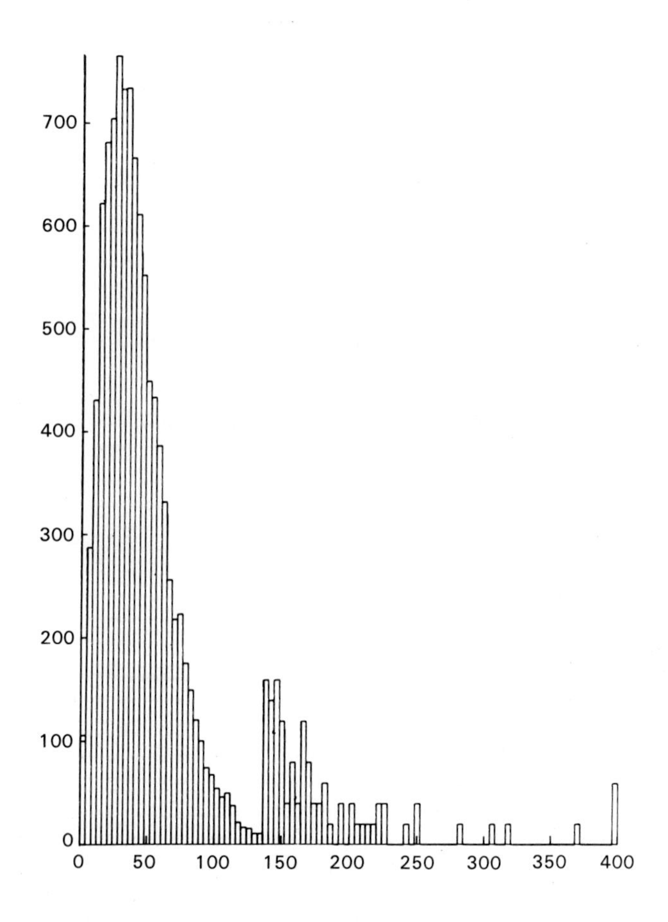

Fig. 4

The probability distribution of the modulus of the received field for a typical SAR image.

Calculating the intensity moments over the complete image we obtain the values 2.9, 20 and 244 for the second, third and fourth normalised intensity moments respectively. From Eq. (20) the corresponding value of ν was found to be 2.22. If the distribution were indeed a K distribution then the predicted values of the third and fourth moments would be 17 and 157. Any discrepancy between the predictions and the measurements stems from two primary causes. Firstly, the experimental results depend dramatically on the threshold value chosen for target identification. Reducing this threshold to 210, for example, gives almost perfect agreement. Secondly, in selecting the value of ν it is assumed that there are no components on a shorter scale than the pixel separation in the image. Since this is obviously a gross over-simplification for real scenes we would expect this approach to lead to an inappropriate choice of ν. Comparison with figure 3 suggests that such an approach could lead to a predicted value which was too small by a factor that could be as large as 50% (Oliver (1984)). In the light of the magnitude of these effects it would be foolish to attempt to obtain a closer match between experiment and theory. Granted these problems, it is apparent that the clutter is reasonably well represented as giving rise to an intensity which has K-distribution statistics.

Having examined the statistics of the received power let us next consider the correlation properties of the signal. In figure 5 we show a measurement of the spatial autocorrelation function of the image in one dimension taken over the entire image. The results both with and without target thresholding are included for comparison. If we consider first the results for clutter alone, denoted by (x), various discrepancies between these and the predictions of the simple model, illustrated in figure 2, are apparent. Firstly, the steep decay between the zero delay and single-pixel delay indicates that the form of the resolution limit is certainly not Gaussian as assumed for the model. Indeed consideration of the SAR processor suggests that a sinc-function weighting would be more appropriate. This would indeed give a much steeper initial gradient to the autocorrelation function for the same nominal resolution. Secondly, it is obviously incorrect to represent the surface in terms of a single exponential decay. From figure 5 one might deduce that a combination of two length scales might adequately represent the surface. In fact, the surface would be expected to contain a variety of length scales. Thus, again, the simple model is not adequate. However, it does indicate that an approach along these lines of interpretation should prove profitable. The corresponding results when all the data are included are denoted by (⊙). The small delay end of the autocorrelation function is now dominated by the returns from a very few scatterers. It is difficult to draw any conclusions, however tentative, in this region. For larger delays the function appears similar to that for clutter alone, suggesting that this decay constant is indeed characteristic of the clutter rather than the strong targets.

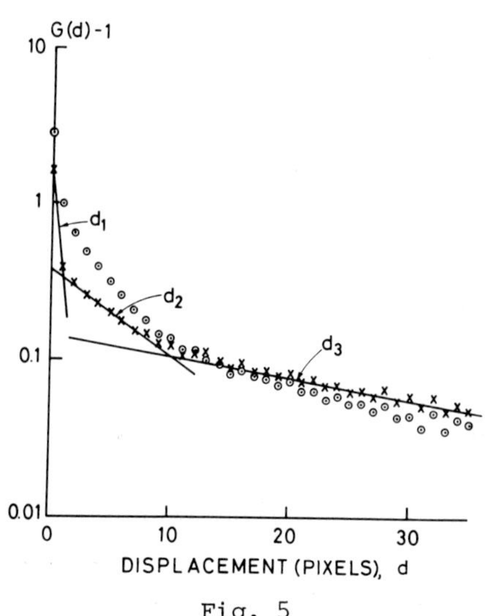

Fig. 5

The autocorrelation function of the detected intensity for the image used for figure 4. Results including all data (⊙) and with clutter alone (x) are shown.

The final comparison is shown in figure 6 in which the values of the normalised intensity moments are plotted as a function of the one-dimensional resolution. The complex image data obtained from the RSRE SAR are summed coherently over numbers of pixels in the azimuth direction. The intensity moments of these new lower-resolution image are then calculated. The results are compared with the predictions of the simple model for a value of ν of 1.67, corresponding to a second moment of 3.2. When the results are combined with measurement, reasonable agreement is obtained for an overall effective line-width given by $1/\Gamma$ = 19 pixels. This is a reasonable compromise with the two decay constants d_2 and d_3 in figure 5 which take the values of 8 and 34 pixels respectively. Thus, while the present analysis is an over-simplification, the description of the observed intensity distribution in terms of an integrated K distribution is certainly promising.

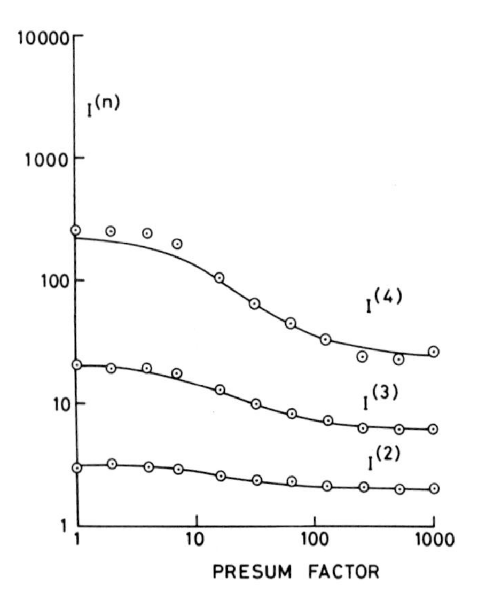

Fig. 6

The dependence of the normalised intensity moments from the image used for figure 4 on the illumination length. The results are compared with the appropriate model predictions.

5. CONCLUSIONS

In this paper we have introduced a simple model to describe clutter. This model treats the local cross-section at a point in the surface as being made up of the incoherent summation of ν complex random walk mechanisms. The statistics of the detected intensity for such a model are shown to be K distributed. In addition, we introduce the correlation properties of the surface directly into the analysis in terms of a single correlation length and period. Based on this model, the behaviour of the received intensity may be calculated for any combination of illumination length, or resolution, and surface scale. Typical results are derived from the model in order to demonstrate its general properties. When we apply the same approach to real radar data it is apparent that, while various discrepancies exist, the fundamental features of the model are generally followed. No detailed comparison can be made at present as the model is too simple and the data too unreliable. However, we expect to extend these studies towards obtaining a useful grasp of the nature of the clutter which forms so much of the basic image data in reconnaissance radar systems.

6. ACKNOWLEDGEMENTS

The author wishes to acknowledge valuable assistance from Mr. P.S. Dombi who performed all the SAR processing computation.

REFERENCES

Bedard, G., Chang, C. and Mandel, L., 1967, "Approximate formulas for photoelectric counting distributions". *Phys. Rev.* **160**, 1496-1500.

Bertollotti, M., Crossignani, B. and di Porto, P., 1970, "On the statistics of Gaussian light scattered by a Gaussian medium". *J. Phys. A* **3**, L37-L38.

Jakeman, E., 1974, in "Photon Correlation and Light-Beating Spectroscopy" (H.Z. Cummins and E.R. Pike, eds.), 75-149, Plenum, New York.

Jakeman, E., 1975, "The effect of wavefront curvature on the coherence properties of laser light scattered by target centres in uniform motion". *J. Phys. A* **8**, L23-L28.

Jakeman, E., 1980a, "On the statistics of K-distributed noise". *J. Phys. A* **13**, 31-48.

Jakeman, E., 1980b, "Statistics of integrated gamma-Lorentzian intensity fluctuations". *Opt. Acta.* **27**, 735-741.

Jakeman, E., Pike, E.R. and Pusey, P.N., 1976, "Photon correlation study of stellar scintillation". *Nature* **263**, 215-217.

Jakeman, E., Pike, E.R., Parry. G. and Saleh, B., 1976, "Speckle patterns in polychromic light". *Opt. Comms.* **19**, 359-364.

Jakeman, E. and Pusey, P.N., 1973, "The statistics of light scattered by a random phase screen". *J. Phys. A* **6**, L88-L92.

Jakeman, E. and Pusey, P.N., 1975, "Non-Gaussian fluctuations in electromagnetic radiation scattered by a random phase screen". *J. Phys. A* **8**, 369-410.

Jakeman, E. and Pusey, P.N., 1976, "A model for non-Rayleigh sea echo". *IEEE Trans AP*-**24**, 806-814.

Jakeman, E. and Pusey, P.N., 1978, "Significance of K distributions in scattering experiments". *Phys. Rev. Letts.* **40**, 546-550.

Oliver, C.J., 1982, "Fundamental properties of high-resolution sideways-looking radar". *IEE Proc. F* **129**, 385-402.

Oliver, C.J., 1983, in "13th European Microwave Conference Proceedings", 552-557 Microwave Exhibitions and Publishers, Tunbridge Wells.

Oliver, C.J., 1984, "A model for non-Rayleigh scattering statistics". *Opt. Acta*, **31**, 701-722.

Parry, G., Pusey, P.N., Jakeman, E. and McWhirter, J.G., 1977, "Focussing by a random phase screen". *Opt. Comms*. **22**, 195-201.

Pusey, P.N., 1977, in "Photon Correlation Spectroscopy and Velocimetry" (H.Z. Cummins and E.R. Pike, eds.), 45-141 Plenum, New York.

Ward, K.D., 1981, "Compound representation of high resolution sea clutter". *Electron. Letts*. **17**, 561-565.

Ward, K.D., 1982, in "Radar 82". IEE Conf. Publ. 216, 203-207.

APPENDIX

Table of values of z, as defined in Eq. (16), used in calculation of the coefficients f, as defined in Eq. (15).

COEFF	Z-VALUE	COEFF	Z-VALUE
f_1	$rd\sqrt{\alpha} + \Gamma/\sqrt{\alpha}$	f_{12}	$\Gamma\sqrt{3/\alpha}$
f_2	$-rd\sqrt{\alpha} + \Gamma/\sqrt{\alpha}$	f_{13}	$\Gamma/2\sqrt{\alpha}$
f_3	$\Gamma/\sqrt{\alpha} - i\beta rd/\sqrt{\alpha}$	f_{14}	$-i\omega/4\sqrt{\alpha}$
f_4	$\Gamma/\sqrt{\alpha} + i\beta rd/\sqrt{\alpha}$	f_{15}	$\Gamma\sqrt{\frac{3}{\alpha}} - \frac{i\omega}{4}\sqrt{\frac{3}{\alpha}}$
f_5	$rd\sqrt{\alpha} + \Gamma/\sqrt{\alpha} - i\omega/2\sqrt{\alpha}$	f_{16}	$\Gamma/2\sqrt{\alpha} - i\omega/4\sqrt{\alpha}$
f_6	$-rd\sqrt{\alpha} + \Gamma/\sqrt{\alpha} + i\omega/2\sqrt{\alpha}$	f_{17}	$\frac{\Gamma}{2}\sqrt{\frac{3}{\alpha}} - \frac{i\omega}{4}\sqrt{\frac{3}{\alpha}}$
f_7	$\Gamma/\sqrt{\alpha} - i(\omega/2\sqrt{\alpha} + \beta rd/\sqrt{\alpha})$	f_{18}	$\Gamma/2\sqrt{\alpha} + i\omega/4\sqrt{\alpha}$
f_8	$\Gamma/\sqrt{\alpha} + i(\omega/2\sqrt{\alpha} + \beta rd/\sqrt{\alpha})$	f_{19}	$\frac{\Gamma}{2}\sqrt{\frac{3}{\alpha}} + \frac{i\omega}{4}\sqrt{\frac{3}{\alpha}}$
f_9	$\Gamma/\sqrt{\alpha}$	f_{20}	$i\omega/4\sqrt{\alpha}$
f_{10}	$\Gamma/\sqrt{\alpha} - i\omega/2\sqrt{\alpha}$	f_{21}	$\Gamma\sqrt{3/\alpha} + \frac{i\omega}{4}\sqrt{\frac{3}{\alpha}}$
f_{11}	$\Gamma/\sqrt{\alpha} + i\omega/2\sqrt{\alpha}$		

ON THE INTERPRETATION OF LASER PROPAGATION EXPERIMENTS

R.H. Clarke
(Imperial College of Science and Technology)

ABSTRACT

The mean intensity formula of Tatarskii is derived from phase-screen theory. Comparison is made with some propagation experiments of a laser beam through a turbulent atmosphere, and the variance and dissipation-scale size of the turbulent refractive index fluctuations are deduced.

1. INTRODUCTION

This paper is written with two distinct but related objectives in mind. One is to interpret the results of experiments on laser propagation through a turbulent atmosphere, such as those of King et al. (1983), in the light of theory. The other objective is to point out how the highly mathematical approach of Tatarskii (1971) and Ishimaru (1978) to the theory can be related to the more physical phase-screen approach of Ratcliffe (1956) and Uscinski (1971) and Flatte (1983). The reason for pursuing this second objective is that phase-screen theory is more naturally suited to the interpretation of experiments.

2. THE PHYSICAL PICTURE

Figure 1 shows the laser beam propagating through a turbulent atmosphere above a curved earth. The solid line depicts the path of the e^{-1} points on the beam intensity profile in the absence of turbulence. The beam centre line, shown chain-dotted, follows a circular path due to the underlying linear trend in the height variation of the refractive index in the first few tens of metres above the ground (see Figure 2).

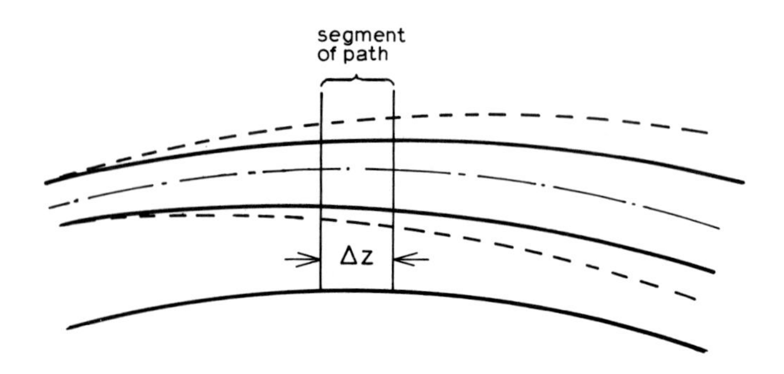

Fig. 1. Beam propagation in the atmosphere;
———— in the absence of turbulence,
- - - - - in the presence of turbulence.

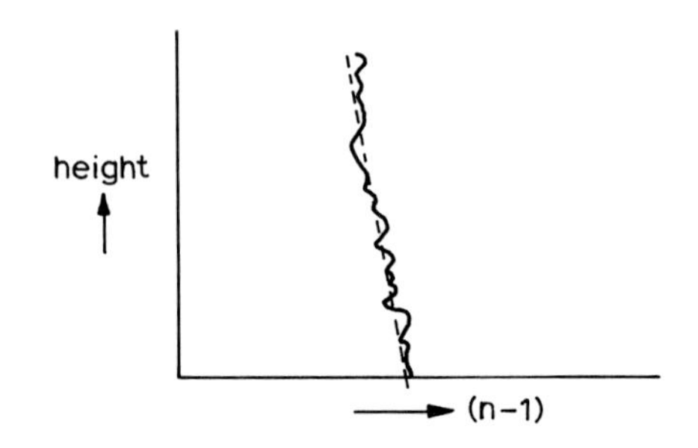

Fig. 2. Refractive index vs. height profile.

(Incidentally, if allowance is made for a second-order parabolic trend in the variation with height of the refractive index, this will produce a continuous but slight defocussing of the beam). If allowance is now made for the turbulence induced fine structure of the refractive index profile this will produce the increased beam spread indicated in Fig. 1 by the dashed curves.

In the experiment of King et al (1983) the width of the laser beam (between e^{-1} amplitude points) is initially about 16 cm, which would mean that after 37 km *in vacuo* the width of the beam would only be about 25 cm. In the real atmosphere the average beam was observed to be approximately Gaussian in shape and 6 m in width. It can be seen from this that the atmosphere induces a significant amount of scatter over 37 km, but that it can be fairly described as forward scatter. This is because a typical "blob" of atmospheric turbulence has a scale size which is very large compared to the laser wavelength of 0.63 μm.

In order to analyse the propagating beam the path will be divided into a series of segments, as indicated in Fig. 1, in a time-honoured manner (Fejer (1953), Bramley (1954), Consortini et al (1963), Chandrasekhar (1952), Lee & Harp (1969). The deterministic linear and parabolic trends in the refractive-index profile will be ignored, since their effects can be restored in an obvious way. A typical path segment is shown in Figure 1. The first segment, with its associated geometry, is shown in Fig. 3 where it is assumed that the beam is launched from the z=o plane. The departure $n_1(x,y,z)$ of the refractive index from its mean value will be assumed to be a locally stationary random process whose magnitude is certainly less than 10^{-6}.

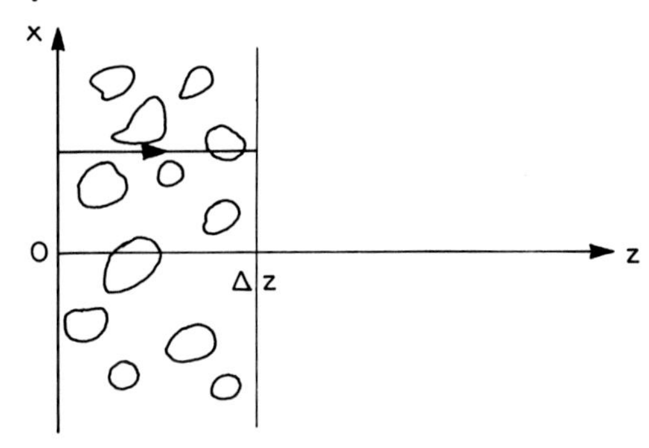

Fig. 3. First segment showing a ray path

3. PHASE-SCREEN THEORY FOR BEAMS

The details of the phase-screen theory for laser beams propagating through a clear but turbulent atmosphere are given in Clarke (1984). The basis of the theory is Papoulis' redefinition (Papoulis, 1974) for optical fields of Woodward's ambiguity function (Woodward, 1953) which was originally defined for radar time-waveforms. The field ambiguity function is simply the Fourier transform of the lateral mutual coherence (i.e., autocorrelation) function of the field. Papoulis showed that this field ambiguity function has the very useful property that, under conditions of Fresnel diffraction, it propagates without changing its form in a uniform medium.

If the direction of propagation is predominantly along the z axis, our principal interest will be in the lateral mutual coherence function

$$\Gamma(x,y;\xi,\eta;z)$$

$$= \langle f^*(x-\xi/2,y-\eta/2,z)f(x+\xi/2,y+\eta/2,z)\rangle \quad - \qquad (1)$$

of the field component $f(x,y,z)$. The asterisk denotes complex conjugate, and the sharp brackets indicate the expectation in a probabilistic sense of the enclosed quantity. Thus the mean intensity is

$$\langle I(x,y,z)\rangle = \langle |f(x,y,z)|^2\rangle$$
$$= \Gamma(x,y;o,o;z). \tag{2}$$

The definition of field ambiguity function, for the field over the plane z, is

$$A(\mu,\nu;\xi,\eta;z) = \frac{1}{\lambda^2}\int_{-\infty}^{\infty}\int_{-\infty}^{\infty}\Gamma(x,y;\xi,\eta;z)$$
$$\exp\{jk(\mu x+\nu y)\}dxdy. \tag{3}$$

It is assumed that the field is strictly monochromatic, of time dependence $\exp(j\omega t)$. Random fluctuations in time are ignored here for greater simplicity of notation, but could be incorporated without difficulty. Papoulis showed (Papoulis, 1974) that under conditions of Fresnel diffraction (interpreted physically this means for narrow beams and/or forward scatter), if the field traverses a uniform medium of width Δz, the ambiguity function over the plane $z + \Delta z$ is given by

$$A(\mu,\nu;\xi,\mu;\ z+\Delta z) = A(\mu,\nu;\ \xi-\mu\Delta z,\ \eta-\nu\Delta z;\ z). \tag{4}$$

This means that the form of the ambiguity function is unchanged in traversing the uniform medium from plane z to plane $z + \Delta z$; merely its argument changes.

It should be noted that what Ratcliffe (1956) called the "field autocorrelation function" was in fact $A(o,o;\xi,\eta;z)$. It can be seen from equation (4) that this is unchanged in both form and argument in traversing from one plane to the next in a uniform medium. This result is very useful if one wishes only to calculate the mean far-field intensity. But the complete ambiguity function must be used to calculate the mean intensity in the near field, or the mutal coherence function of the field, whether it be near or far.

Since the magnitude of the refractive-index fluctuations is very small, their effect can be treated as a perturbation. Indeed it can be shown, starting from the parabolic approximation to the wave equation (Tatarski (1971)) (for which Fresnel diffraction is appropriate), that propagation through such a tenuous random medium can be artifically divided into two stages. One stage treats diffraction while ignoring scattering; and the next suppresses diffraction while allowing for the presence of turbulence in terms of phase integrals along parallel ray paths. This is precisely the "beam propagation" method of Van Roey et al (1981) and the numerical "split-step Fourier" technique of Tappert (1977). Physically it is tantamount to assuming that the irregularities are swept on to the face of each segment, where their effect is concentrated, leaving diffraction to occur in the now-uniform segment interiors.

Thus the field ambiguity function, once calculated just to the right of the entrance plane of each segment, is now envisaged as traversing the segment unchanged. Then when it reaches the exit plane the field ambiguity function changes abruptly in response to the accumulated random phase which it encounters there.

For essentially forward-scatter propagation it is reasonable to calculate the cumulative phase variation along horizontal ray paths. Thus for the first segment this phase variation, along the ray path shown in Fig. 3, is

$$\Phi(x,y) = -k \int_{0}^{\Delta z} n_1(x,y,z)\,dz, \tag{5}$$

where k is the mean propagation constant. The phase variance will be of the order of

$$\sigma_\Phi^2 = k^2 \sigma_{n_1}^2 \, a \, \Delta z, \tag{6}$$

where $\sigma_{n_1}^2$ is the variance of the refractive index fluctuations, and a is its scale size, provided that

$$\Delta z >> a. \tag{7}$$

This condition incidentally ensures that $\Phi(x,y)$ is a Gaussian random process, by the Central Limit Theorem. However, given that this condition must be met it is best to keep Δz as small as possible so that

$$\sigma_\Phi^2 << 1. \tag{8}$$

It will be noticed that the various physical restrictions imposed are often encountered in more rigorous analyses as necessary approximations to enable a solution to be found. In particular the condition that $\Delta z >> a$ means that the effects due to adjacent segments are almost entirely statistically independent, which is the well-known Markov approximation.

Applying the above method to each segment in turn, the field ambiguity function at the entrance face to the first segment can be calculated from the known field distribution at the output of the laser transmitter as, say,

$$A(\mu,\nu;\xi,\eta;o) = A_o(\mu;\nu;\xi,\eta) \tag{9}$$

After traversing the supposedly uniform segment between $z = o$ and $z = \Delta z$, at which point equation (4) gives the field ambiguity function as $A_o(\mu,\nu;\ \xi-\mu\Delta z,\ \eta-\nu\Delta z)$, the field encounters the aggregated phase fluctuation of equation (5). Making the reasonable assumption that the refractive index is statistically stationary across the beam, the ambiguity function of the field entering the second segment is

$$A(\mu,\nu;\xi,\eta;\ \Delta z) = A_o(\mu,\nu;\ \xi - \mu\Delta z,\ \eta - \nu\Delta z)$$

$$\langle \exp\{j[\Phi(x+\xi,y+\eta) - \Phi(x,y)]\}\rangle \tag{10}$$

and since $\Phi(x,y)$ is Gaussian it follows that

$$A(\mu,\nu;\ \xi,\eta;\ \Delta z) = A_o(\mu,\nu;\ \xi-\mu\Delta z,\eta-\nu\Delta z)\ \exp\{-\sigma_\Phi^2\,[1-\rho(\xi,\eta)]\} \tag{11}$$

where σ_Φ^2 is the phase variance of equation (6) and $\rho(\xi,\eta)$ is the normalised lateral autocovariance function of the refractive index fluctuations.

The argument can be continued out to a distance

$$z = \sum_{i=1}^{N} \Delta z_i \tag{12}$$

provided only that essentially forward scatter is involved. The ambiguity function over the plane z is then

$$A(\mu,\nu;\ \xi,\eta;\ z) = A_o(\mu,\nu;\ \xi-\mu z,\eta-\nu z)$$

$$\exp\{-\sum_{n=1}^{N} \sigma_{\Phi_n}^2 [1-\rho(\xi-\mu\{z-\sum_{i=1}^{n}\Delta z_i\},\ \eta-\nu\{z-\sum_{i=1}^{n}\Delta z_i\})]\} \tag{13}$$

in which

$$\sigma_{\Phi_n}^2 = k^2\sigma_{n_1}^2\ \zeta_o\Delta z_n \tag{14}$$

is the phase variance for the nth section.

Since the Δz_i can be taken to be of the order of meters, equation (13) can be written in integral form as

$$A(\mu,\nu;\xi,\eta;\ z) = A_o(\mu,\nu;\ \xi-\mu z,\eta-\nu z)$$
$$\exp\{-k^2\sigma_{n_1}^2\ \zeta_o[\ z-\int_o^z \rho(\xi-\mu z',\eta-\nu z']\} \qquad (15)$$

if it is assumed that the turbulence is statistically uniform throughout the length of the path.

Taking the Fourier transform of this last equation yields the lateral mutual coherence function

$$\Gamma(x,y;\ \xi,\eta;\ z) = \int_{-\infty}^{\infty}\int_{-\infty}^{\infty} A(\mu,\nu;\xi,\eta;z)\ \exp\{-jk(\mu x+\nu y)\}\ d\mu d\nu \qquad (16)$$

with equation (15) substituted. The result is identical to Tatarski's solution of the partial differential equation for the mutual coherence function, as quoted by Ishimaru (1978). It yields the mean intensity, as indicated in equation (2), by setting ξ and η equal to zero.

4. APPLICATION TO A LASER BEAM PROPAGATING THROUGH A TURBULENT ATMOSPHERE

If the field emerging from the laser transmitting assembly is the fundamental Gaussian mode

$$f(x,y) = \exp\ \{-\ \frac{x^2+\ y^2}{W_o^{\ 2}}\} \qquad (17)$$

where W_o is the beam waist size, then equations (1), (3) and (9) give

$$A_o(\mu,\nu;\xi,\eta) = \frac{\pi W_o^2}{2\lambda^2}\ \exp\{-\ \frac{\pi^2 W_o^{\ 2}}{2\lambda^2}(\mu^2+\nu^2)\}\ \exp\{-\ \frac{\xi^2+\eta^2}{2W_o^{\ 2}}\} \qquad (18)$$

Equations (2), (16) and (18) then give the mean intensity at the plane z as

$$<I(x,y,z)> = \frac{\pi W_o^{\ 2}}{2\lambda^2}\int_{-\infty}^{\infty}\int_{-\infty}^{\infty}$$
$$\exp\{-(\ \frac{\pi^2 W_o^2}{2\lambda^2}\ +\ \frac{z^2}{2W_o^{\ 2}})(\mu^2+\nu^2)\}$$
$$\exp\{-k^2\sigma_{n_1}^{\ 2}\ \zeta_o[\ z\ -\int_o^z \rho(-\mu z',\ -\ \nu z')dz']\}$$
$$\exp\{-jk(\mu x+\nu y\}\ d\mu d\nu. \qquad (19)$$

Approximating the inner integral by $z\rho(-\frac{\mu z}{2}, -\frac{\nu z}{2})$ and replacing $-\mu z$, $-\nu z$ by ξ,η throughout, one has

$$\langle I(x,y,z)\rangle = \frac{\pi W_o^2}{2\lambda^2 z^2}\int_{-\infty}^{\infty}\int_{-\infty}^{\infty}$$

$$\exp\{-(\frac{\pi^2 W_o^2}{2\lambda^2 z^2} + \frac{1}{2W_o^2})(\xi^2+\eta^2)\}$$

$$\exp\{-\sigma_{\Phi T}^2 [1-\rho(\xi/2, \eta/2)]\} \exp\{j\frac{k(\xi x+\eta y)}{z}\} d\xi d\eta, \qquad (20)$$

where

$$\sigma_{\Phi T}^2 = k^2 \sigma_{n_1}^2 \zeta_o z \qquad (21)$$

is the total phase fluctuation variance along the path.

The intensity formula of equation (20) can be used to reveal a great deal of interesting detail about a beam propagating in a turbulent medium. For present purposes I want merely to highlight a few aspects. The first one is that, putting $\sigma_{\Phi T} = 0$, it can be checked that the intensity corresponds to the formulas for a fundamental-mode laser beam propagating in a uniform medium (Kogelnik and Li (1966)). The next point to notice is that the mean field is a specific fraction $\exp(-\sigma_{\Phi T}^2/2)$ of the undisturbed field, which is why it is referred to as the coherent part. Since the cumulative randomisation of the phase of the propagating beam is a loss-free mechanism, the power lost from the coherent part is converted into incoherent scatter. It is clear from equation (21) that there will come a distance at which the coherent part has become negligibly small. This is conveniently taken as occurring when $\sigma_{\Phi T}^2 = 1$. Figure 4 shows the evolution of the laser beam through turbulence, over a fairly short distance, on the assumption that $W_o \gg a \gg \lambda$.

It is assumed that the turbulence is homogeneous and fully developed then the inner and outer scales of turbulence, broadly defining the inertial subrange, are of the order of millimetres and metres respectively. From a physical point of view it is the smaller, dissipation scale which is going to contribute significantly to the break-up and broadening of the beam. Let this dissipation scale size be a, so that in the neighbourhood of zero separation (Tatarski (1961))

$$\rho(\xi,\eta) = 1 - \frac{\xi^2+\eta^2}{a^2} . \qquad (22)$$

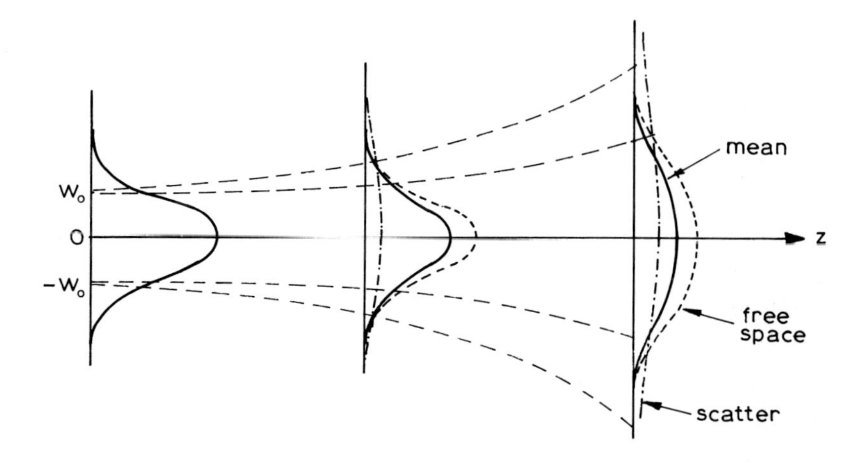

Fig. 4. The changing beam profile

This parabolic behaviour will dominate in equation (20) when $\sigma_{\Phi T}^2 >> 1$, that is, at the longer path lengths. So substituting equation (22) into equation (20) and integrating, the average intensity at these longer path lengths, making the reasonable assumption that $W_o >> a$, is

$$\langle I(x,y,z)\rangle = \frac{W_o^2}{W^2} \exp \{- \frac{2(x^2+y^2)}{W^2}\} \tag{23}$$

where

$$W = \frac{\sqrt{2}\lambda\sigma_{\Phi T}z}{2\pi a} = \frac{\sqrt{2}\sigma_{n_1}}{\sqrt{a}} z^{3/2} \tag{24}$$

Thus the expected shape of the average intensity profile, at distances large enough to make $\sigma_{\Phi T}^2$ large in comparison with unity, is Gaussian with a beam waist size which depends on the three-halves power of the distance. The shape and distance dependence of the width of the average intensity profile are the same as those obtained by Ishimaru (1978) in his mathematically more complex analysis. The same shape and distance dependence would be expected from the diffusion equation as pointed out by Weston.

5. SOME DEDUCTIONS FROM EXPERIMENT

If the experimentally observed value from King et al (1983) of w = 3m at a distance of z = 37 km, is substituted into equation (24) we get

$$\sigma_{n_1}/\sqrt{a} = 3 \times 10^{-7}. \tag{25}$$

We can attempt to resolve the ambiguity here by incorporating a generally established, if rather approximate, fact - that when a helium-neon laser beam propagates through the atmosphere it has effectively lost its coherence after about 1 km (Hufnagel and Stanley (1964)). This will occur when the phase variance $\sigma^2_{\Phi T}$ of equation (21) is unity, with the result that

$$\sigma_{n_1}\sqrt{a} = 3.2 \times 10^{-9} \tag{26}$$

Combining equations (25) and (26) yields

$$\sigma_{n_1} = 3 \times 10^{-8} \text{ and } a = 1 \text{ cm}.$$

The values given in equations (25) and (26) come from completely different experiments, performed at different times and in radically different environments. (The latter is derived from an experiment in which the laser beam is propagating above the hot surface of a car park). But the fact that the values obtained for σ_{n_1} and a are not unreasonable suggests than an experiment specifically designed to elicit these values could be valuable, particularly in view of the difficulty of measuring σ_{n_1} and a directly.

6. CONCLUSIONS

It has been shown that the intensity formula of turbulent scattering theory can be applied to laser propagation experiments to elicit turbulence parameters which are well nigh impossible to measure directly. In the process it has been demonstrated that Fourier-transform/phase-screen theory is perfectly compatible with the differential equation approach for finding the lateral mutual coherence function, of which the mean intensity is a special case.

7. ACKNOWLEDGEMENTS

I am grateful to Dr. H.E. Rowe and to A.T. & T. Bell Laboratories for encouragement and support.

8. REFERENCES

Bramley, E.N., 1954, "The Diffraction of Waves by an Irregular Refracting Medium" *Proc.Roy.Soc.* **A 225**, 515-518.

Chandrasekhar, S., 1952, "A Statistical Basis for the Theory of Stellar Scintillation", *Mon.Not.Roy.As.Soc.* **112**, 475-483.

Clarke, R.H., 1984, "Analysis of Laser Beam Propagation in a Turbulent Atmosphere", A.T. & Bell Laboratories Technical Journal. To appear in August, 1985.

Consortini, A., Ronchi, L., Scheggi, A.M. and Toraldo di Francia, G., 1963, "Influence of the Atmospheric Turbulence on the Space Coherence of a Laser Beam", *Alta Frequenza*. **32**, 790-794.

Fejer, J.A., 1953, "The Diffraction of Waves in Passing through an Irregular Refracting Medium". *Proc.Roy.Soc.* **A 220**, 455-471.

Flatte, S.M., 1983, "Wave Propagation through Random Media: Contributions from Ocean Acoustics". *Proc. IEEE,* **71(11)**, 1267-1294.

Hufnagel, R.E., and Stanley, N.R., 1964, "Modulation Transfer Function Associated with Image Transmission through Turbulent Media", *J.Opt.Soc.Am.* **54**, 52-61.

Ishimaru, A., 1978, "The Beam Wave Case and Remote Sensing". In "Laser Beam Propagation in the Atmosphere" (J.W. Strohbehn, Ed.). Springer-Verlag, Berlin.

King, B.G., Fitzgerald, P.J., and Stein, H.A., 1983, "An Experimental Study of Atmospheric Optical Transmission". *Bell System Tech. J.,* **62(3)**, 607-629.

Kogelnik, H. and Li, T., 1966, "Laser Beams and Resonators". *Applied Optics*. **5(10)**, 1550-1567.

Lee, R.W. and Harp, J.C., 1969, "Weak Scattering in Random Media, with Applications to Remote Probing". *Proc. IEEE.* **57**, 375-406.

Papoulis, A., 1974, "Ambiguity Function in Fourier Optics", *J.Opt.Soc.Am.,* **64** (**6**), 779-788.

Ratcliffe, J.A., 1956, "Some Aspects of Diffraction Theory and their Application to the Ionosphere". Reports on Progr. in Phys. **19**, 188-267.

Tappert, F.D., 1977, "The Parabolic Approximation Method". In "Wave Propagation in Underwater Acoustics". (Keller, J.B., and Papadakis, J., eds.) Springer-Verlag, Berlin.

Tatarskii, V.I., 1971, "The Effects of the Turbulent Atmosphere on Wave Propagation". Israel Program for Scientific Translations.

Tatarskii, V.I., 1961, "Wave Propagation in a Turbulent Medium". McGraw-Hill, N.Y.

Uscinski, B.J., 1977, "The Elements of Wave Propagation in Random Media" McGraw-Hill, London.

Van Roey, J., van der Donk, J., and Lagasse, P.E., 1981, "Beam Propagation Method: Analysis and Assessment". *J.Opt.Soc.Am.* **71**, 803-810.

Weston, D.E., Private Communication.

Woodward, P.M., 1953, "Probability and Information Theory with Applications to Radar", Pergamon, London.

RADIO WAVE SCATTERING IN THE INTERPLANETARY AND INTERSTELLAR MEDIA

P.J. Duffett-Smith
(Cavendish Laboratory, University of Cambridge)

ABSTRACT

Scattering in irregular celestial plasmas provides a powerful tool in metre-wave radio astronomy both for studying conditions within the plasmas themselves and for measuring the sizes of radio sources with high resolving power. Three investigations are described briefly in which the scattering properties of the interplanetary and interstellar media have been investigated.

1. INTRODUCTION

1.1 Radio astronomers' bathroom windows

When a radio astronomer observes the Universe, he has to do so through several ionised regions. These contain random irregularities of density, and hence refractive index, so that incoming radio wavefronts suffer random phase perturbations. There is generally relative motion between the ionised region and the line of sight to the radio source, with the result that the random diffraction pattern on the ground moves across the radio telescope. The radio source then appears to scintillate if it has a small enough angular size and if the bandwidth of the receiver is sufficiently narrow; otherwise its apparent angular size is broadened by the scattering.

The closest scattering screen is the ionosphere at a height of about 200 km. The spectrum of the density irregularities roughly follows a power law, but those irregularities which are larger than the Fresnel first half-period zone do not contribute to intensity fluctuations (although they do of course affect the phase) so that the scintillation pattern appears to have a characteristic scale size and the fluctuations of intensity a characteristic time scale. Typical values at 81.5 MHz when the scattering is weak are 5 km for the scale size and 10 s for the time scale. There are also quasiperiodic travelling ionospheric disturbances with scales of about 100km and about 10 minutes. These do not affect the scintillation pattern, but cause the dominant phase-shifts in interferometric experiments at metre wavelengths.

Next out from the Earth comes the interplanetary medium, or solar wind - the plasma which fills the space of the solar system. The scattering distance is about 10^8 km, with apparent pattern scales of about 200 km and 1 s. The strength of the scattering varies with the direction of the line of sight through the medium, being largest close to the Sun where the plasma density is greatest. At 81.5 MHz the scattering becomes strong within about 35° of the Sun.

Going further still from the Earth, the next scattering screen is the interstellar medium which fills the space between the stars in our Galaxy. Here the scattering distance might typically be 3.10^{15} km and the pattern scales 10^6 km and 10 minutes. The scattering is very strong at metre wavelengths.

Outside the Galaxy, there is undoubtedly an intergalactic medium whose mean density increases with redshift. Remarkably little is known about it and scattering within it has yet to be detected.

1.2 The uses of scattering

One's attitude to the scattering depends very much on one's experimental point of view. It may simply be a nuisance, degrading the observations. On the other hand, if you think you already know about the radio source you are observing through the plasma, you can use the scattering to probe the medium along the line of sight to the radio source, gaining knowledge of the fluctuations in electron density.

Alternatively, if you think you know about the medium, you can use the scattering to measure the radio source. Generally, large sources do not scintillate while point sources scintillate most strongly. Intermediate sources scintillate to some extent which can be measured and compared with models to derive angular sizes. The technique is useful because of the very high resolving power available in the interplanetary and interstellar media. For example, the interplanetary medium resolves sources in the range 0.1 - 2 arcsec, equivalent to a radio telescope with a 6000-km baseline at 81.5 MHz. Such instruments are not normally available. The interstellar

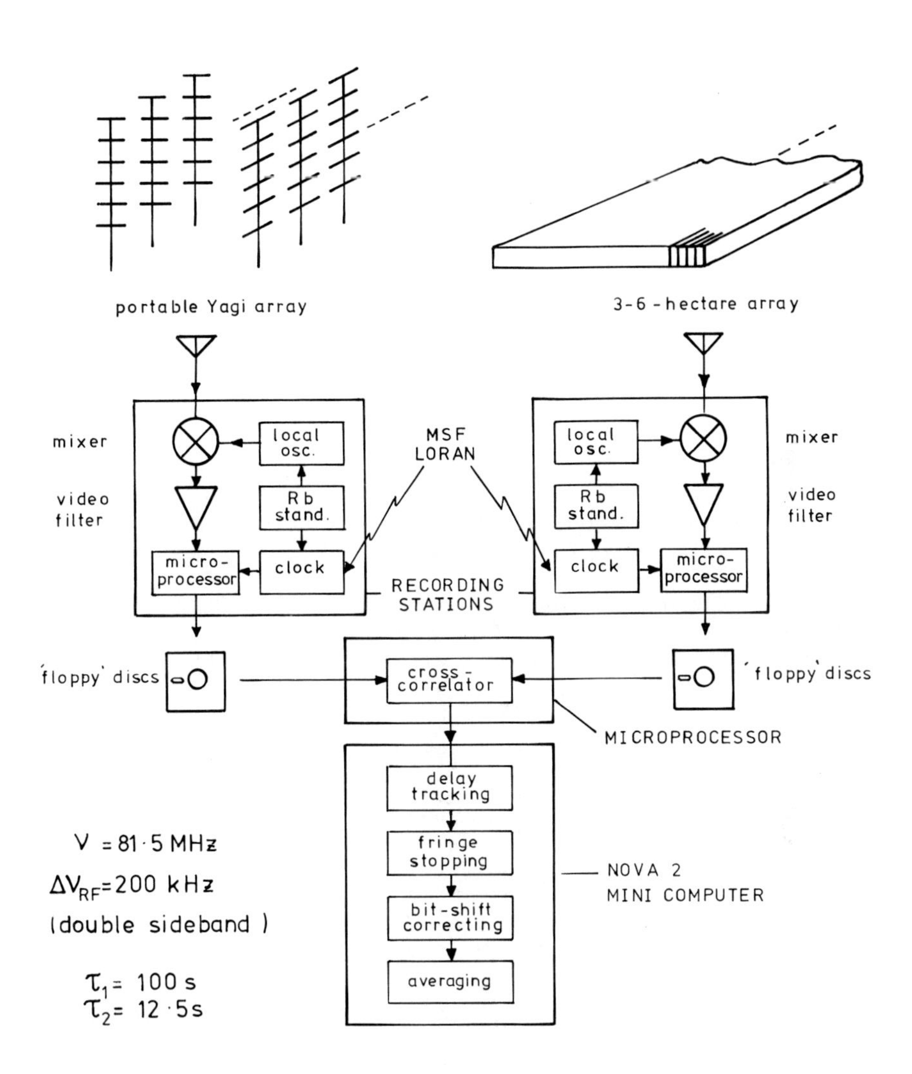

Fig. 1 A schematic diagram of the long-baseline interferometer used by Rees & Duffett-Smith to measure the phase power spectrum of the solar wind.

medium has a resolution in the microarcsecond range.

In this contribution, I outline briefly the techniques that have been used in three investigations of ionised regions, two of the interplanetary medium and one of the interstellar medium.

2. THE PHASE-POWER SPECTRUM OF THE SOLAR WIND

2.1 Experimental details

Rees and Duffett-Smith (1984) have measured the phase-power spectrum of the solar wind at 81.5 MHz using a radio interferometer with a baseline of 130 km. They observed the effect of the scattering on the apparent fringe visibilities of two unresolved sources, 3C48 and 3C147. The medium introduced random phase fluctuations across the wavefront which reduced the apparent visibility to

$$\gamma = \gamma_o \exp(-\phi_o^2/2)$$

where ϕ_o^2 was the variance of the phase difference between the two aerials and γ_o the visibility which would have been measured in the absence of the scattering.

A schematic diagram of the experiment is shown in Figure 1. The radio antenna used at Cambridge was the 36,000-m^2 array, consisting of 4096 full-wave dipoles phased to provide full-gain beams along the local meridian. The other end of the interferometer consisted of two identical rows of 6-element yagi antennas set to observe in orthogonal polarisations to allow for differential Faraday rotation in the ionosphere between the two ends of the interferometer. Eight yagi antennas spaced 1 wavelength apart were used in each row, and the whole array was set up on a rugby training field at Christ's Hospital, near Horsham in Sussex. The radio receiving, timing, and recording equipment was housed in a motor caravan which also served as hostel accommodation during the 2 months of observing.

The radio signals, of total bandwidth 200 kHz, were digitised and recorded on to "floppy discs" using standard microcomputers, and later correlated using the same micro-computers. To achieve coherence, the local oscillators of the receivers at each end were phase-locked to rubidium atomic frequency standards with stabilities of several parts in 10^{11}. The data acquisition at each end was synchronised to within a few microseconds using the coded radio signals on 60 kHz from MSF Rugby, and on 100 kHz from the LORAN-C navigation signals at Sylt in Germany. The correlated data were transferred to a more-powerful computer for further analysis, finally yielding two numbers for each observation: the mean-square phase deviation for coherent integration over 12.5 s, and over 100 s (the total length of any one observation). Observations were made every day for two months while the lines of sight to the radio sources covered the solar-elongation range 60° - 20°. (The solar elongation is the angle between the Sun and the radio source).

2.2 The experimental filters

To understand this experiment further, it is necessary to appreciate the two filters which applied (see Figure 2). The first of these may be called

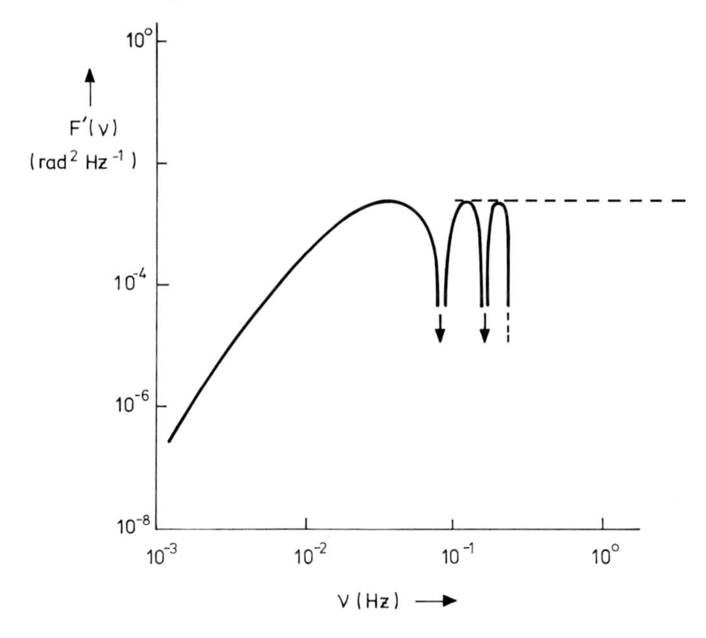

Fig. 2 The experimental filter. The effective phase power spectrum may be regarded as the true spectrum multiplied by the solid curve.

the baseline filter and it acted to remove the effects of some pattern scales. For example, the scale whose length was exactly the same as that of the baseline would have had no effect on the interferometer phase when moving parallel to the baseline, while one exactly half as long would have had maximum effect. The data can be thought of as being multiplied by the filter, although its effect is really through a two-dimensional strip integral.

The second filter may be called the integration-time filter. Fluctuations much longer than the coherent integration time have little effect, while those much shorter have maximum effect. The solid curve in Figure 2 shows the difference between the combined effects of the two filters for the two averaging periods of 12.5 and 100 s. We may regard the effective spectrum as the real spectrum multiplied by this curve.

2.3 Models of the spectrum

Power-law spectra, known to be good representations from spacecraft measurements at low spatial frequencies, were fitted to the experimental data. The results are shown in Figure 3 together with those of an ISEE propagation experiment. Fresnel diffraction, which acts to remove the effects of scale sizes larger than about the first half-period zone, enables the form of the spectrum to be modelled by a Gaussian curve for the purposes of calculating interplanetary scintillation. A recent model of this sort by Readhead, Kemp & Hewish (1978) is also shown in Figure 3. It is consistent with the data, but may have slightly too little phase power associated with it.

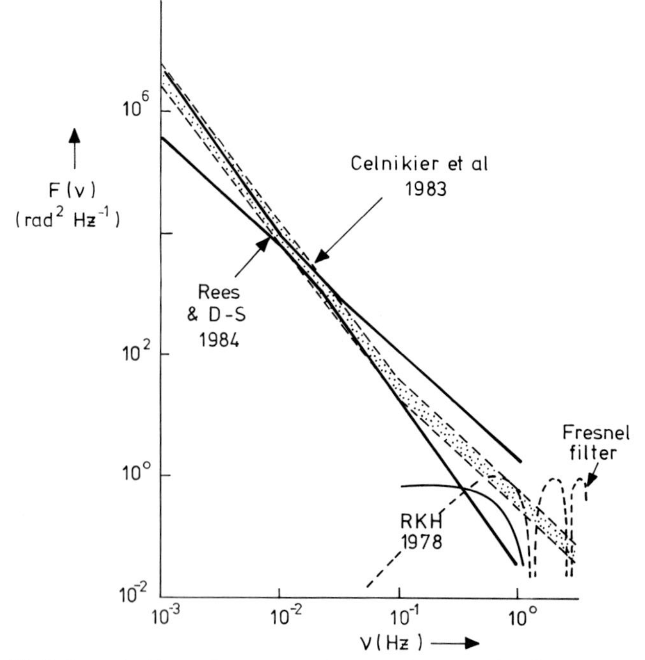

Fig. 3 Models of the phase power spectrum of the solar wind. The range of models fitting the results of Rees & Duffett-Smith lies between the solid curves, while that fitting the results of the ISEE propagation experiment by Celnikier et al. is shown by stippled shading. The experimental filter introduced by Fresnel diffraction effects for intensity scintillation is shown by the dashed curve.

3. INTERPLANETARY WEATHER

3.1 Experimental details

When the interplanetary scintillation (IPS) index of a radio source is measured as a function of solar elongation, it is observed to vary in a systematic manner depending on the angular size of the radio source (see Figure 4). Superimposed on this systematic trend

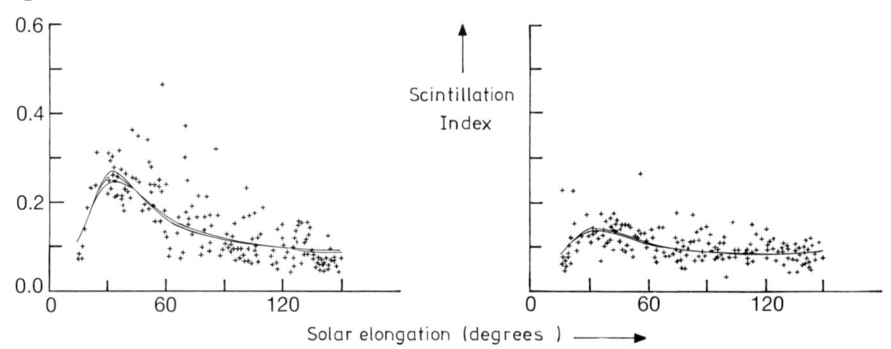

Fig. 4 The variation of interplanetary scintillation index with solar elongation. Individual measurements at 81.5 MHz are shown by '+' for two radio sources of different angular sizes, 0.3 arcsecond on the left and 1.1 arcseconds on the right. Mean trends are indicated by the solid curves.

are random variations which may cause the observed index to be up to a factor of 3 times greater or smaller than the expected value. These random variations are caused by changes within the solar wind itself, of density integrated along the line of sight through the medium. Such variations are often found to be correlated over large areas of the sky and might then be termed the interplanetary weather.

Tappin & Hewish (1984) have used the 36,000-m^2 array at Cambridge to monitor the weather. First, they selected a more-or-less uniform grid of 2500 scintillating radio sources covering most of the sky visible from Cambridge. They made daily observations of these radio sources over the course of a year to establish the mean trend for each of them. It is known that the average properties of the interplanetary medium are remarkably constant as far as IPS is concerned so that the same mean curve might be assumed to apply for several years. Next, the daily value of the scintillation index, m, of each source was compared with that expected from the mean curve, m_o, to give an enhancement factor

$$g = m/m_o.$$

This factor reflected the integrated state of the solar wind along the line of sight to the radio source. Tappin & Hewish plotted the enhancement factor for each source every day for 400 days to get detailed weather maps of the interplanetary medium.

3.2 A weather event

Figure 5(a-j) shows the progress of one event which lasted for 8 days. A "puff" of high density emerged from the Sun and travelled outward, enveloping the Earth, followed by a region of low density travelling outward in the same way. Figure 5 displays the daily weather maps in sun-centred coordinates. The circle centred on the Sun represents the locus of points which have a solar elongation of 90°; points outside the circle lie outside the Earth's orbit while points inside the circle lie inside the Earth's orbit. Enhanced scattering ($g > 1$) is shown by lighter shading, decreased scattering ($g < 1$) by darker shading.

The importance of these weather events lies partly in the fact that they are responsible for geomagnetic disturbances, causing loss of HF radio links and fluctuations in magnetometer readings. The IPS method presents the only viable method of studying the events on a large scale and, since they can be detected several days in advance of their striking the Earth, the method could be used for interplanetary weather forecasting. It is relatively inexpensive, effective, and might assume some importance with the increase of extra-vehicular activity (EVA) in space missions.

4. INTERSTELLAR SCATTERING

4.1 Experimental details

Rees & Duffett-Smith (in preparation) have used the Cambridge 36,000-m^2 array and the method of IPS to map the scattering in the interstellar medium. They used the mean variation of IPS index with solar elongation to determine the apparent angular sizes of about 3200 sources in the area of sky between declinations 20° and 60° and between right ascensions 0^{h} and 12^{h}. Other investigations (e.g. Duffett-Smith, Purvis & Hewish 1980) have shown that the IPS 'mean' angular size in this region is about 0.6 arcsec while the interstellar scattering angle at 81.5 MHz is approximately given by

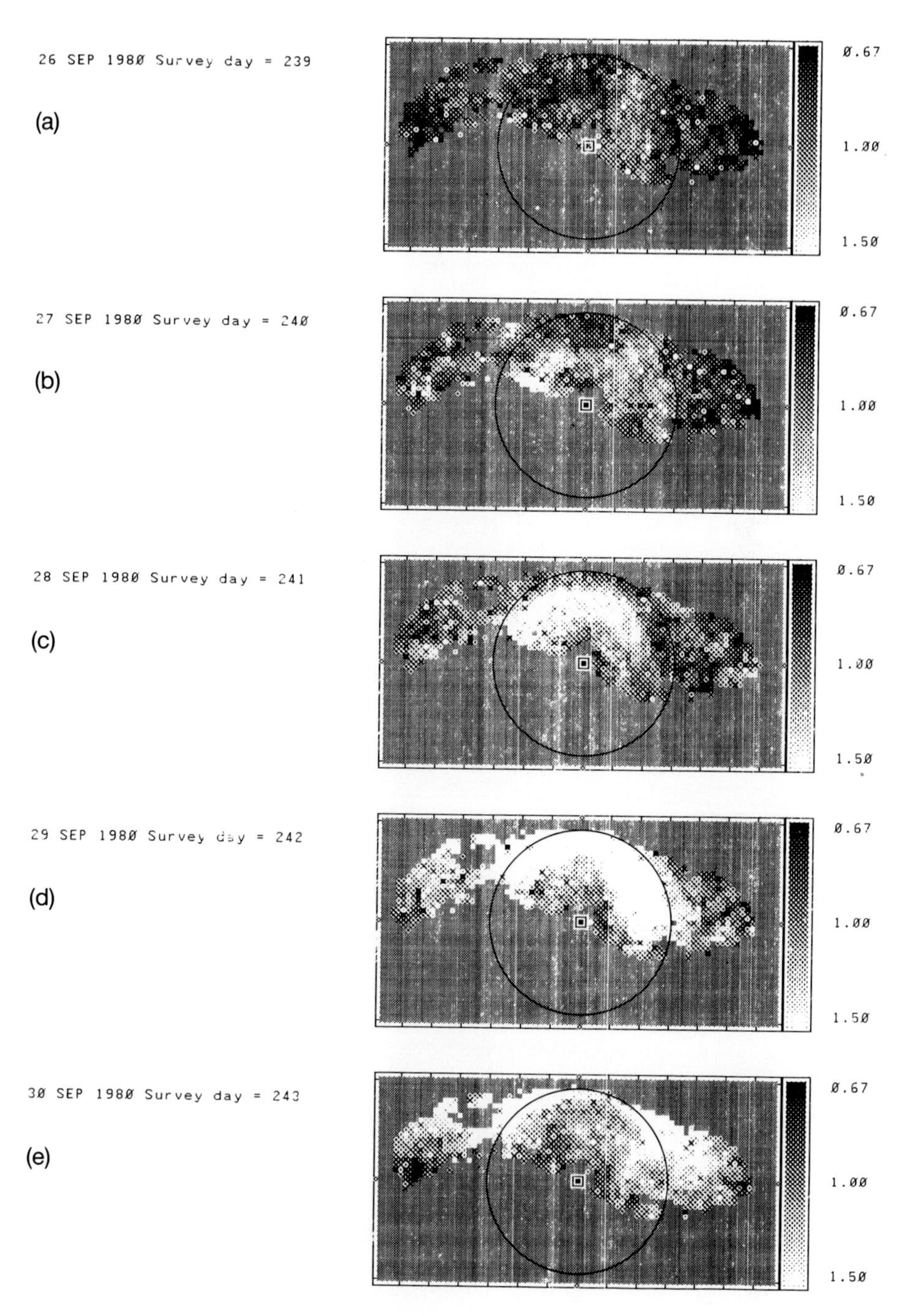
26 SEP 198Ø Survey day = 239
(a)
Ø.67
1.ØØ
1.5Ø
27 SEP 198Ø Survey day = 24Ø
(b)
Ø.67
1.ØØ
1.5Ø
28 SEP 198Ø Survey day = 241
(c)
Ø.67
1.ØØ
1.5Ø
29 SEP 198Ø Survey day = 242
(d)
Ø.67
1.ØØ
1.5Ø
3Ø SEP 198Ø Survey day = 243
(e)
Ø.67
1.ØØ
1.5Ø

Fig. 5

Ø1 OCT 198Ø Survey day = 244

(f)

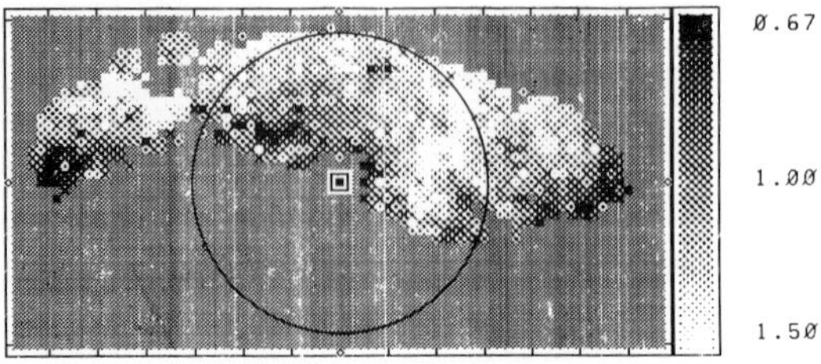

Ø.67
1.ØØ
1.5Ø

Ø2 OCT 198Ø Survey day = 245

(g)

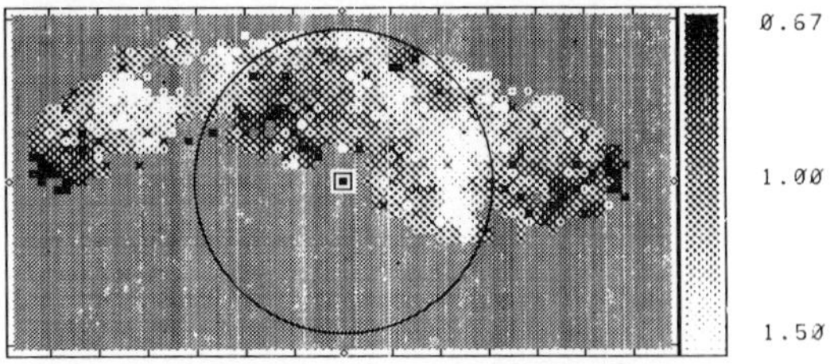

Ø.67
1.ØØ
1.5Ø

Ø3 OCT 198Ø Survey day = 246

(h)

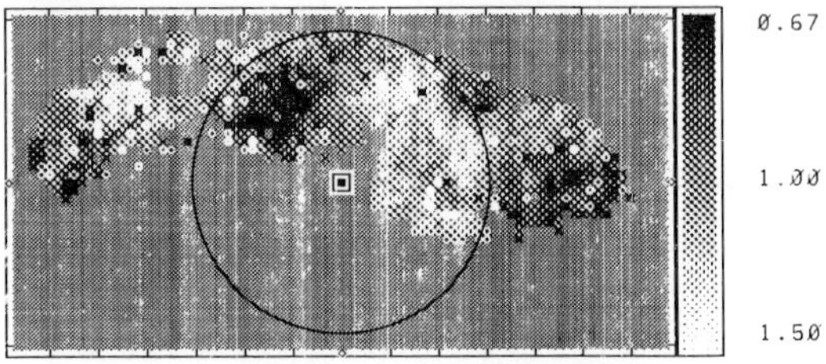

Ø.67
1.ØØ
1.5Ø

Ø4 OCT 198Ø Survey day = 247

(i)

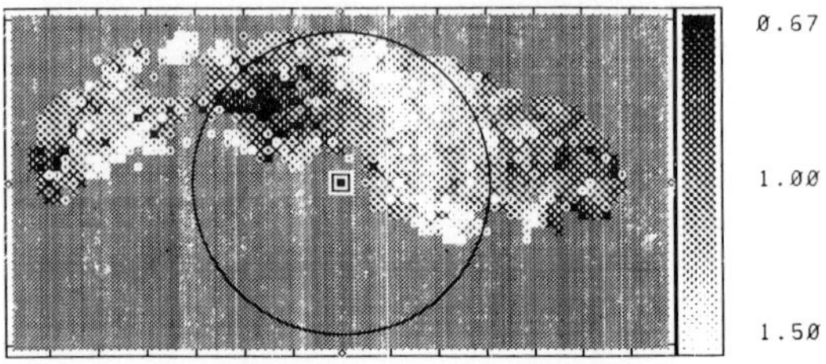

Ø.67
1.ØØ
1.5Ø

Ø5 OCT 198Ø Survey day = 248

(j)

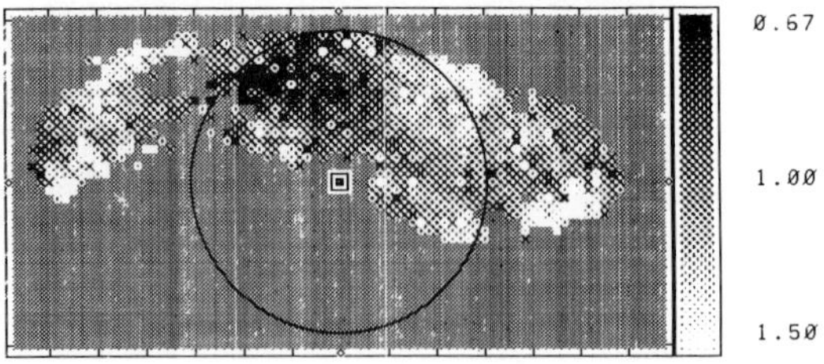

Ø.67
1.ØØ
1.5Ø

Fig. 5 (a-j) An interplanetary weather event. Each map shows the enhancement factor measured on the date indicated. Darker shading indicates less scattering, lighter shading indicates more scattering. The sun is in the centre of the map and the line of 90° solar elongation is shown by the circle. Uniform grey shading indicates the region of sky which is not seen by the 3.6-hectare array at Cambridge. The points are plotted in heliocentric ecliptic coordinates using Mollweide's equal area projection. East is to the left and north to the top.

$$\theta_s = 0.15/(\sin|b|)^{\frac{1}{2}} \text{ arcsec}$$

where b is the galactic latitude (Duffett-Smith & Readhead, 1976). The apparent angular size of a radio source, measured by IPS and viewed through the interstellar medium, is given by

$$\theta_A^2 = \theta_o^2 + \theta_s^2$$

where θ_o is the intrinsic size of the source. The scattering becomes comparable to θ_o, and hence θ_A is significantly changed, for $|b| \lesssim 10^o$.

4.2 A map of interstellar scattering

The apparent angular sizes of the radio sources were averaged in areas $5^o \times 5^o$ and plotted in celestial coordinates to obtain a map of the variation of interstellar scattering. Preliminary results are shown in Figure 6 together with contours of constant galactic latitude. It is apparent that the scattering increases markedly towards the galactic plane as expected, but that it is patchy. There is also an unexpected region of enhanced scattering in the lower right-hand corner of the map, but the reality of this feature has yet to be established firmly. Further refinements need to be made to the technique to improve the accuracy of the maps, but it is already clear that it provides a powerful tool in the exploration of the Galaxy.

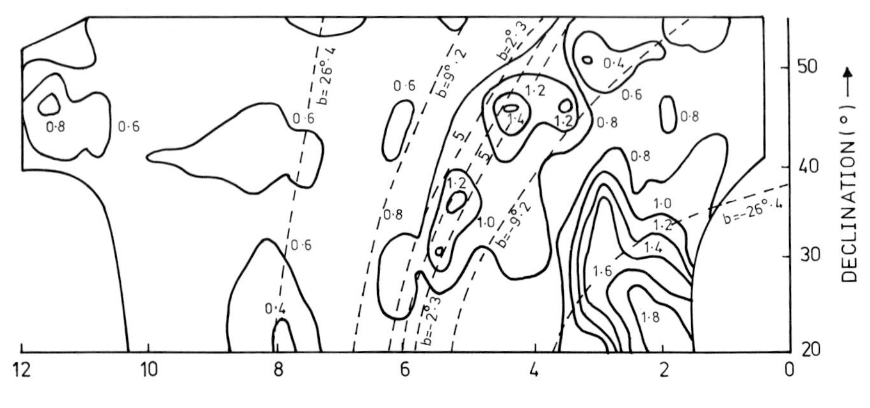

Fig. 6 A map of interstellar scattering at 81.5 MHz between 0^h < right ascension < 12^h and 20° < declination < 60°. The contours of scattering are labelled in units of arcseconds. Contours of constant galactic latitude are indicated by the dashed curves. The reality of the scattering feature in the lower right-hand corner of the map has yet to be verified.

REFERENCES

Celnikier, L.M., Harvey, C.G., Jegou, R., Kemp, M. & Moricet, P., 1983, "A determination of the electron density fluctuation spectrum in the solar wind, using the ISEE propagation experiment". *Atron. Astrophys.*, **126**, 293.

Duffett-Smith, P.J., Purvis, A. & Hewish, A., 1980, "A statistical study of faint radio sources at 81.5 MHz - 1. The data". *Mon.Not. R. astr. Soc.*, **190**, 891.

Duffett-Smith, P.J. & Readhead, A.C.S., 1976, "The angular broadening of radio sources by scattering in the interstellar medium". *Mon.Not. R. astr. Soc.*, **174**, 7.

Rees, W.G., & Duffett-Smith, P.J., 1984, "The phase power spectrum of the solar wind measured by long-baseline interferometry at 81.5 MHz". *Mon.Not. R. astr. Soc.*, **212**, 463.

Rees, W.G. & Duffett-Smith, P.J. 1984, "A map of interstellar scattering in the region 0^h < RA < 12^h and $20^o < \delta < 60^o$". In preparation for *Mon.Not. R. astr. Soc.*, (publication 1985).

Readhead, A.C.S., Kemp, M.C. & Hewish, A., 1978, "The spectrum of small-scale density fluctuations in the solar wind". *Mon.Not. R. astr. Soc.,* **185**, 207.

Tappin, S.J. & Hewish, A., 1984, "Images of interplanetary disturbances" (provisional title). In preparation for *Sol. Phys.* (publication 1985).

FREQUENCY DRIFT IN PULSAR SCINTILLATION

F.G. Smith
(Jodrell Bank, Macclesfield, Cheshire)

The dynamic spectrum of scintillation in a pulsar often showns organised patterns, in which peaks of intensity consistently drift in frequency at a definite rate. We have measured this drift rate for 32 pulsars. The results and their interpretation in terms of the spectrum of turbulence in the interstellar medium are set out in a paper by Smith and Wright, recently submitted for publication.

Out of a sample of 63 pulsars, selected partly as a set in which the scintillation bandwidth was known to be between 100 kHz and 5 MHz at our observing frequency of 408 MHz, and supplemented by other pulsars with dispersion measure less than 60 pc cm^{-3}, we were able to measure a rate of frequency drift in 25 and obtain a lower limit for drift rate in a further 7. We also measured the characteristic bandwidth and fading rate for these 32 pulsars. Most of the others gave too low a signal-to-noise ratio for any such measurements to be made.

Our analysis set out first to establish that frequency drifting is related to turbulence distributed along the whole of the path to the pulsar, rather than to an isolated cloud. We therefore examined the relation of the drift rate $\frac{d\nu}{dt}$ to the pulsar distance Z and the velocity V_s of the scintillation pattern, and found that the large scale turbulence responsible for frequency drifting is closely related to the smaller scale turbulence responsible for scintillation.

We then find the relation between the angles of scattering θ_s involved in scintillation and the angle of refraction θ_r

involved in frequency drifting. The ratio $m = \theta_r/\theta_s$ depends on the spectrum of irregularities in the interstellar medium. Our observations give $m\cos\phi$, where ϕ is the angle between the direction of frequency dispersion and the direction of pattern velocity. The theoretical relation is

$$m\cos\phi = 8^{-\frac{1}{2}} B \tau^{-1} \frac{d\nu}{dt}^{-1} .$$

Our data give a mean value $\bar{m} = 0.35$, (assuming that ϕ is randomly distributed).

Assuming a power law spectrum for irregularities $P(q) = A(qL)^{-\alpha}$, with $\alpha < 4$, the angles θ_s and θ_r are determined by two scales of irregularity a_{min} and a_{max}, which are the limits pointed out by Scheuer for deep scintillation and multi-path propagation respectively. Following Roberts and Ables (1982), we expect

$$\theta_s \propto a_{min}^{\frac{\alpha-4}{2}} \text{ and } \theta_r \propto a_{max}^{\frac{\alpha-4}{2}} .$$

The ratio $\frac{a_{max}}{a_{min}}$ is approximately $B\nu^{-1}$, which for our observations ranged from 200 to 400.

For a Kolmogorov spectrum we therefore expect m to lie between 0.25 and 0.45, nicely bracketting the observed value of 0.35.

We therefore regard our observations as strong support for the Kolmogorov spectrum. We do not speculate, however, on the physical processes which generate such a spectrum.

REFERENCES

Smith, F.G. and Wright N.C., 1985, "Frequency drift in pulsar scintillation", Mon. Not. R. Astr. Soc. (in press).

Roberts, J.H. and Ables, J.G., 1982, M.N.R.A.S. **201**, 1119.

QUASI-PERIODIC SCINTILLATION PATTERNS

A. Hewish
(Cavendish Laboratory, Cambridge)

ABSTRACT

The dynamic spectra of naturally occurring scintillation patterns sometimes display significant quasi-periodic variations. The nature of this phenomenon is discussed and it is shown how these patterns may be used to investigate the power spectrum of the irregularities in the random medium which causes scintillation. Application of the method to pulsar scintillation indicates that Kolmogorov turbulence is often a poor representation of the interstellar plasma.

1. INTRODUCTION

Scintillation techniques are of particular value for the remote-sensing of astrophysical plasmas which are inaccessible to direct measurement. They have been widely used to study the interplanetary medium where suitably coherent illumination of the plasma is provided by extragalactic radio sources of small angular size, notably quasars and hot-spots in the outer lobes of powerful radio galaxies. Similar methods were later applied to the interstellar plasma, following the discovery of pulsars and the realisation that their very small physical size gave rise to radiation of even greater spatial coherence. The distances and sizes of the interstellar clouds responsible for pulsar scintillation are such that the characteristic timescales tend to be rather long. Thus correspondingly larger observation times are necessary to determine ensemble-average parameters of the wavefield. This disadvantage is compensated by the opportunity provided for the study of deterministic phenomena related to the particular arrangement of a given sample of irregularities along the line of sight.

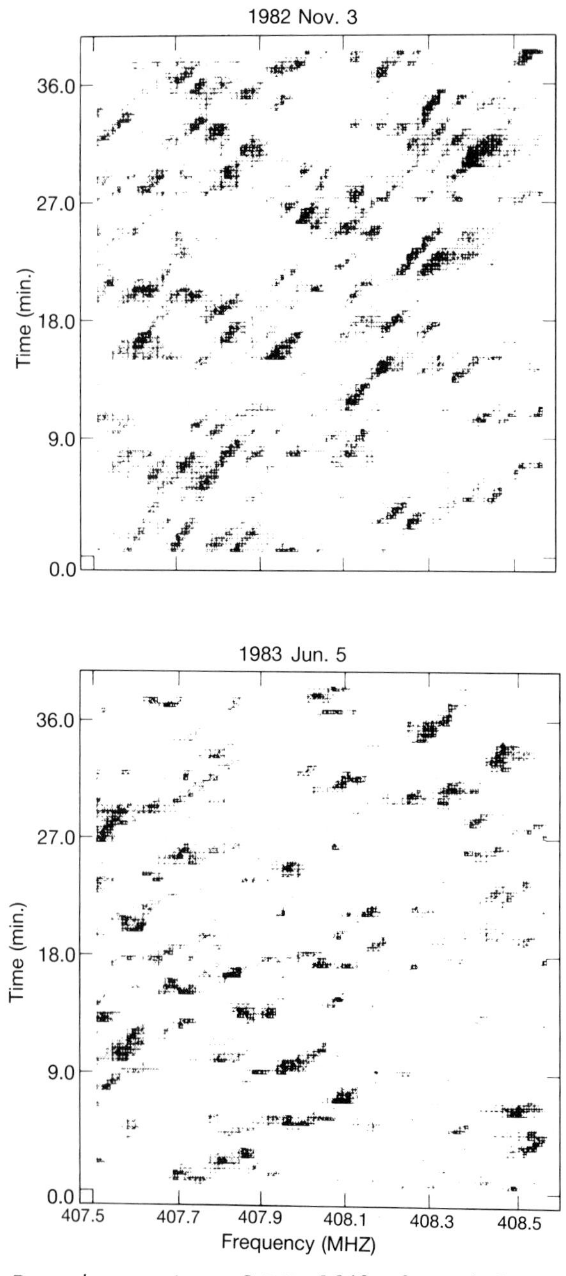

Fig. 1 Dynamic spectra of PSR 1642-03 at 408 MHz. Eight grey-scale levels above the e^{-1} threshold are displayed.

One notable deterministic feature of pulsar scintillation is the presence of quasi-periodic fringes in the dynamic spectra of the wavefield, an example of which is illustrated in Fig. 1. Similar effects have been seen in many other cases and it is likely that this phenomenon is typical of pulsar scintillation. The conditions under which quasi-periodic scintillation could arise were considered by Hewish (1980) who discussed a model combining random diffraction from a large number of small-scale irregularities with more systematic refraction caused by much larger irregularities. It was shown that quasi-periodic fringes would occur when typical angles of refraction θ_r were greater than the width of the angular spectrum θ_s causing random diffraction. Roughly speaking, the small-scale irregularities generate a speckle pattern which is refracted, or "steered", by the large-scale irregularities. When conditions are such that a point in the observing plane is intersected by a small number of refractively steered beams the random pattern must be systematically modulated by, for example, simple two or three - source interference effects. Thus, while the random pattern may change on a relatively short timescale, the systematic interference effects will persist for so long as the geometry of the larger scale refracted beams remains more or less unchanged.

Scintillation patterns exhibiting quasi-periodicity contain information on the relative magnitudes of θ_r and θ_s which may in turn be used to investigate the form of the wavenumber spectrum of the irregularity spectrum on two scales. No adequate theory yet exists for an analytical treatment of these deterministic phenomena which involve strong scattering in an extended medium. The considerations in this paper represent a physical, rather than mathematical approach. Nevertheless, even at this level, observations of pulsar scintillation are often found to be inconsistent with Kolmogorov models of turbulence in the interstellar gas. This result contrasts with deductions based on ensemble-average analysis of the radio-frequency autocorrelation function (eg. Armstrong and Rickett, 1981). The discrepancy may reflect the fact that information about θ_r comes from regions of the spectrum at smaller wavenumbers than those contributing to the speckle pattern of intensity and that a simple power law model is inappropriate on an extended range of scales.

2. THEORY

In the following analysis we show how θ_r and θ_s are related to the spectrum of the irregularities and examples will be given of different types of dynamic spectra that should be seen when $\theta_r > \theta_s$. Emphasis will be placed on physical arguments rather than on an exact treatment. Most theoretical work germane to this problem has concentrated on the solution of the equation for the fourth moment of the wave field. Such an approach is less useful here since we are concerned more with the recognition and analysis of deterministic interference phenomena which modulate the stochastic, diffractive scattering process, than with ensemble average properties.

2.1 The dependence of θ_r and θ_s on the medium

We assume for simplicity that the extended medium may be regarded as a thin screen midway between the source and the observer. Let $P_N(K_x, K_y, K_z)$ be the wavenumber power spectrum of the density variations in the medium where the spatial wavelength is $2\pi K^{-1} = 2\pi(K_x^2 + K_y^2 + K_z^2)^{-1/2}$. In what follows it is helpful to consider the spectrum as divided into two portions, although in reality the distinction becomes blurred at the interface. Let K_F define a critical wavenumber such that spectral components for which $K > K_F$ cause random intensity variations at the observer due to the interference of waves scattered into an angular spectrum of width θ_s, while components with $K < K_F$, considered in isolation, would give rise to tilted wavefronts on a scale such that no interference was possible. For a strongly scattering screen at distance z from the observer, geometrical optics shows that irregularities causing angular tilts of magnitude θ_r on a scale L will not contribute to intensity variations if $L > z\,\theta_r$. Hence $K_F \sim 2\pi L^{-1}$.

To estimate the magnitude of the phase gradients imposed upon wavefronts at the screen by different portions of the spectrum consider the phase modulation $\Phi(x,y)$ across the plane. By a well known result the wavenumber spectrum of $\Phi(x,y)$ is

$$P_\Phi(K_x, K_y) = AP_N(K_x, K_y, 0) \text{ where A is constant}$$

Hence the mean square gradient in some direction (which can be along x without loss of generality) imposed by irregularity components in the range K_x to $K_x + \Delta K_x$ is

$$<(\delta\Phi/\delta x)^2> = \frac{K_x^2}{2} \Delta K_x \int_{-\infty}^{\infty} P_\Phi (K_x, K_y)\, dK_y \qquad (1)$$

This may be expressed in terms of the angular tilts across the wavefront by putting

$$\delta\theta_r = \frac{\lambda}{2\pi} \quad \frac{\delta\Phi}{\delta x}$$

where λ is the radio wavelength.

The magnitudes of θ_r and θ_s are therefore given by

$$<\theta_r^2> = \frac{\lambda}{2\pi} \int_0^{K_F} \frac{K_x^2}{2} \left[\int_{-\infty}^{\infty} P_\Phi(K_x, K_y)\, dK_y \right] dK_x \qquad (2)$$

$$<\theta_{rs}^2> = \frac{\lambda}{2\pi} \int_{K_F}^{\infty} \frac{K_x^2}{2} \left[\int_{-\infty}^{\infty} P_\Phi(K_x, K_y)\, dK_y \right] dK_x \qquad (3)$$

For power law spectra of the type

$$P_N (K_x, K_y, K_z) \propto (K_x^2 + K_y^2 + K_z^2)^{-\alpha/2}$$

The evaluation of (2) will depend upon fixing some lower bound, K_e, on K, as $K \to 0$, if $\alpha > 4$. The evaluation of (3) will similarly depend upon fixing an upper bound, K_u, for $K > K_F$ if $\alpha \leq 4$; provided that $\alpha > 2$, so that Φ^2 does not diverge at large K, K_u is effectively the limit above which only weak scattering occurs. Hence

$$\int_{K_U}^{\infty} \int_{-\infty}^{\infty} P_\Phi \, dK_y \, dK_x \geq 1 \text{ rad.}$$

The general behaviour of θ_r and θ_s is best illustrated by returning to expression (1). Consider for example a Kolmogorov spectrum for which $\alpha = 11/3$. In this case components in the range ΔK_x produce angular tilts given by

$$\langle \delta\theta_r^2 \rangle \propto K_x^{-2/3} \Delta K_x .$$

Putting $\Delta K_x = \beta K_x$, where $\beta < 1$ to define an appropriate range, we obtain $\langle \delta\theta_r^2 \rangle \propto K_x^{1/3}$. This is the familiar result that angular tilts for Kolmogorov turbulence vary as $L^{-1/6}$, where L is the scale over which tilts are considered. Clearly the smaller scales (Larger K) dominate the angular tilts in a Kolmogorov spectrum so that in this case $\theta_r > \theta_s$ demands $\alpha > 4$. In order to investigate the constraints imposed upon spectra consistent with the observed quasi-periodic structure, we now turn to a closer examination of the types of dynamic spectra expected when $\theta_r > \theta_s$.

2.2 The generation of quasi-periodic patterns

When plane waves enter a uniform random medium the diffraction pattern after propagation through a distance z is characterised by intensity fluctuations which have a correlation time $t_c \sim c/\pi z\theta_s^2$. In these relations θ_s is a measure of angular spectrum of waves scattered by a large number of irregularities for which $K > K_F$, V is the speed at which the pattern is convected past the observer and f_c is the range beyond which correlation falls to a low value (Lee 1976). In what follows we take z to be the distance of the pulsar from the observer, and assume that the same scattering occurs in a thin screen at distance z/2. The appropriate relations then become $t_c \sim \lambda/\pi\theta_s V$, $f_c \sim 4c/\pi z\theta_s^2$. The dynamic spectrum of such a pattern is sketched in Fig. 2(a).

If, in addition to the angular scattering θ_s, the wavefront is tilted through an angle θ_r due to the presence of large-scale irregularities for which $K < K_F$, the dynamic spectrum is sheared by dispersion in the medium to produce systematic gradients of magnitude

$$\frac{df}{dt} \sim \frac{V_x f}{z\theta_r}$$

where V_x is the component of V along the direction of refractive "steering" of the pattern (Hewish 1980).

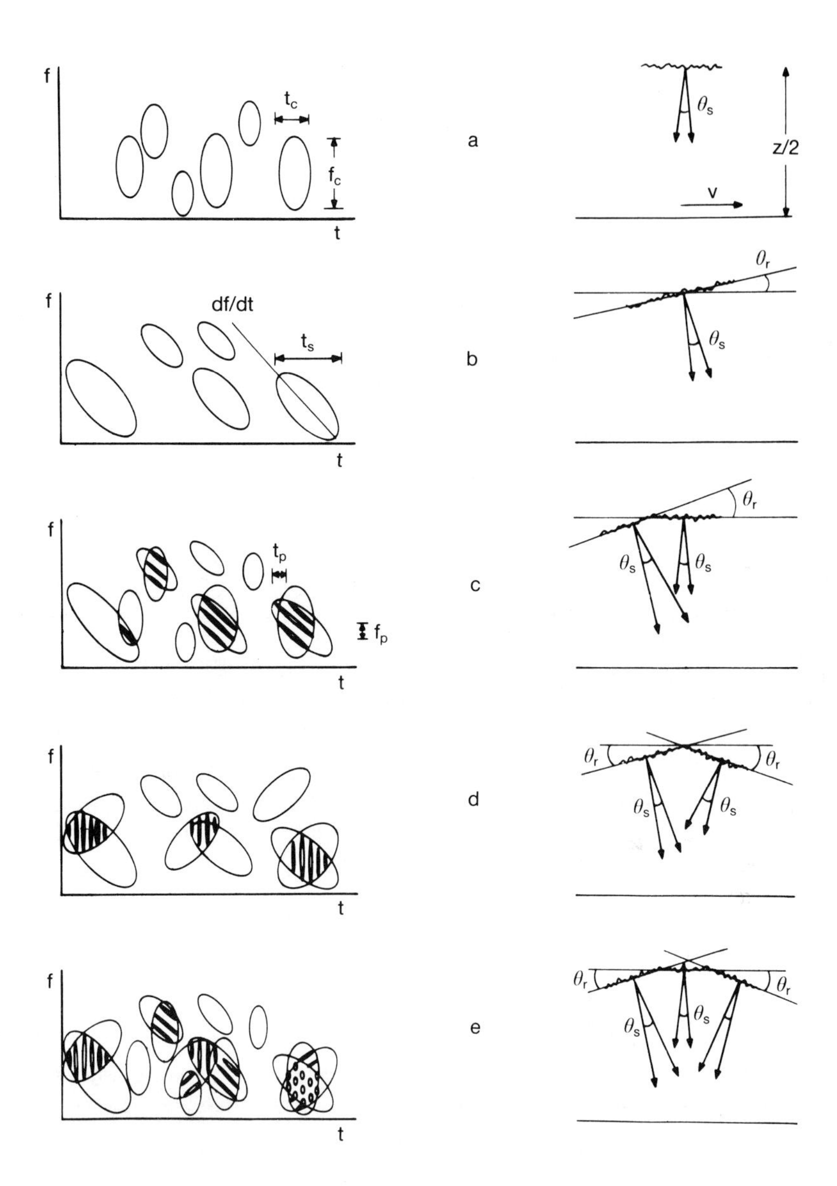

Fig. 2 Schematic dynamic spectra of pulsar scintillation for various geometries of large-scale refraction.

This effect is shown in Fig. 2(b). The total shear t_s across the correlation bandwidth f_c is

$$t_s = f_c \frac{dt}{df} = \frac{f_c z\theta_r}{V_x f} \tag{4}$$

Hence $t_s/t_c = 4\theta_r/\theta_s$

Equation (4) may provide a useful measure of θ_r/θ_s but the method is of limited value since the case $t_s/t_c > 4$, corresponding to $\theta_r > \theta_s$, demands good alignment of V_x and V. When the latter does not apply, both the correlation bandwidth and the total shear are reduced by the factor θ_s/θ_r when $\theta_r > \theta_s$(Hewish 1980). Hence the method underestimates θ_r/θ_s in general. This modification of the correlation bandwidth due to the presence of large-scale irregularities shows that analysis of the frequency correlation function to obtain the irregularity spectrum (e.g. Armstrong and Rickett 1981) can lead to errors if $\theta_r > \theta_s$. It should be noted that the absence of systematic gradients df/dt in the dynamic spectrum does not necessarily imply $\theta_r < \theta_s$ since V_x may be small.

When $\theta_r > \theta_s$ the intensity pattern may result not only from the superposition of a large number of waves scattered through angles θ_s, but also from the interference of a small number of wavefronts tilted through angles θ_r. Simple geometrical optics suggests that this could occur if θ_r is caused by refraction by irregularities of wavenumber $K \sim 2\pi\ (z\theta_r)^{-1}$. On the thin screen model that we have adopted this means that the net effect of these large-scale irregularities is to produce rough focussing at a distance somewhat less than that of the observer from the screen. The random pattern characterised by f_c and t_c will then be modulated by interference fringes on a finer scale provided that the field incident upon the large irregularities is coherent over distances $\sim z\theta_r$. Some examples of dynamic spectra that could result from the interference of a few tilted wavefronts, combined with small-angle scattering are sketched in Fig. 2. The spatial scale of the interference fringes is of order $2\lambda/\theta_r$ which gives a temporal period $t_p \sim 2\lambda/\theta_r V_x$. The corresponding periodicity in radio frequency results from typical path differences ~ $z\theta_r^2/4$ between tilted wavefronts, which gives a separation in frequency $f_p \sim 4c/z\theta_r^2$. Roberts and Ables (1982) derived similar relationships; our magnitudes follow from the thin screen assumption discussed at the beginning of this section.

An important feature of dynamic spectra such as those illustrated in Fig. 2 (c-c) is that they are characterised by three distinct scales. The smallest scale is the fringe scale t_p and f_p. Next there is the larger scale t_c and f_c which governs amplitudes and phases of the fields across the tilted wavefronts; thus the fringe groups can only maintain coherent fringe-phase, and fringe visibility, over scales ~ t_c and f_c. This effect can produce rapid changes in the fringe patterns, while the overall arrangement of large-scale tilted wavefronts is relatively constant. Finally there is the longest time-scale $t_r \sim 2\pi(K_oV)^{-1}$ where K_o is the wavenumber of the irregularities giving rise to systematic tilts; this is the time required for the largest irregularities to be replaced.

The simple considerations that we have outlined show how dynamic spectra may be used to make quantitative estimates of θ_r/θ_s and this theory is applied to observations of PSR 1642-03 in the next section. As a simple diagnostic we note that $t_c/t_p \sim \theta_r/2\pi\theta_s$ and $f_c/f_p \sim \theta_r^2/\pi\theta_s^2$, so that simply counting the number of fringes within a coherent group gives an approximate measure of θ_r/θ_s. These relations also show that quasi-periodicity may be more evident along the frequency axis than along the time axis owing to the squared dependence on θ_r and θ_s.

3. APPLICATION OF THE METHOD

A good example of quasi-periodic scintillation is provided by observations of pulsar PSR 1652-03 at frequencies in the neighbourhood of 408 MHz as shown in Fig. 1. (I am indebted to A. Wolszczan and D.A. Graham for these measurements which were made with the 100 m radio telescope of the Max Planck Institut fur Radioastronomie, Bonn). The dynamic spectrum obtained in 1982 November 3 shows a random speckle pattern with a coherence time $t_c \sim 8$ min and coherence bandwidth $f_c \sim 0.24$ MHz. Superimposed on this random pattern may be seen diagonally-running interference fringes on a smaller scale. The interference fringes persist for longer than the sample shown and have a timescale $\geq$ 2 days. While the persistence of the fringes is obvious, the fact that the fringe phase coheres for only ~ 8 min may be demonstrated by tilting the page and viewing diagonally along the pattern. The disappearance of the fringes when viewed in this fashion immediately shows the lack of long-term coherence in frequency or time. The later observations of 1983 June 5 reveal some fringe effects but at a barely significant level.

When a correlation analysis is carried out on the data for 1982 November 3 we obtain $f_p = 60 \pm 5$ kHz, $f_c = 240 \pm 10$ kHz. The ratio θ_r/θ_s may be derived from the simple relation $\theta_r/\theta_s = (\pi f_c/f_p)^{1/2} = 3.5 \pm 0.2$. If we now assume a simple power law of index α for the wavenumber spectrum of the irregularities we obtain from (2) and (3) the relations:

$$<\theta_s^2> \propto \int_{K_f}^{\infty} K_x^{-\alpha+3} \, dK_x \propto K_f^{-\alpha+4} \tag{5}$$

$$<\theta_r^2> \propto \int_{K_o}^{K_F} K_x^{-\alpha+3} \, dK_x \propto K_o^{-\alpha+4} \quad \text{for } K_o << K_F \tag{6}$$

where K_F and K_o correspond to the wavenumbers which characterise the speckle pattern and the refractive steering respectively.

Thus we have

$$\frac{<\theta_r^2>}{<\theta_s^2>} = (K_F/K_o)^{\alpha-4} \tag{7}$$

which can be used to derive α when K_F and K_o are known. We identify K_F with the scale of the irregularities responsible for the speckle pattern and adopt the value $K_F \sim 2\pi/10^9$ m^{-1}. This is of the same order as the Fresnel wavenumber $2\pi(\lambda z/2\pi)^{1/2}$ and evidence in support of this value has been summarised by Armstrong, Cordes and Rickett (1981).

Obtaining a good estimate for K_o is difficult because consecutive observations over many days are required to sample the slow timescale t_r adequately. Limited evidence based on data like that in Fig. 1 suggest $t_r \sim 2$ days and K_o is then obtained from the relation $K_o = 2\pi(t_r V)^{-1}$ where V is the velocity (36 km s^{-1}) derived from the known distance and proper motion of PSR 0642-03. Substitution of these values in (7) leads to $\alpha = 5.35 \pm 0.05$ which differs significantly from the Kolmogorov index $\alpha = 3.67$.

4. CONCLUSION

A theory has been presented which shows that useful information may be obtained from deterministic features of the dynamic spectra of scintillation patterns generated by propagation in random media. The method is in contrast to conventional techniques based upon ensemble averages which would remove the important quasi-periodic features. A more sophisticated analysis is required for accurate quantitative work. Application of the method to pulsar scintillation shows that the commonly-assumed Kolmogorov model of turbulence in the interstellar medium can be incorrect at small wavenumbers.

REFERENCES

Armstrong, J.W., Cordes, J.M. and Rickett, B.J., 1981, "Density Power Spectrum in the local Interstellar Medium" *Nature*, **291**, 561-563.

Armstrong, J.W. and Rickett, B.J., 1981, "Power spectrum of small-scale density irregularities in the interstellar medium" *Mon. Not. R. astr. Soc.*, **194**, 623-638.

Hewish, A., 1980, "Frequency-time structure of pulsar scintillation" *Mon. Not. R. astr. Soc.*, **192**, 799-804.

Lee, L.C., 1976, "Strong scintillations in astrophysics IV. Cross-correlation between different frequencies and finite bandwidth effects" *Astrophys. J.*, **206**, 744-752.

THE IDENTIFICATION OF TWO SCALES IN STRONG INTERSTELLAR SCINTILLATION

B.J. Rickett
(University of California, San Diego)

ABSTRACT

The asymptotic theory of strong scintillations predicts large fractional intensity fluctuations on two spatial scales. The small scale is essentially diffractive in nature, while the large scale is refractive. Measurements of laser propagation in the atmosphere show both scales in approximate agreement with the theory.

The radio signals from pulsars also vary on two scales (minutes and days). The faster variations are decorrelated over a narrow bandwidth and are caused by diffractive scattering in the interstellar medium. The slower variations are correlated over an octave in frequency and have recently been identified as the refractive counterpart of the faster variations. The two scales vary in opposite ways with both distance and with radio frequency and their geometric mean approximates the Fresnel scale as expected. Only pulsars are small enough to show diffractive scintillation, but some compact extra-galactic sources may also be small enough to show refractive variability over days to months.

1. INTRODUCTION

The random intensity variations caused by propagation in an irregular weakly refracting medium are characterised as strong scintillations when the fractional intensity variations are of the order one or greater. The general theoretical solution for the statistics of the intensity variations $\Delta I(x,y)$ is not available; however, asymptotic results in strong scintillation predict two spatial scales for ΔI (e.g. Prokhorov et al. 1975). Recent measurements of laser propagation through atmospheric turbulence give experimental confirmation of the two scales (Coles and

Frehlich, 1982). This paper presents observations of pulsar intensity variations which also conform to the two scales of strong scintillation but are here caused in the inhomogeneous interstellar medium.

2. LASER SCINTILLATION AND ASYMPTOTIC THEORY

Figure 1 shows the spatial correlation of the intensity of a laser beam after propagation through 1km of turbulent air above a flat desert floor (Coles and Frehlich, 1982). The auto-correlation function can best be described as the sum of narrow (r_l) and wide (r_L) components, as indicated. These measurements are true spatial correlations, not relying on a velocity to map time to space; care was also taken to diverge the laser beam by enough to eliminate beam wander as a contribution to ΔI.

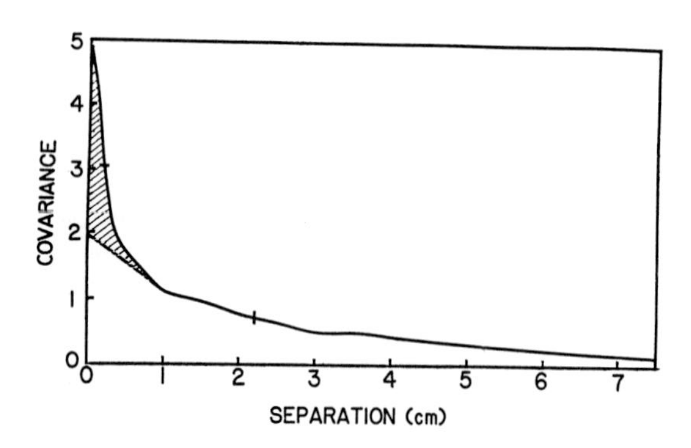

Fig. 1. The observed intensity covariance for a spherically diverging laser beam after propagating 1 km horizontally at a height of 1 m above a hot dry lake bed. The Fresnel scale r_F = 11 mm. The scintillations were strong with total variance of 5. The large and small components are shaded for clarity; the variance in the large component is 2. The two scales have been marked on the curve and their product is very close to the expected value of $r_F^2/2$.

The strength of the turbulence was estimated simultaneously by telescopic recordings of the laser's scattered angular spectrum along the same path. The angular spectrum is the Fourier transform of the second moment of the field, for which there is a general analytical relation to the phase structure function (see Coles and Frehlich for details). Thus for each intensity correlation measured there were also estimates of the structure function $D_\phi(r)$:

$$D_\phi(r) = < [\phi(\underline{r}') - \phi(\underline{r}'+\underline{r})]^2 >$$

where the phase is a line integral of the refractive index deviations n at wavelength $2\pi/k$.

$$\phi(x,y) = \int_0^L k\ n(x,y,z')\ dz'$$

The strength of the scattering at distance L is defined by:

$$u = D_\phi(r_F) \quad ; \quad r_F = (L/k)^{0.5} \qquad (1)$$

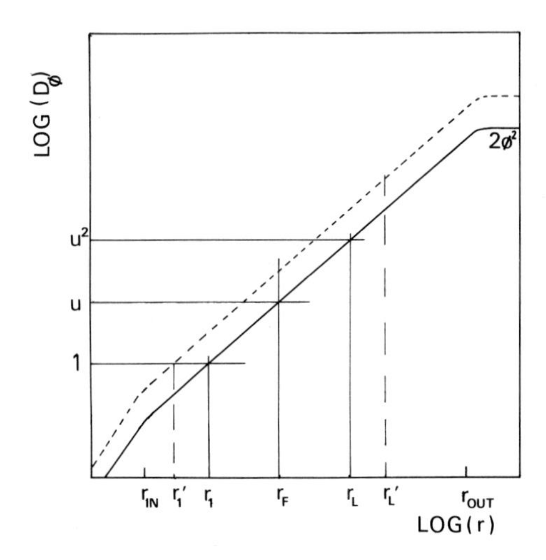

Fig. 2. Schematic form of the phase structure function, for a medium with a power law spectrum ($2<\alpha<4$) and $r_{out} >> r_{in}$. The geometric mean of the large and small scales is equal to the Fresnel scale, providing that $r_{out} > r_L > r_F > r_1 > r_{in}$.

The strength of the scintillation is characterised by $u=D_\phi(r_F)$.

The dashed lines show the effect of increasing the strength of the index irregularities keeping the inner and outer and Fresnel scales constant.

Strong scintillation is for $u > 1$ and the data of Figure 1 correspond to u approximately 100. The asymptotic theory of Prokhorov et al. (1975) predicts two scales given by:

$$D_\phi(r_1) = 1 \quad ; \quad r_L = r_f^2 \, r_1 \tag{2}$$

The phase structure function is shown schematically in Figure 2 indicating the various scales. A physical interpretation is that r_1 is the diffractive scale over which a radian of phase difference typically exists, from which we can define the width of the angular spectrum as $\theta_s = (kr_1)^{-1}$. The large scale is then $r_L = L\theta_s$, which is the radius of the scattering disc, the largest region that can influence the field received at a particular point. For media with a power law spectrum of inhomogeneities there is increasing power with larger scales; thus the phase deviations of size r_L can cause the greatest refractive displacements of small scale diffractive pattern, resulting in regions of enhanced or diminished intensity on a scale of r_L.

Figure 3 shows the results for the laser scintillations discussed above. The plus symbols represent the scales from data similar to that in Figure 1 and are plotted against u derived from the simultaneous angular spectrum measurements. The general behaviour is clear that the two scales diverge as the scintillations become stronger. The dots show the scales expected from (2) and the observed $D_\phi(r)$; they do not agree in detail. In addition the scintillation index observed was larger than that expected from the asymptotic theory. Coles and Frehlich suggest that the discrepancies are caused by the presence of an inner scale causing the spectrum to deviate from the Kolmogorov form, which is consistent with the rest of their results. In the present context we are more concerned that strong laser scintillations do indeed show two scales each contributing a variance of the order of the square of the mean.

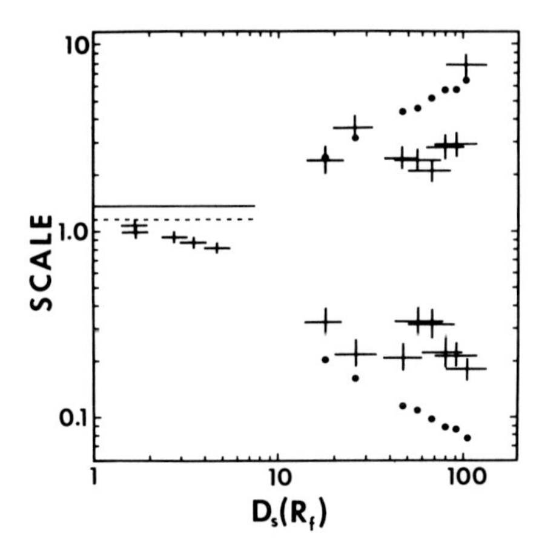

Fig. 3

Fig. 3. The observed spatial scale of intensity in units of the Fresnel scale (11 mm for these data). The data are plotted as crossed error bars. The two scales expected from the asymptotic theory are plotted as small circles in strong scintillation. The weak scintillation scale is plotted as a solid line for the measured structure function and as a dashed line for a Kolmogorov spectrum with no inner scale.

3. PULSAR SCINTILLATION IN THE INTERSTELLAR MEDIUM

Pulsars emit radio waves from about 30 MHz to 10 GHz with a characteristic period of a few seconds or less. When averaged over 1-2 minutes the intrinsic pulse amplitude variation is largely removed and the remaining variability on time scales of minutes to weeks or months is now known to be due to interstellar scintillation. The pulsing nature of the signals is not relevant to the following discussion; pulsars are to be regarded as steady point sources in an extended inhomogeneous plasma. Typical distances are several thousand light years (10^{19} m). Other cosmic radio sources have sufficient angular extent to almost entirely smooth out the scintillations, in the same way that planets barely twinkle when compared with stars. An interesting exception to this statement is discussed as a conclusion of the work reported here.

Soon after the discovery of pulsars in 1967 their variability over minutes to hours was recognised as diffractive scintillation (Scheuer, 1968). The hallmark of these variations is their 100% scintillation index (rms intensity/mean) and their decorrelation over a small fractional bandwidth. For a given pulsar and centre frequency observations yield a characteristic time scale (Δt) and characteristic frequency difference (Δf) for decorrelation to say half of the zero lag value. Typical values are $\Delta t \sim$ 1-100 minutes and $\Delta f \sim$ 10-10000 KHz at frequencies of 100-1000 MHz, where pulsars are most readily studied. The steep decrease of Δf with pulsar distance was the first evidence that this variability was indeed a scintillation effect (Rickett, 1969). Another aspect of the same diffractive scintillation is that for very distant pulsars their pulses are broadened in time by $(2\pi\Delta f)^{-1}$ (Rankin et al. 1970).

Since that time diffractive interstellar scintillation (ISS) has been much studied (see review by Rickett, 1977 and recent comprehensive study by Cordes et al. 1984). It is recognised from this work that the interstellar electron density spectrum is a power law function of wavenumber with an exponent near the Kolmogorov value 11/3; the spatial distribution is however not very uniform, especially enhanced toward the galactic centre. Figures 4a) and b) show the typical diffractive scintillation time Δt plotted against centre frequency and integrated electron density toward the pulsar (dispersion measure DM, which is a reasonable estimate of distance $L \sim DM/0.03$ in parsecs). The straight lines are the expected behaviour for a uniform distribution of Kolmogorov scattering irregularities.

Typical ISS observations span several hours sufficient to include many samples of Δt and Δf. However, the intensity averaged over such a span is not stable and itself varies over days, weeks or months. This was recognised early in pulsar observations (e.g. Hewish et al. 1968) and was well studied by Helfand et al (1977). These slow variations were agreed to be intrinsic to the pulsar radiation process, until Sieber (1982) noticed a relationship with pulsar distance. He showed that the characteristic fading time for slow variations increased markedly with pulsar distance and with wavelength. He concluded that a propagation process was responsible but noted that the form of the relationships were exactly the reverse of those shown in figure 4 for the well-known diffractive ISS. Rickett et al. (1984) identified the slow variations as the refractive component of strong scintillations similar to that seen in the laser measurements.

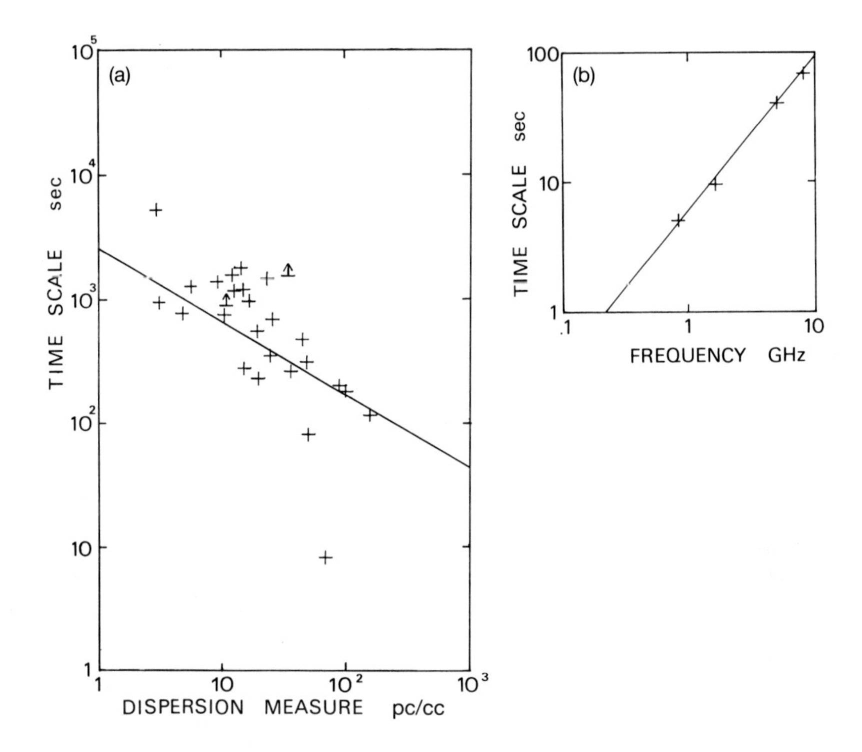

Fig. 4a. Diffractive time scale measured at 400 MHz by Backer (1975) for various pulsars versus their dispersion measure. Fig. 4b. Diffractive time scale measured versus frequency for PSRO833-45 (Backer, 1974).

Helfand et al. (1977) recorded daily mean pulsar intensity at 156 MHz over a four year period. They formed the auto-correlation function of intensity and defined a time scale t_1 such that $\rho(t_1/2) = \rho(1)/2$. The effects of residual diffractive ISS and estimation error were removed by using ρ at 1 day lag to estimate the variance. Figure 5a) shows their values plotted against the integrated electron density toward each pulsar (its dispersion measure). Fig 5b) shows the fading times for a single pulsar (in the crab nebula) plotted against frequency. Data are taken from Sieber's analysis of observations by Rankin et al. (1974), except that the 74 MHz value is derived from Rickett and Seiradakis (1982). We now compare the results of figures 3, 4, 5, with the asymptotic theory.

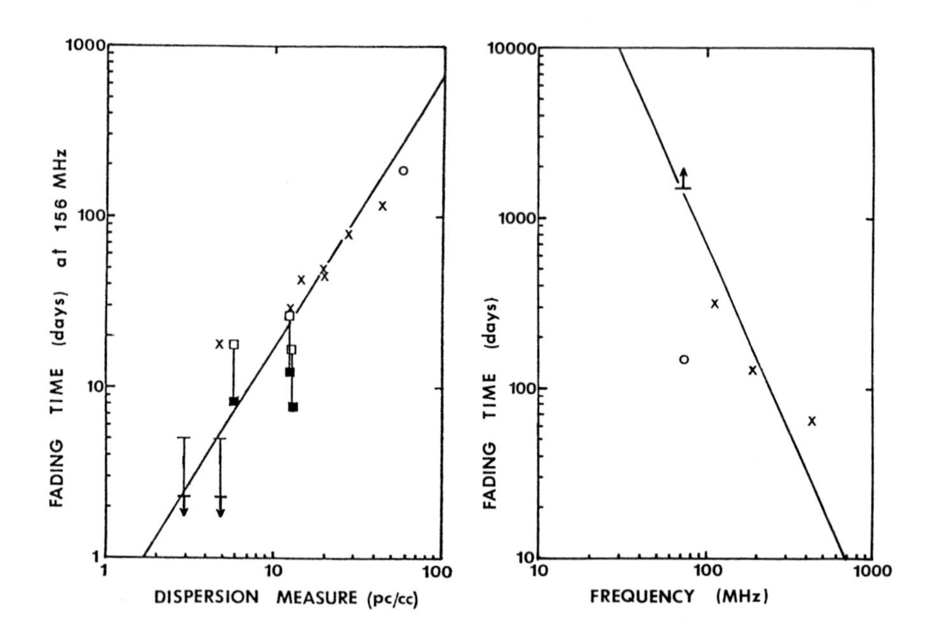

Fig. 5a. Pulsar fading times at 156 MHz plotted against dispersion measure. The different symbols represent different observers as discussed by Rickett et al (1984). The line is from the asymptotic theory as discussed in the text. Fig. 5b. The fading time of the Crab pulsar PSRO531+21 versus frequency. The different symbols again represent different observers as discussed by Rickett et al (1984).

Equations (1) and (2) define the small and large scales in terms of distance L and the structure function. If we assume a power law spectrum with a three-dimensional exponent of 11/3 in accordance with observations, we expect the small scale $r_1 \propto L^{-0.6} \lambda^{-1.2}$ for an ionised medium well above the local plasma frequency. Thus if also the velocity v is independent of distance we expect $t_1 \propto DM^{-0.6} f^{1.2} v^{-1}$. The line in Figure 4a) has the predicted slope and the multiplicative constant adjusted to make it fit the average trend of the data. This equation is:

$$t_1 = 8000\ DM^{-0.6}\ f_G^{1.2} \quad \text{seconds}$$

where DM is the dispersion measure in units of parsec. cm^{-3} and f_G is the frequency in GHz. We use DM to estimate the pulsar distance, as discussed above, and obtain $r_L = L/kr_1 = L/kvt_1$, giving finally:

$$t_L = r_L/v = DM^{1.6}\ f_G^{-2.2}\ 400\ (v_k/100)^{-2} \quad \text{seconds.}$$

This line plotted in Figure 5a) gives a reasonable fit to the observations when $v_k = 100$ km/s is chosen. Pulsars have large proper motions and 100 km/s is a typical value (see Lyne and Smith, 1983 for a discussion). The same equation with $v_k = 100$ and DM chosen for the Crab pulsar (PO531+21) is plotted in Figure 5b); again the agreement is convincing evidence that refraction by large scale inhomogeneities from a power law spectrum in the cause of slow pulsar variability.

The success of the explanation above leads to the prediction of other effects of irregular refraction in the interstellar medium. The dynamic spectra of diffractive ISS are discussed by Hewish and by Smith (this volume pages 203 and 201). Both papers support the view that the drift of scintillation features in such spectra is due to the frequency dependence of the refractive displacement of the diffractive pattern. Their interpretations differ in what this implies about the wavenumber spectrum in the medium. I have also examined this question and conclude that the presently available observations can be explained as a spectrum with an exponent near or greater than the critical value of 4 (this includes the Kolmogorov value of 11/3), and I have also suggested an observational test of this question (Rickett, 1985).

Pulsars are the only radio sources small enough and bright enough to show diffractive ISS. Indeed the absence of ISS in other sources has been used to set limits on their structure at the microarcsecond level (Dennison and Condon, 1981). The diameter threshold for diffractive ISS is approximately:

$$\theta_{source} \lesssim r_1/L \ ,$$

which is typically a microarcsecond. For refractive ISS the threshold is less stringent;

$$\theta_{source} \lesssim r_L/L \sim \theta_s$$

which is typically a milli-arcsecond. From very long baseline interferometry it is known that such structure is not uncommon, particularly in flat spectrum extra-galactic radio sources. Thus intensity variations on time scales of days to months are to be expected from small diameter components of such sources. As suggested by Rickett, Coles and Bourgois (1984) this may well explain the metre wavelength variability of some sources and other variations which have proved hard to explain as intrinsic source variability.

In summary, the existence of two spatial scales of intensity variation as predicted by asymptotic theories of strong scintillation, are seen in both laser propagation through a turbulent atmosphere and in radio propagation through the ionised interstellar medium. The small diffractive scale is well known, but the large refractive scale has been largely overlooked. The latter leads to the prediction of interstellar modulation effects which should be detectable in some extra-galactic radio sources.

4. REFERENCES

Backer, D.C. 1974, "Interstellar scattering of the Vela pulsar", *Astrophys. J.*, **190**, 667-671.

Backer, D.C. 1975, "Interstellar scattering of pulsar radiation", *Astron. and Astrophys.*, **43**, 395-404.

Coles, W.A. and Frehlich, R.G. 1982, "Simultaneous measurements of angular scattering and intensity scintillation in the atmosphere", *J. Opt. Soc. Am.*, **72**, 1042-1048.

Cordes, J.M., Weisberg, J.M. and Boriakoff, V. 1984, "Small scale density turbulence in this interstellar medium", *Astrophys., J.*, in press.

Dennison, B. and Condon, J.J. 1981, "A search for interstellar scintillations in a large sample of low-frequency variable sources", *Astrophys. J.*, **246**, 91-99.

Helfand, D.J., Fowler, L.A. and Kuhlman, J.V. 1977, "Pulsar flux density observations: long term intensity and spectral variables", *Astron. J.*, **82**, 701-705.

Hewish, A., Bell, S.J., Pilkington, J.D.H., Scott, P.F. and Collins, R.A. 1968, "Observation of a rapidly pulsating radio source", *Nature (Lond.)*, **217**, 709-713.

Lyne, A.G. and Smith, F.G. 1982, "Interstellar scintillation and pulsar velocities", *Nature (Lond.)*, **298**, 825-827.

Prokhorov, A.M., Bunkin, F.V., Gochelashvily, K.S. and Shishov, V.I. 1975, "Laser irradiance propagation in turbulent media", *Proc. IEE.*, **63**, 790-811.

Rankin, J.M., Comella, J.M., Craft, H.D., Richards, D.W., Campbell, D.B. and Counselman, C.C. 1970, "Radio pulse shapes, flux densities and dispersion measure of pulsar NP0532", *Astrophys. J.*, **162**, 707-725.

Rankin, J.M., Payne, R.R. and Campbell, D.B. 1974, "The Crab nebula pulsar: radio frequency spectral variability", *Astrophys. J.*, **193**, L71-L74.

Rickett, B.J. 1969, "Frequency structure of pulsar intensity variations", *Nature (Lond.)*, **221**, 158-159.

Rickett, B.J. 1977, "Interstellar scattering and scintillation of radio waves", *Ann. Rev. Astron. Astrophys.*, **15**, 479-504.

Rickett, B.J. 1985, "Refractive interstellar scintillation and radio source variability", Preprint, submitted to *Astrophys. J.*

Rickett, B.J. and Seiradakis, J.H. 1982, "The flux of the Crab pulsar at 74MHz from 1971 to 1981", *Astrophys. J.*, **256**, 612-616.

Rickett, B.J., Coles, M.A. and Bourgois, G. 1984, "Slow scintillation in the interstellar medium", *Astron. Astrophys.*, in press.

Scheuer, P.A.G. 1968, "Amplitude variations in pulsed radio sources", *Nature (Lond.)*, **218**, 920-922.

Sieber, W. 1982, "Casual relationship between pulsar long-term intensity variations and the interstellar medium", *Astron. Astrophys.*, **113**, 311-313.

TESTING THE THEORIES OF FORWARD VOLUME SCATTERING

T.E. Ewart
(Applied Physics Laboratory and School of Oceanography, University of Washington)

ABSTRACT

Recently there have been many theoretical developments in the area of forward scattering of waves propagating in random media. Here, it is contended that our lack of understanding of the detailed autocorrelation or autospectral function of the medium is likely to be the limiting factor in testing the theories. An overview of some relevant research that supports this contention, including numerical simulations and an ocean experiment, is presented. If this point of view is valid, future research may need redirection. Areas for further study are indicated.

1. THE CURRENT STATUS

To predict the statistical behaviour of waves propagating in a random medium, the stochastic properties of the index of refraction structure of that medium must be well understood. If one assumes the underlying processes producing the irregularities to be Gaussian, the medium autocorrelation function, $\rho(\delta x, \delta y, \delta z, \delta t)$, is required as input to any proposed scattering theory ($\delta x, \delta y, \delta z$, and δt are the x, y, z, and t coordinate lags). For the case discussed here, medium isotropy reduces the scope of that requirement and cylindrically symmetric propagation is assumed to take place in the x, z plane. Wave propagation is in the x direction. Our understanding of the forward scattering of wavefields propagating in random media has reached the point where knowledge of the detailed structure of the index of refraction of the medium is the most probable limiting factor in testing theories. This limitation has made it difficult to compare theoretical predictions with available experimental data accurately.

A solution of the parabolic fourth moment equation by Uscinski (1982) for plane wave propagation in a medium with random irregularities is a seminal contribution to the field. Subsequently, Uscinski *et al.* (1983) obtained an approximate evaluation of a solution for propagation with point source initial condition. An expression is given for the correlations of intensity in a propagating wave as a function of δz and δt. The corresponding Fourier transform, $P_{I/<I>}(\beta,f)$, is the power spectral density of the intensity fluctuations as a function of wavenumber, β and time frequency, f. That evaluation is denoted here as $m_{IV}^{(0)}$. Macaskill (1983) has obtained a result for m_{IV} that is a more precise evaluation of the fundamental solution. Predictions based on Macaskill's solution are denoted here as $m_{IV}^{(1)}$. The superscripts indicate the precision in the evaluation of the basic solution for m_{IV}. The inputs to both evaluations of the solution are the medium autocorrelation function, the scattering parameter Γ, and the scaled range X, where Γ is the Fresnel length divided by the scattering length and X is the range scaled by the Fresnel length. It is important to note that the solutions of Uscinski *et al.* and Macaskill apply to all ranges where the parabolic approximation holds, and over wide ranges of Γ and X. In practice the parameters are obtained from models or derived from measurements. The testing of these theories is a necessary next step.

1.1 Numerical Experiments

Numerical work by Macaskill and Ewart (1984) demonstrates that the form of the medium autocorrelation function strongly affects the intensity correlations of a wave propagating in that medium. In that work, simulated index of refraction data obtained by Monte Carlo methods are input to a numerical solution of the parabolic wave equation in order to study the behaviour of waves propagating in media with specific forms for ρ. Using that technique to simulate waves propagating in a medium with ρ, Γ, and X specified, the theories can be tested in the following sense. The numerical experiments provide realisations of the complex amplitude of a wave where the propagation satisfies the parabolic wave equation. If the intensity correlations computed from the simulated results are in agreement with those obtained from the analytic solutions of the parabolic moment equations, then one gains confidence that both the theory and the simulation techniques are correct.

Examples of realisations of the normalised intensity, $I/<I>$, for plane wave propagation through a medium described by two different autocorrelation functions are shown in Fig. 1.

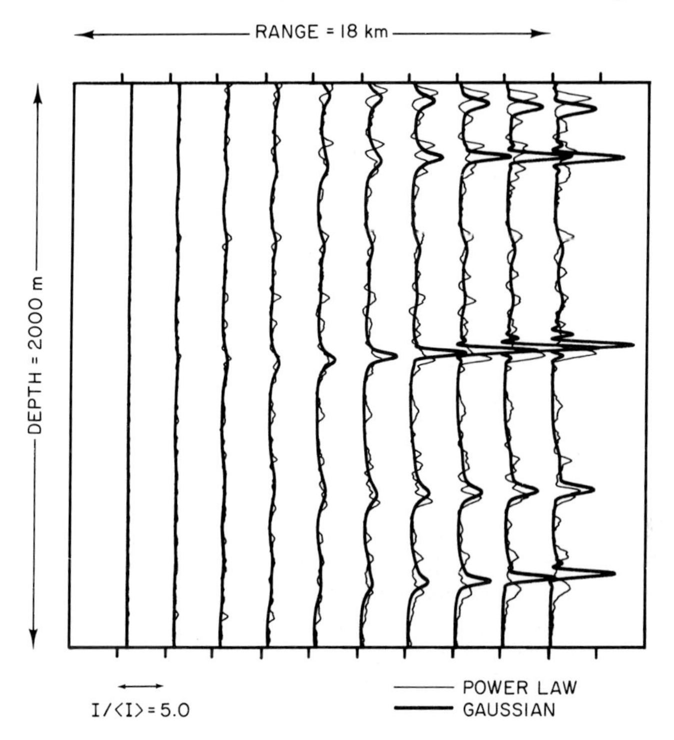

Fig. 1. Intensity realisations of plane wave propagation in a medium with Gaussian or power law autocorrelations in z.

The input index of refraction data used to define the medium for each realisation is generated with Gaussian and with power law (β^{-4} power spectral density) correlation functions; the variance of the phase deviations is representative of observed open ocean internal wave fluctuations, and the scaled range and depth spans correspond to typical open ocean values (2000 m depth, 18 km range) for acoustic wave propagation at 4 kHz. Time correlations are neglected and $\rho(\delta z)$ is input. The index of refraction fluctuations are assumed to be delta function correlated in the range direction in an integral sense. (The delta function correlation, or strong Markov assumption, raises some questions about the scattering for both the theory and the numerical simulations; nevertheless its validity is assumed, and the topic will not be discussed.) The ocean processes are simulated as filtered white noise, and the underlying white noise process is the same for both realisations. It is clear from the plots that major phase defects produce local focuses in the same regions for either form of $\rho(\delta z)$. However, the detailed character of the intensity fluctuations is quite different. This reflects scattering from the "single scale" of the Gaussian $\rho(\delta z)$ versus scattering from the "range of scales" of the power law. As expected, the normalised intensity realisation produced from a power law $\rho(\delta z)$ contains far more high wavenumber energy.

Macaskill and Ewart compare the results of the simulations with theoretical predictions of both the scintillation index of the intensity fluctuations observed transverse to the direction of propagation and its spectral decomposition. This comparison is seen in Figs. 2 and 3 where results computed from many realisations, produced using the Gaussian form for $\rho(\delta z)$ as in Fig. 1, are averaged. Those results are the scintillation index S_I^2 as a function of scaled range X and the power spectral density of intensity as a function of vertical wavenumber $P_{I/<I>}(\beta)$. The input parameter Γ is 1000 for both computations, and $P_{I/<I>}(\beta)$ is shown for X =0.1 (corresponding to a range near the medium focus for the Gaussian case). Four theoretical predicions are shown in Fig. 2. Predictions based on the well known Rytov approximation and far-field asymptotic expansion techniques are included as well as $m_{IV}^{(0)}$ and $m_{IV}^{(1)}$. The $m_{IV}^{(1)}$ evaluation provides considerably more precise predictions than the others. Similar results obtain for a wide range of Γ and X. Notably, the predictions of $m_{IV}^{(0)}$ fit the form of S_I^2 and $P_{I/<I>}(\beta)$ well, but clearly underpredict the observations. This is important, in what follows here, because the solution for $m_{IV}^{(1)}$ has not yet been evaluated for arbitrary medium autocorrelation functions such as those proposed for the ocean transmission case.

These comparisons provide evidence that the theoretical predictions are correct for a wide range of scattering parameters and medium correlation functions. As presented, the Monte Carlo solutions test the theoretical predictions of two-dimensional plane wave propagation in the transverse dimension only. The $m_{IV}^{(0)}$ and $m_{IV}^{(1)}$ predictions, however, include both the transverse and the temporal intensity fluctuations. Temporal intensity fluctuations have been determined experimentally, and those results can be used for further testing of the theories.

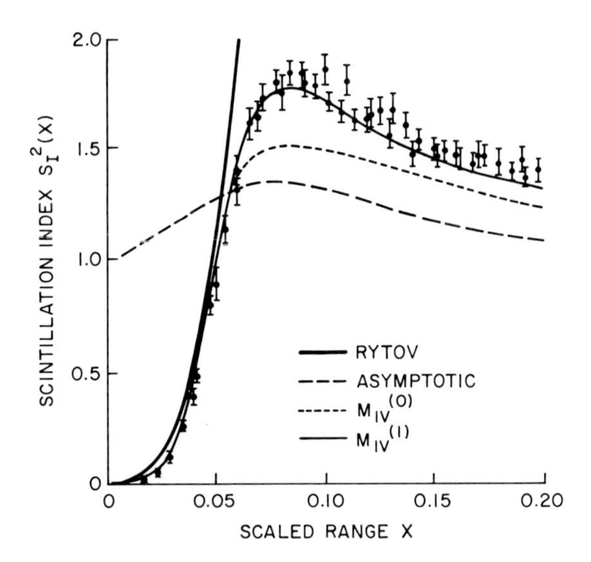

Fig. 2. Scintillation index as a function of scaled range for simulated plane wave propagation.

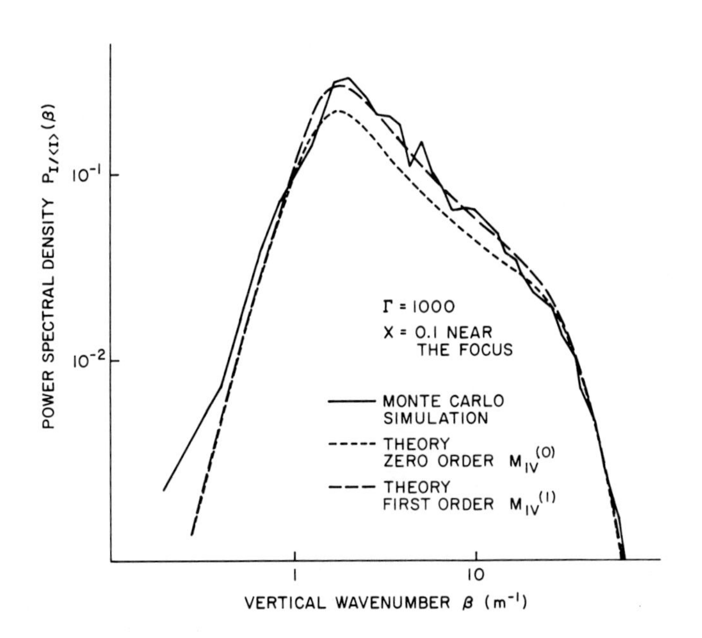

Fig. 3. Spectrum of the normalised intensity versus wavenumber.

1.2. Experiment-Theory Comparisons

Simultaneous measurements of the complex amplitude of acoustic waves propagating through a random medium, and the index of refraction structure of that medium were made in the Mid-Ocean Acoustic Transmission Experiment, MATE. The experiment is discussed by Ewart and Reynolds (1984), and only an overview is presented here. An important difference between MATE and other experiments is that extensive measurements of the medium index of refraction were made simultaneously with the acoustic measurements. Measurements of the sound velocity structure were determined well enough to make estimates of $\rho(\delta x, \delta y, \delta z, \delta t)$.

An artist's drawing of the experiment layout is shown in Fig. 4. In the acoustic measurements, narrowband and wideband pulses at four acoustic frequencies spanning 2.5 octaves from 2 to 13 kHz were transmitted over an 18 km wholly refracted Fermat path. Four transducers located at the corners of a rectangle oriented perpendicular to the transmission path (3 m vertical and 150 m horizontal) received the acoustic pulses. Six days of data recorded at two receivers separated 3 m vertically have been processed. The results are discussed by Ewart and Reynolds (1984), and some are repeated here.

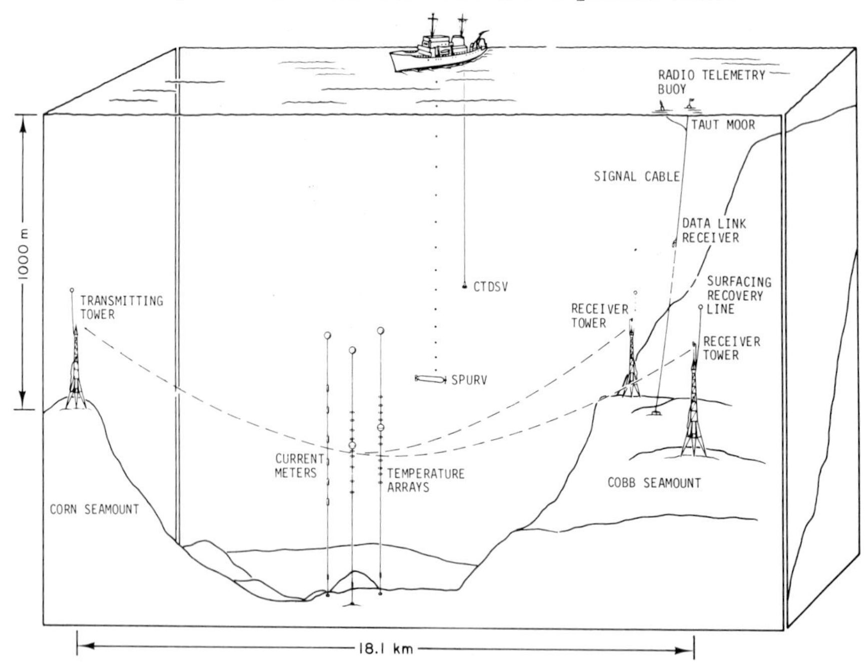

Fig. 4. The MATE experimental layout.

To determine the form of $\rho(\delta x, \delta y, \delta z, \delta t)$ for the sound velocity fluctuations encountered during MATE, measurements of the index of refraction fluctuations in three dimensions and time are required. Figure 5 illustrates samples of the spatial and temporal measurements made during MATE. It is the intent here to give the reader a feeling for the type and quantity of data that go into a determination of $\rho(\delta x, \delta y, \delta z, \delta t)$. The measurements of temperature, conductivity, pressure, and sound velocity made by the CTDSV determine the medium autocorrelation function for a large range of δz, a lesser range in δt, and some values of δx and δy. The moorings give estimates of $\rho(\delta x, \delta y, \delta z, \delta t)$ for a wide range in δt, several values of δz, and three range values. The Self-Propelled Underwater Research Vehicle, SPURV, samples a large range in δx or δy, and a single value of δz. ("Top" and "bottom" in Fig. 5C refer to two temperature sensors on SPURV spaced 1 m vertically.)

The parameters of various models of internal waves and finestructure have been determined from the oceanographic data. Simply stated, finestructure is defined as those fluctuations that do not arise from internal wave processes. The random internal wave and finestructure processes produce the space-time modulations of the mean sound velocity profile. Over the propagation depths of the single Fermat path isolated in the MATE analysis, the mean sound velocity profile is linear; this feature of the MATE environment eliminates to first order the necessity to account for profile effects. Using the form of $\rho(\delta x, \delta y, \delta z, \delta t)$ determined from the models, predictions of the power spectra of phase and intensity can be tested against the results of the acoustic measurements.

The scattering regimes of MATE span the broad region where the acoustic field has undergone multiple scatter but the wave has not travelled far enough to approach saturation. The broad range of regimes can be seen by studying Fig. 6, in which phasor plots of the complex amplitude measured during approximately one inertial cycle (16.4 h) at a single receiver are plotted for each frequency. The character of the scattering changes markedly, from 2 kHz where there are large phase wraps and weak amplitude fluctuations, to 13 kHz where the scattering appears to be more nearly saturated. In this region neither the Rytov approximation for the near field nor the far field asymptotic expansion solutions are applicable. From Fig. 2 it is clear that a propagating wavefield generally converges slowly to the saturation value of $S_I^2 = 1.0$.

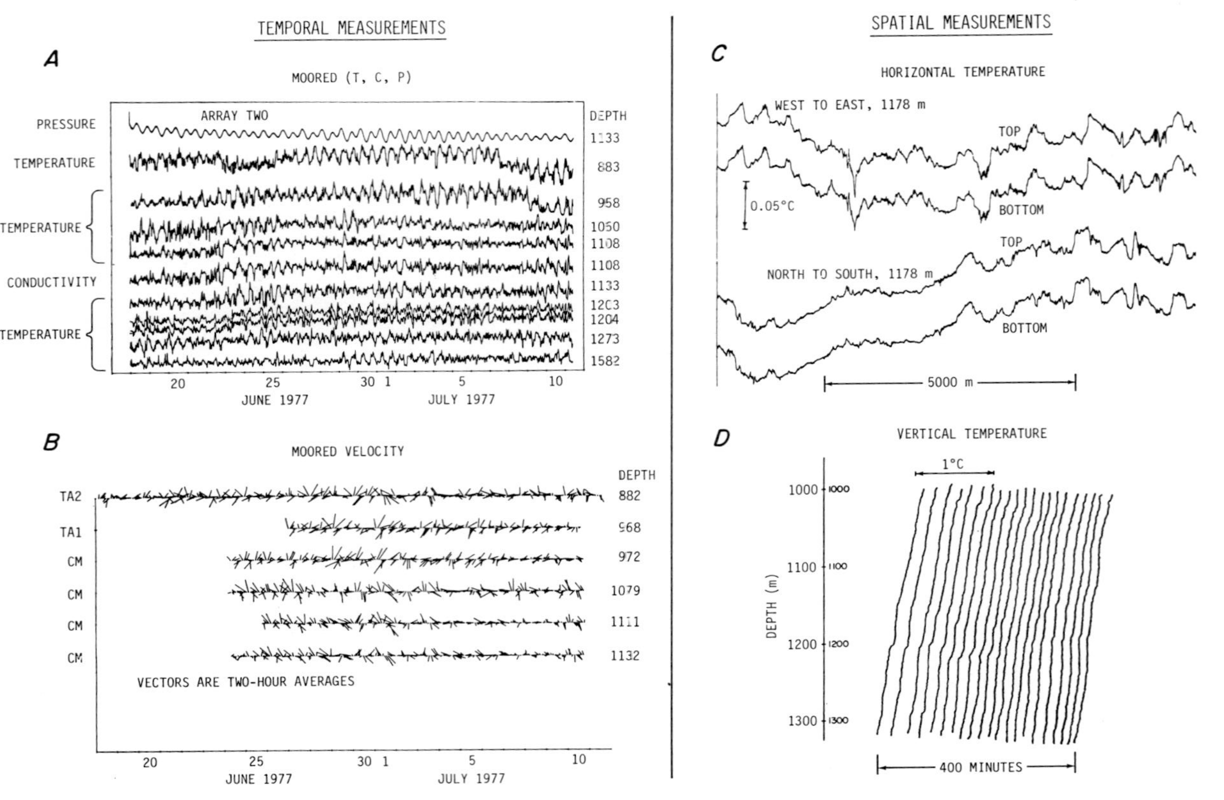

Fig. 5. Selected MATE environmental data.

For the ocean, away from boundaries and major current features, statistical isotropy in the horizontal is generally assumed. This assumption is shown to be valid by Ewart and Reynolds for the MATE environment, and they drop the δy dependence. They also assume that the stochastic properties of the index of refraction result from stationary processes. This assumption must be questioned, as its validity is at the root of current research on oceanographic processes. In their studies $\rho(\delta x, \delta y, \delta z, \delta t)$ is obtained by fitting the oceanographic data to various models for linear internal waves. Details of the modelling are given by Levine and Irish (1981), and by Ewart and Reynolds (1984). For the purposes of this paper only the results of the acoustic predictions are presented.

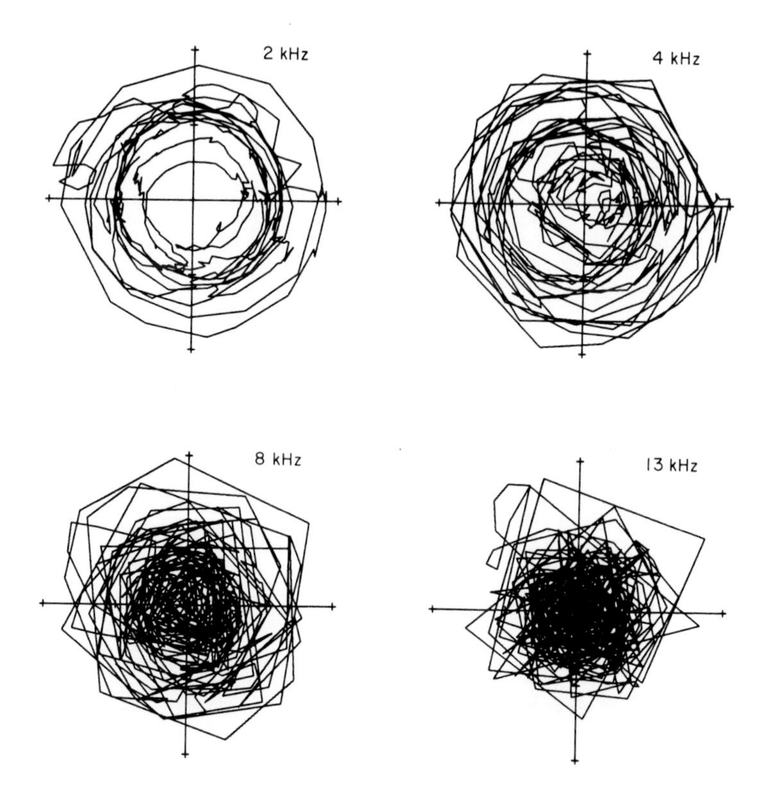

Fig. 6. Phasor plots of the MATE complex amplitudes measured during a single inertial period.

Several investigators have tried to predict the results from an earlier experiment at the MATE site reported by Ewart (1976). Only 4 and 8 kHz results were available from that experiment. Those theoretical predictions are compared here with the MATE results. In Fig. 7 the values of the power spectral density of the log intensity for the MATE measurements are compared with the predictions of Desaubies (1978), where the scattering parameters and $\rho(\delta x,\delta y,\delta z,\delta t)$ were obtained for linear internal waves as previously discussed. [The predictions of Desaubies are virtually identical to those of Munk and Zachariasen (1976).] Clearly, both the form and the acoustic frequency scaling of geometric acoustics (the variance scales as the acoustic wavenumber) is incorrect.

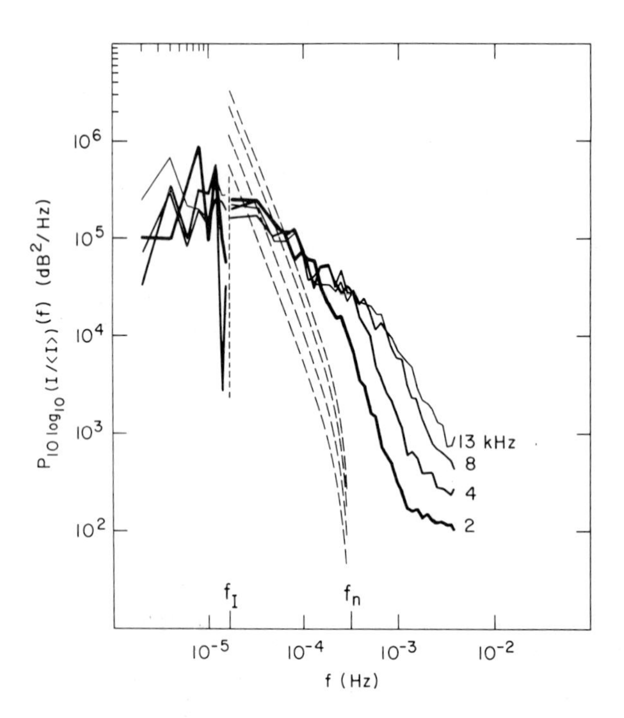

Fig. 7. Measured and predicted spectra of the log intensity. Theory from Desaubies (1978).

Flatte *et al.* (1980) [reported also in Flatte (1983)] obtain a modification to $\rho(\delta x,\delta y,\delta z,\delta t)$ by assuming that the smaller scale internal waves are advected by larger ones. They parameterise this advection by a velocity constant. A discussion of the merit of this modification is beyond the scope of this paper, but it is noted that their prediction of the log intensity spectrum based on their form for

$\rho(\delta x,\delta y,\delta z,\delta t)$ and the Rytov approximation appears to agree with the experimental results at 4 kHz. A computation of their predictions of the spectrum of the log intensity for the four frequencies of MATE is plotted with the experimental results in Fig. 8. It is clear that while the prediction for 4 kHz agrees reasonably well out to the buoyancy frequency, f_n, the acoustic frequency scaling of geometric acoustics is wrong as it is in Fig. 7.

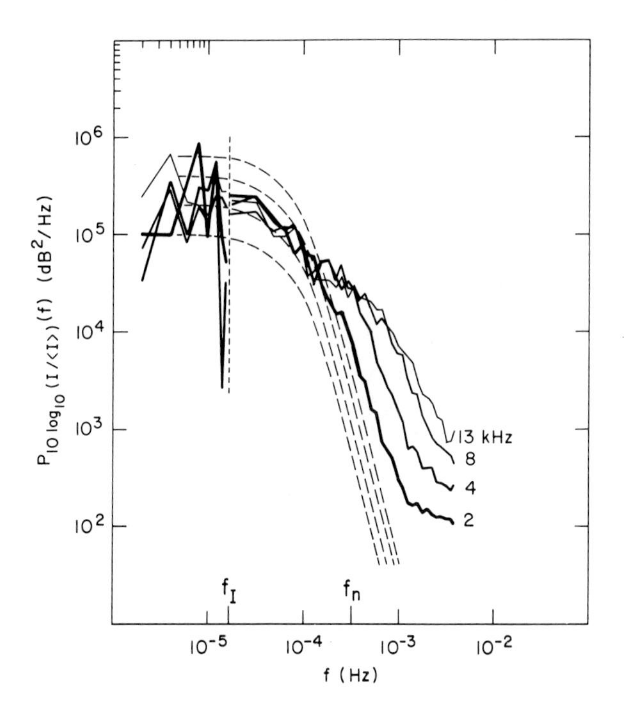

Fig. 8. Measured and predicted spectra of the log intensity. Theory from Flatte *et al.* (1980).

The evaluation of $m_{IV}^{(0)}$ by Uscinski *et al.* (1983) predicts the power spectral density of the normalised intensity when the scattering parameters and $\rho(\delta x,\delta y,\delta z,\delta t)$ are input. This prediction is shown in Fig. 9 for four different linear internal wave models for $\rho(\delta x,\delta y,\delta z,\delta t)$. The MATE index of refraction measurements are used to determine each model and its normalisation. In this plot the frequencies above 2 kHz are each offset by a decade. The models used are those of Garrett and Munk (1979), Desaubies (1976), and Levine and Irish (1981); Q designates the power law of the temporal frequency dependence in the models. Clearly the differences in the four predictions are too small to permit a ranking of the models.

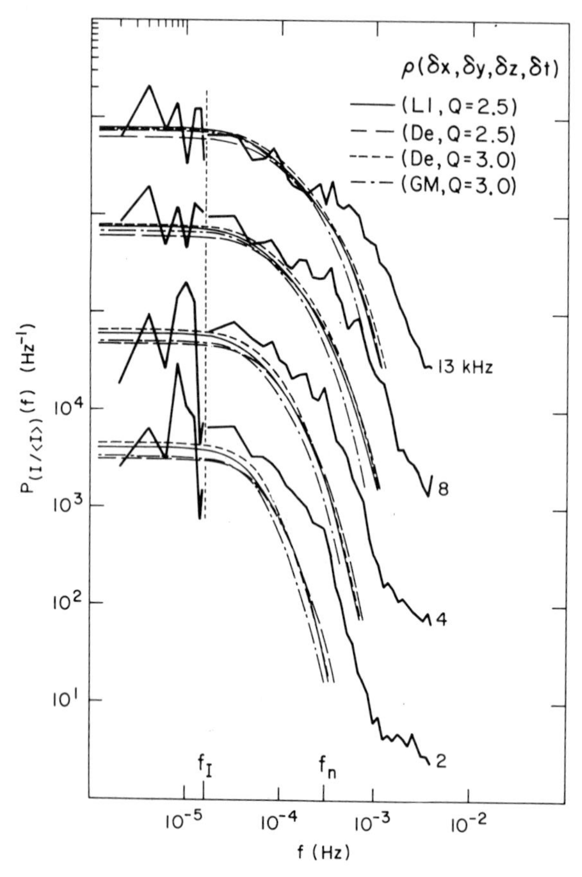

Fig. 9. Measured and predicted spectra of intensity. Theory from Uscinski *et al.* (1983).

The predicted values of $S_I^{\;2}$ for 2 to 13 kHz from the Levine and Irish model are 0.26, 0.51, 0.88, and 1.09 versus 0.73, 1.18, 1.95, and 1.85 for the measured values. The Uscinski *et al.* predictions tend to show geometric acoustics frequency scaling at the lower acoustic frequencies. This is expected, as the values of $S_I^{\;2}$ are quite underpredicted. Two points can be made. First, as mentioned earlier, evaluation of $m_{IV}^{\;(1)}$ by Macaskill has not been included, and examination of Fig. 2 shows that this correction provides a significant improvement in the prediction of S_I^2. Second, Levine and Irish show that one half of the variance of the oceanic fluctuations arises from the noninternal wave processes termed finestructure. In more recent work by Ewart *et al.* (1984), inclusion of a model of finestructure as well as of internal waves improves the predictions of the four values of S_I^2, giving 0.56, 0.83, 1.17, and 1.27. These values will be increased further when evaluation of $m_{IV}^{(1)}$ is included as in Fig. 2.

2. SUMMARY AND CONCLUSIONS

Propagation in a random inhomogeneous medium with a specified correlation function and a wide range of scattering parameters has been studied by using Monte Carlo simulations of the random field as input to numerical solutions of the parabolic wave equation. The results indicate that the available solutions of the parabolic fourth moment equation predict the outcome of these numerical experiments quite precisely for the case of spatial fluctuations of the wavefield in z. The results from MATE provide a more precise test of theories with experiment than hitherto possible; similar comparisons of theory with measured temporal fluctuations give general agreement, but lack precision. It is clear from the experimental studies that obtaining the form and normalisation of $\rho(\delta x,\delta y,\delta z,\delta t)$ is a major obstacle. This is due to our lack of understanding of the statistical behaviour of finestructure (including the stationarity of the processes). Our lack of knowledge of the index of refraction field probably limits our ability to predict the acoustic fluctuations from the measured environmental parameters. The vast amount of data taken during MATE is an indication of how formidable this task is for a real world autocorrelation function. Some improvements are suggested by these studies.

The first improvement is to measure the spatial and the temporal acoustic fluctuations simultaneously to obtain $I(z,t)$, since the predictions $m_{IV}^{(0)}$ and $m_{IV}^{(1)}$ yield the space-time correlations. Another way to test the theories more precisely is to include the acoustic frequency cross correlations of intensity. Because the contribution to the intensity variance arises from different index of refraction wavenumber bands in β for each acoustic frequency, the cross correlations may prove to be more sensitive to the precise form of $\rho(\delta x,\delta y,\delta z,\delta t)$. A more robust method of obtaining the medium autocorrelation function would be to find the best form for $\rho(\delta x,\delta y,\delta z,\delta t)$ that derives from all of the index of refraction data in a minimum mean square error sense. This should be carried out with minimal recourse to models. All of these improvements are planned in the future.

3. REFERENCES

Desaubies, Y.J.F., 1976, *J. Phys. Ocean.* **6**, 976.

Desaubies, Y.J.F., 1978, *J. Acoust. Soc. Am.* **64**, 1460.

Ewart, T.E., 1976, *J. Acoust. Soc. Am.* **60**, 46.

Ewart, T.E. and Reynolds, S.A., 1984, *J. Acoust. Soc. Am.* **75**, 785.

Ewart, T.E., Macaskill, C. and Uscinski, B.J., 1985, *J. Acoust. Soc. Am.* 77 (5), 1732.

Flatte, S.M., Leung, R. and Lee, S.Y., 1980, *J. Acoust. Soc. Am.* **68**, 1773.

Flatte, S.M., 1983, *Proc. IEEE.* **71** (11), 1267.

Garrett, C. and Munk, W., 1979, *Ann. Rev. of Fluid Mech.* **11**, 339.

Levine, M.D. and Irish, J.D., 1981, *J. Phys. Oceanogr.* **11**, 676.

Macaskill, C., 1983, *Proc. Roy. Soc.* **386**, 461.

Macaskill, C. and Ewart, T.E., 1984, *IMA J. Appl. Math.* **33**, 1.

Munk, W., and Zachariasen, F., 1976, *J. Acoustic. Soc. Am.* **59**, 818.

Uscinski, B.J., 1982, *Proc. Roy. Soc.* **380**, 137.

Uscinski, B.J., Macaskill, C. and Ewart, T.E., 1983, *J. Acoust. Soc. Am.* **74**, 1474.

OPTICAL SCATTERING EXPERIMENTS

E. Jakeman
(Royal Signals and Radar Establishment, Malvern)

ABSTRACT

A review of data acquired in light scattering experiments designed to investigate non-Gaussian fluctuations will be given. Systems investigated include dynamic scattering in liquid crystals, laboratory generated thermal plumes, mixing layers, rippled liquid surfaces and specially prepared solid diffusers. Comparison with theoretical predictions will be made where possible.

1. INTRODUCTION

Since the advent of the laser there has been a wealth of coherent light scattering experiments from a very wide range of scattering systems. Many of these experiments were designed to investigate or exploit the most familiar laser light scattering phenomena: "speckle", the pattern of bright and dark regions formed when coherent light is scattered by almost any kind of rough surface Fig 1. Speckle is the optical analogue of Gaussian noise and arises when randomly phased contributions from many independent scattering elements add together coherently. The central limit theorem predicts that in this situation the resultant field vector is a circular complex Gaussian process with a Rayleigh distributed amplitude and negative exponential distribution of intensity fluctuations (Dainty, 1975). Gaussian speckle is now a thoroughly investigated and fully understood phenomenon which has found application in many areas of pratical measurement such as surface roughness and anemometry. Many naturally occurring optical phenomena, however, are broadband (white light) effects visible to the naked eye and do not owe their existence to the coherence of the source. These are non-Gaussian effects and their statistical and correlation properties are on the whole not well characterised or understood (Jakeman, 1984). In order to start to rectify this situation a programme of non-Gaussian light scattering experiments was begun at RSRE some ten years ago. Since then we have investigated a wide variety of simple scattering systems with the principle aim of gaining some insight into the statistical nature of non-Gaussian intensity patterns for noise modelling purposes.

Fig. 1 Laser light scattered by a large area of ground glass: a Fraunhofer region Gaussian speckle pattern.

Many of these systems were generic i.e. typical of scattering systems encountered in practice, rather than artificially contrived to enable their characteristics to be measured by other techniques. However, a few specially prepared and characterised scatterers have been investigated and it is important to carry out more such experiments in the future as the aim of the programme shifts towards the development of remote sensing techniques.

In recent years a number of other groups have carried out non-Gaussian scattering experiments, in particular Professor Dainty's team at Rochester (now at Imperial College) (Levine and Dainty, 1983; Chandley and Escamilla, 1979) and Professor Asakura's Group in Japan (Ohtsubo and Asakura, 1978). However in this paper I shall restrict my historical survey to our own work on diffusing layers and surfaces at RSRE with which I am most familiar, and I would like to acknowledge at this point the invaluable support and encouragement of my experimentalist colleagues over the years, particularly Dr. P.N. Pusey, Dr. G. Parry and more recently Drs. J.G. Walker, D.L. Jordan and R.C. Hollins. Although many of the experimental results have been interpreted in the light of predictions made by solving Maxwell's equations for specific theoretical models, a more empirical approach: the two dimensional random walk model (Pusey, 1977) has often provided much needed insight and has led to one of the most useful outputs from this programme of research: a non-Gaussian noise model with wide applicability, namely, the class of K-distributions (Jakeman and Pusey, 1976). Thus in the next section we briefly review the predictions of the random walk model. Sections 3 and 4 are devoted to scattering by fluid and solid scattering systems respectively with some concluding comments in section 5.

2. THE RANDOM WALK MODEL

In the simplest random walk model the scattered wave field is represented as the sum of N randomly phased vectors of variable length:

$$\varepsilon = \sum_{n=1}^{N} a_n e^{i\phi_n} \tag{1}$$

where the $\{\phi_n\}$ are uniformly distributed and independent of the $\{a_n\}$ which are also statistically identical and independent from each other. The mean intensity and second normalised intensity moment are then given by

$$\langle I \rangle = N\langle a^2 \rangle$$

$$\frac{\langle I^2 \rangle}{\langle I \rangle^2} = 2 - \frac{2}{N} + \frac{\langle a^4 \rangle}{N\langle a^2 \rangle^2} \tag{2}$$

When N >> 1 the second moment reduces to the Gaussian speckle value of two as predicted by the central limit theorem. When N is finite there is a deviation from this value which is inversely proportional to N. If equation (1) is used to model rough surface or diffuser scattering then it is reasonable to expect $N \propto$ illuminated area so that the deviation from Gaussian statistics obtained by illuminating a small area should be inversely proportional to this area. Note that the random phase assumptions made in equation (1) presumes that the scatterer introduces phase fluctuations greater than 2, which is the case of most interest. Also when N is finite equation (2) depends on the detailed statistical behaviour of the form-factors $\{a_n\}$ ascribed to the individual scattering centres. Thus the model will be highly sensitive to the detailed nature of the scatterer. It is usually found in practice (except for certain simple particle scattering systems) that the final term in equation (2) is much larger than the second term so that enhanced fluctuations are observed. (Jakeman and Pusey, 1975; Pusey and Jakeman 1975) The pattern generated by a scatterer in such a non-Gaussian configuration may have a highly complicated structure as indicated in Figure 2 and, in the case of a moving pattern, the detected intensity will have a spikey structure quite distinct from Gaussian noise.

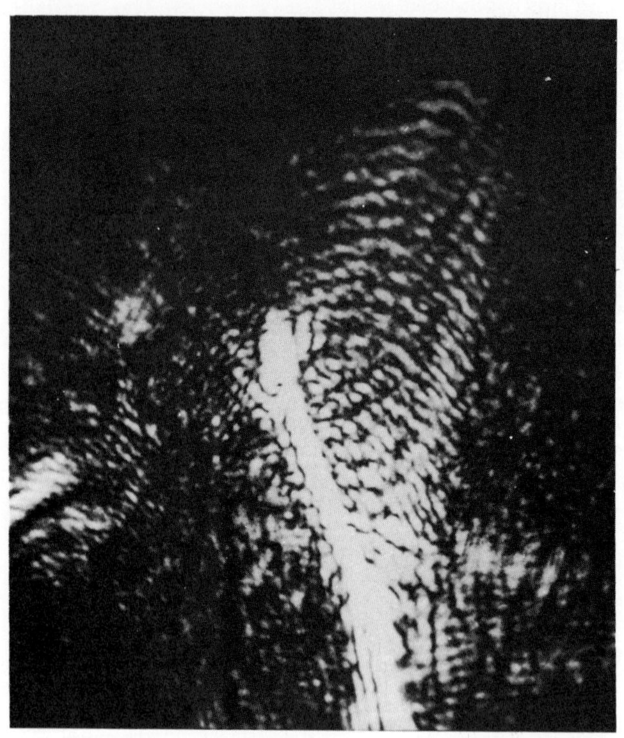

Fig. 2 Laser light scattered by a small area of ground glass: a Fraunhofer region non-Gaussian intensity pattern.

The origin of the correlated regions of high intensity evident in Figure 2 can also be interpreted in terms of the random walk model through the coherence function

$$\frac{\langle I(x)I(x')\rangle}{\langle I(x)\rangle\langle I(x')\rangle} = \left(1 - \frac{1}{N}\right)\left[1 + \frac{|\varepsilon(x)\varepsilon^*(x')|^2}{\langle I(x)\rangle\langle I(x')\rangle}\right] + \frac{\langle a^2(x)a^2(x')\rangle}{N\langle a^2(x)\rangle\langle a^2(x')\rangle} \tag{3}$$

where x and x' represent two space-time points. When N is large equation (3) reduces to the factorisation property of a complex Gaussian process and the second term in the square brackets represents the effect of interference or speckle in the pattern. The associated coherence time reflects the relative motion of scatterers whilst the coherence length will be just inversely proportional to the aperture (illuminated area) size. When N is finite, however, the final term becomes important. This is a single scatterer term characterised by the time constant of the motion of individual scatterers and by one or more length scales associated with their diffraction and focussing properties. It is evident that all of these scales will be longer than those characterising optical frequency speckle, as indeed is observed in Figure 2. Because the single scatterer contributions may be geometrical in origin it may well be present with broadband (white light) illumination when the interference term vanishes.

Summarising then, the random walk model leads us to expect that when only a small area of the scatterer (comparable to the largest scale size present) contributes to the intensity at the detector we expect the second normalised moment of the intensity fluctuation distribution to exhibit a deviation from Gaussian statistics inversely proportional to the illuminated area. We expect the intensity coherence function to exhibit more than one length scale and more than one time scale and we expect residual fluctuations even in white light illumination.

The random walk model (1) can be solved exactly in a formal sense for the distribution of amplitude fluctuations ($A = \sqrt{I}$)

$$P_N(A) = A\int_0^\infty u\,du\,J_0(uA)\,\langle J_0(ua)\rangle^N \tag{4}$$

However this result is not particularly useful as a noise model.

It is well known that

$$\lim_{N\to\infty} P_N(A) \sim 2Ae^{-A^2} \quad \text{for all } p(a)$$

and that

$$\underline{\text{if}}\ \ p(a) \sim 2ae^{-a^2} \quad \underline{\text{then}}\ \ P_N(A) \sim 2Ae^{-A^2} \ \text{for all } N$$

Evidently neither of these results is useful in non-Gaussian noise modelling. However it is not difficult to show that (Jakeman and Pusey, 1976)

$$\underline{\text{if}}\ \ p(a) \sim a^{\nu} K_{\nu-1}(a) \quad \underline{\text{then}}\ \ P(A) \sim A^{\nu N} K_{\nu N-1}(A) \ \text{for all } N \tag{5}$$

where K_{ν} is the modified Bessel function of the second kind. These K-distributions have moments which lie between Rayleigh and log-normal and clearly have a number of attractive features. For example the coherent addition of vectors whose amplitude is K-distributed leads to a resultant whose amplitude is distributed according to a different member of the same class: no new parameter is introduced into the model.

Although K-distributions were introduced as an empirical class of distributions satisfying (4) they have proved to provide an excellent model for data from a wide range of scattering systems, particularly multiscale systems (Jakeman, 1980). It has been conjectured that the modulation of small scales by an underlying larger scale structure leads to clustering or bunching of the scattering centres which results in non-Gaussian fluctuations even in the high scatterer density limit. It is not difficult to show that the negative binomial distribution

$$P(N) = \binom{N - \nu + 1}{N} \frac{(\bar{N}/\nu)^N}{(1 + \bar{N}/\nu)^{N+\nu}} \tag{6}$$

leads to K-distributed amplitude fluctuations in this limit (Jakeman and Pusey, 1978) i.e.

$$\lim_{\bar{N}\to\infty} \langle P_N(A)\rangle \sim A^{\nu} K_{\nu-1}(A)$$

where $\bar{N} = \langle N\rangle$ and $P_N(A)$ is given by equation (4). Thus the K-distribution model can arise as the combined result of scatterer density fluctuations and interference effects. Although the negative binomial model is only one particular cluster model, some kind of clustering property is more likely to be common to a large number of systems than the detailed behaviour of individual scattering centres i.e. p(a). It has been shown that the cluster-interference interpretation is indeed correct for one scattering model which leads exactly to K-distributed intensity fluctuations (Jakeman, 1982).

3. EXPERIMENTS ON DIFFUSING FLUIDS

3.1 Liquid Crystals Experiments

The first quantitative non-Gaussian scattering experiments were carried out on 25 μm layers of the negative nematic liquid crystal MBBA used in early liquid crystal watch displays etc (Jakeman and Pusey, 1975; Pusey and Jakeman, 1975). When driven into a turbulent state by the application of a small potential difference this layer takes on the appearance of ground glass and under the appropriate polarisation conditions behaves as a random phase changing screen. The scattered intensity pattern was investigated in a far field configuration as a function of illuminated area and scattering angle. Fig 3 shows the appearance of the pattern.

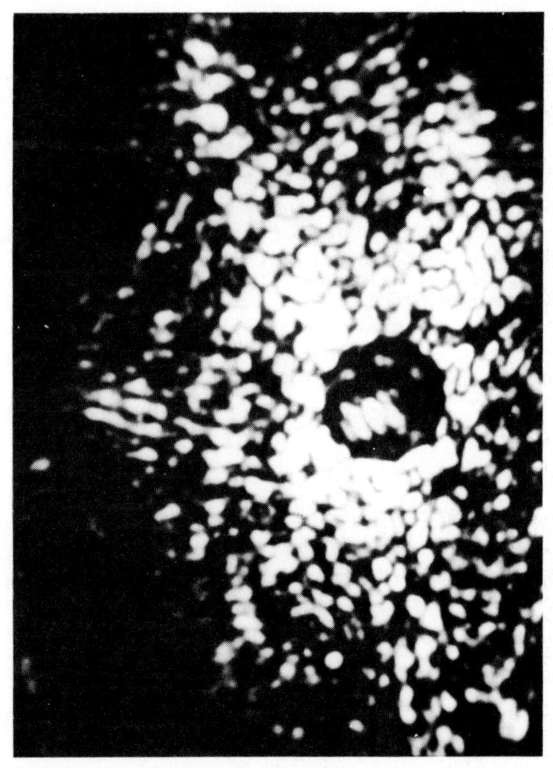

Fig. 3 Laser light scattered by a thin layer of nematic liquid crystal in turbulent motion: a Fraunhofer region non-Gaussian intensity pattern.

Figure 4 shows the second normalised intensity moment as a function of illuminated area and confirms the $(\text{area})^{-1}$ behaviour predicted by the random walk model. However it was estimated that the effective number of scatterers was still relatively large in these experiments (>16).

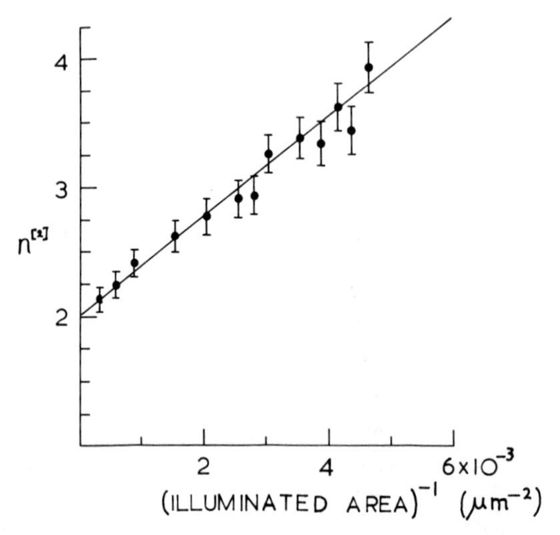

Fig. 4 Normalised second intensity moment ($n^{[2]}$) of laser light scattered by nematic liquid crystal into the Fraunhofer region as a function of illuminated area (Pusey and Jakeman, 1975)

Fig. 5 shows a plot of the spatial coherence function of the intensity fluctuations and exhibits two well defined length scales as expected. By increasing the applied potential difference the size of the scattering structures could be reduced and it was then found that only the central rapid fall-off was preserved, indicating that this length scale related to interference effects or speckle.

A number of other measurements were made on this system including the dependence of the second normalised intensity moment on angle and the temporal coherence properties of the intensity. These could all be explained satisfactorily on the basis of the random walk model and are not reproduced here. An adequate model for the higher order moments of the intensity fluctuation distributions could not be found at the time of the original experiments but it was subsequently found that the K-distribution model was an excellent fit to the data (Fig. 6).

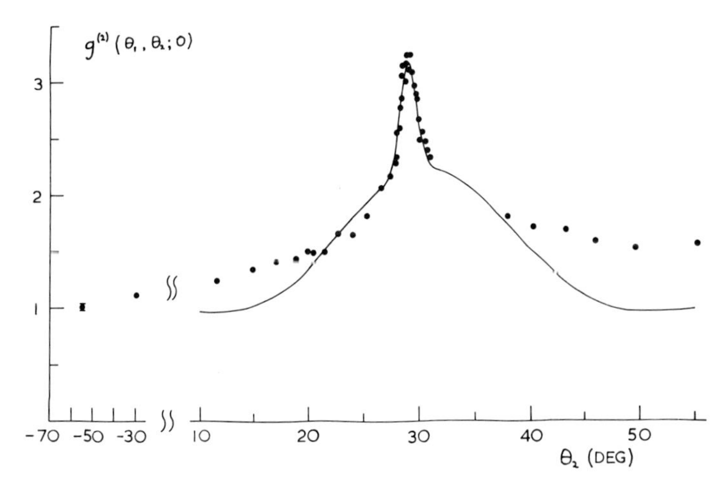

Fig. 5 Spatial coherence function of intensity fluctuations of laser-light scattered by nematic liquid crystal into the Fraunhofer region as a function of detector separation angle (··experiment-theory) (Pusey and Jakeman, 1975)).

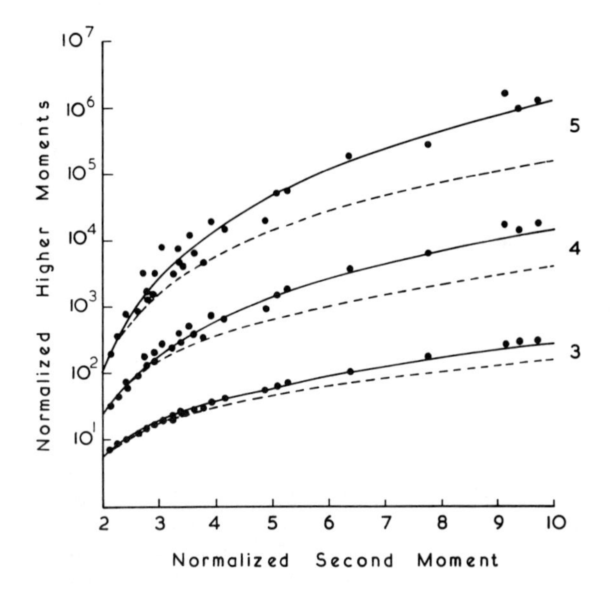

Fig. 6 Comparison of liquid crystal intensity fluctuation data with the higher normalised moments of the K-distribution class as a function of second normalised moment (··data -theory) (Pusey and Jakeman, 1978)).

3.2 Thermal Plume Experiments

Theoretical predictions for the intensity fluctuations in the Fresnel region of a phase changing screen appeared in the literature during the 1950's and 1960's. A number of simple experiments on thermal plumes and mixing layers were devised at RSRE during the mid 70's in order to test these predictions and also to gain insight into the more difficult but important problem of propagation through extended media. Figure 7 shows the focussing curve or scintillation plot for light scattered into the Fresnel region by a thermal plume generated by a heater between two baffles (Parry, Pusey, Jakeman and McWhirter, 1977). Note that enhanced non-Gaussian fluctuations occur in this scattering geometry as a result of the limited area of the scatterer which contributes to the intensity at the detector when this is sufficiently close. The limitation is typically caused by geometric considerations eg. the finite tilts of the scattered wavefront, although diffraction plays the dominant role in the case of fractal scatterers.

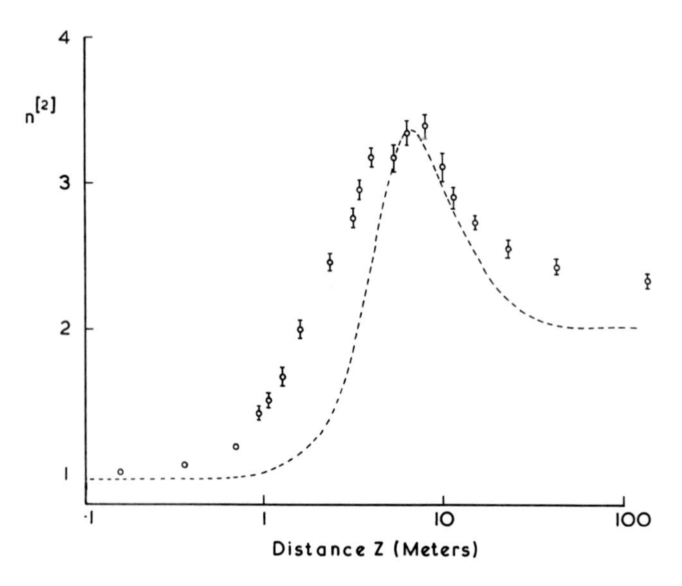

Fig. 7 Second normalised intensity moment of laser light scattered by a thermal plume as a function of propagation distance in the Fresnel region (Φ data, ---- smooth single scale model)(Parry, Pusey, Jakeman and McWhirter, 1977)

Comparison of the data with theoretical curves calculated assuming a joint Gaussian phase function with Gaussian autocorrelation function indicates that the predicted focussing peak is too narrow, probably because a range of scale sizes focussing at different distances are present. This data was subsequently compared with the predictions of a two scale model (Jakeman and McWhirter, 1981) and more recently with the predictions of a sub-fractal model as shown in Fig 8 (Jakeman and Jefferson, 1984). Much better aggreement with the data is obtained using the latter model, the apparent deviation at large propagation distances being attributable to inadequate modelling of outer scale effects.

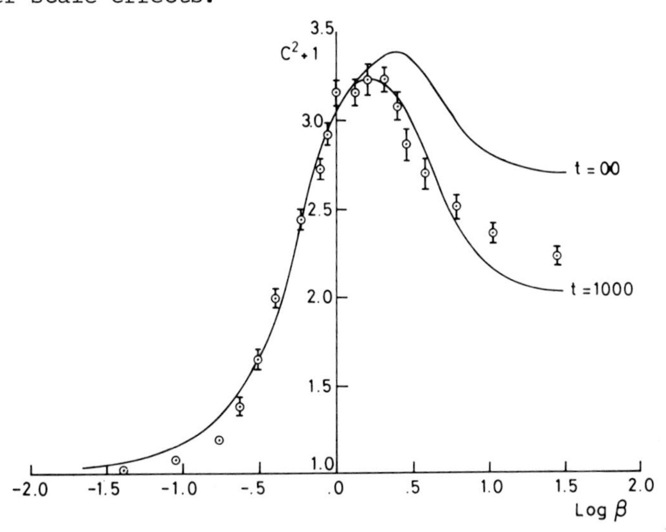

Fig. 8 Data as for Figure 7 compared with the predictions of the sub-fractal model, β is proportional to the propagation distance and t is an outer scale cut-off parameter.

It was found experimentally that different focussing curves were obtained for plumes of different scattering strengths, thus ruling out a simple fractal description of the system. Indeed, geometrical effects are clearly visible in the scattered intensity pattern (Fig 9).

Several other properties of the scattered intensity, including the spatial, temporal and spatio-temporal coherence functions have been measured for this system and the reader is referred to the literature for a full description of the experiments (Parry, Pusey, Jakeman and McWhirter, 1978). One interesting and important observation made was that close to the plume the intensity fluctuations were greater than log-normal (ie the higher moments were greater for a given second moment). At larger propagation distances a narrow region of log-normal behaviour was found but in the focussing peak region and beyond the intensity fluctuations were K-distributed. In a subsequent measurement of non-Gaussian stellar scintillation it was deduced on this basis that the observation was being made in a region preceding the focussing peak (Jakeman, Parry and Pusey, 1978).

Fig. 9 Intensity pattern in the focussing region, thermal plume experiment (Parry, Pusey, Jakeman and McWhirter, 1977).

A number of measurements similar to the above have been made on mixing layers of hot and cold water and qualitatively similar results have been obtained. Strikingly different effects have been observed in the case of mixing layers of brine and water, possibly as a result of the smaller diffusion coefficient leading to smaller scale structure (Walker and Jakeman, 1984). For example, the focussing curve in this latter system exhibited little or no peak despite the obvious strength of the scattering. Far field measurements indicate that geometrical optics effects are still important, however, and a sub-fractal model appears to give good agreement with data (Fig 10).

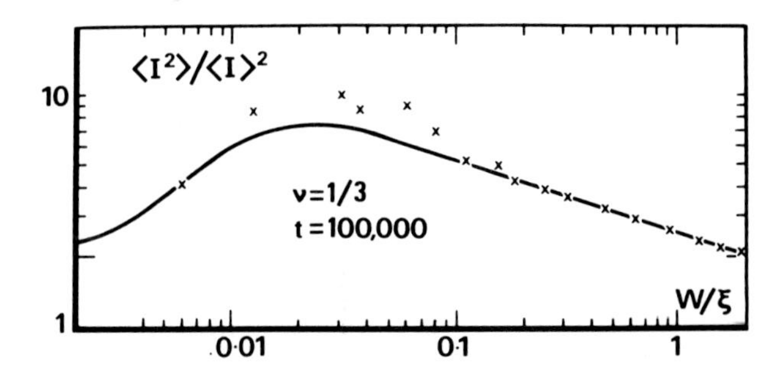

Fig. 10 Laser light scattered into the Fraunhofer region by a turbulently mixing brine/water layer: comparison of experimental data with theoretical predictions based on a sub-fractal model; W is the width of the illuminated region, ν the index of the slope structure function and t an outer scale cut-off parameter (Walker and Jakeman, 1984).

3.3 Rippled Water Surface Experiments

In these experiments a laser beam was propagated up through the surface of a wind-ruffled water surface (Walker and Jakeman, 1982). The ripples so generated were typically less than 1 cm in size so that only the capilliary wave spectrum was excited. The increments of the intensity fluctuation distribution were measured as a function of illuminated area for different "wind" strengths. Figure 11 shows a comparison of data with theoretical predictions based on a joint-Gaussian phase function with Gaussian autocorrelation function. Good agreement is found showing that this scatterer is well described by a smoothly varying, single-scale model. The intensity fluctuations were not accurately K-distributed.

4. EXPERIMENTS ON SOLID DIFFUSING SURFACES

4.1 Prepared Gaussian Diffusers

A number of experiments have been carried out in recent years on specially prepared and characterised diffusing surfaces. Figure 12 shows the focussing curve obtained with one such surface made by exposing photoresist to a superposition of many speckle patterns. Measurements showed that the resulting surface height was approximatelyGaussian with Gaussian autocorrelation function (Jakeman, 1984). Comparison with theory does not show perfect agreement, however. The intensity fluctuations were not accurately K-distributed in any scattering regime.

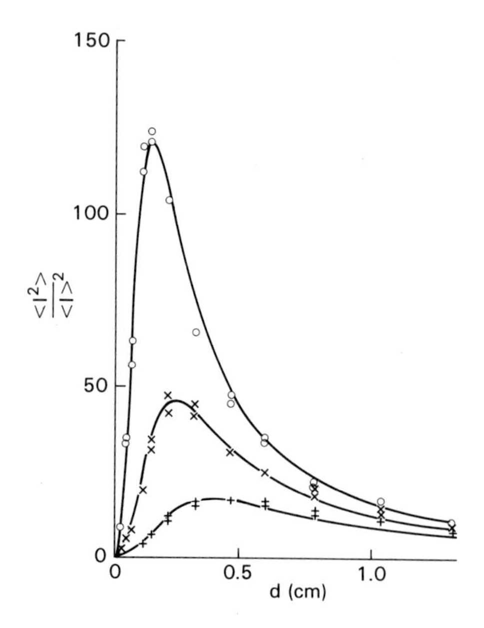

Fig. 11 Scattering by a rippled water surface:comparison of experimental data with the predictions of a smooth single scale model for different roughnesses, d is the width of the illuminated spot (Walker and Jakeman, 1982)

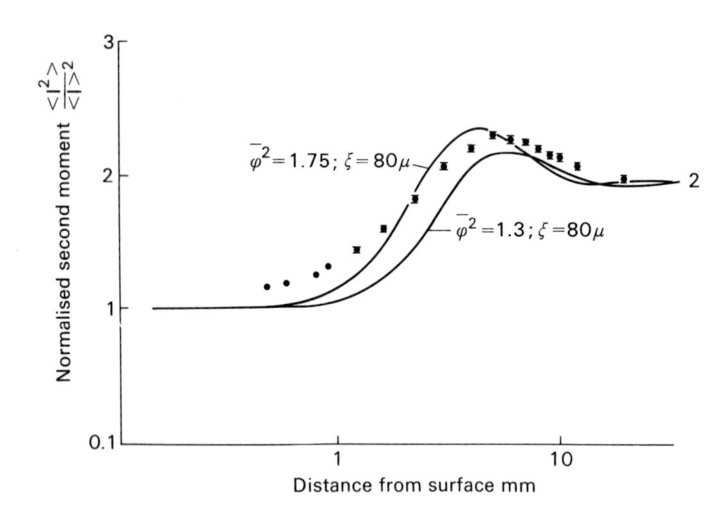

Fig. 12 Fresnel region focussing curve for a prepared rigid Gaussian diffuser compared with the theoretical predictions based on a smooth-single scale model (Jakeman, 1984).

4.2 IR Scattering from a Fractal Surface

Fig. 13 shows the height profile for a randomly engraved germanium surface measured using a "Talysurf" (Jordan, Hollins and Jakeman, 1983; Jordan, Hollins and Jakeman, 1984). Statistical analysis indicates a scaling behaviour over several decades and taken together with the observed Gaussian distribution of height differences suggest that this surface is fractal over almost three decades. The distribution of scattered intensity in the far field is very broad and in spite of height fluctuations of tens of microns only weak non-Gaussian peaks were observed in IR scattering experiments in both Fresnel and Fraunhofer configurations.

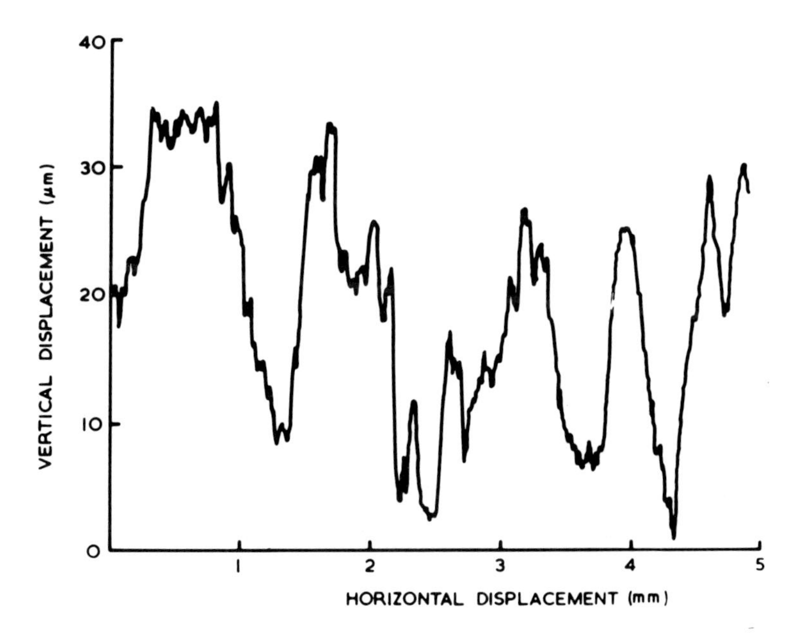

Fig. 13 Topography of a randomly engraved germanium surface (Jordan, Hollins and Jakeman, 1983;1984)

Figure 14 shows a typical focussing curve exhibiting a peak only a little in excess of the Gaussian speckle value. Fig 15 shows the observed intensity pattern in the peak region of the Fresnel configuration visualised using a pyroelectric vidicon tube. Although the picture quality is poor the pattern is evidently very different from those generated by "smoother" diffusers such as thermal plumes (Figure 9) showing little structure of the geometrical optics type.

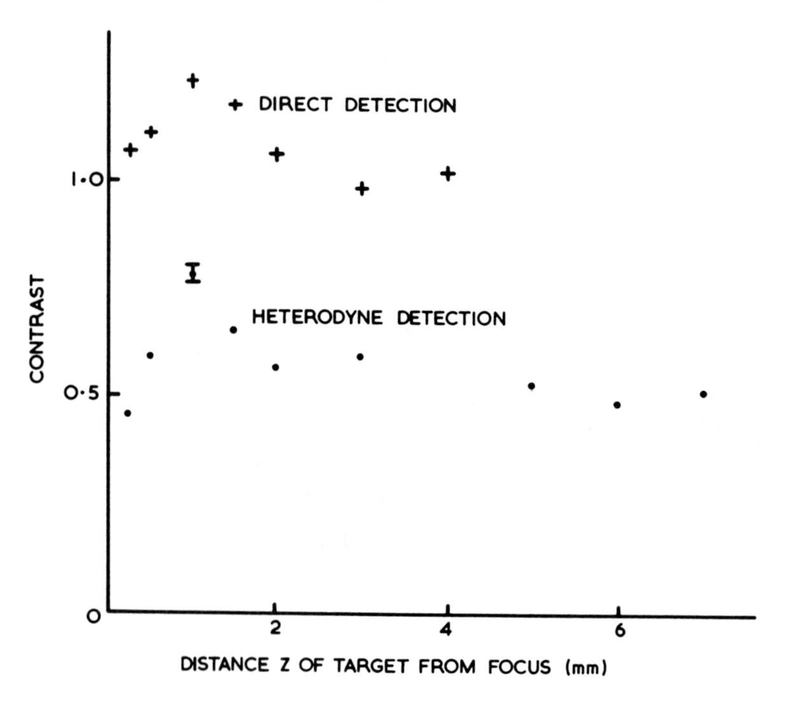

Fig 14 Fresnel region focussing curves for ir radiation scattered by the surface illustrated in Figure 13 (Jordan, Hollis and Jakeman, 1983; 1984).

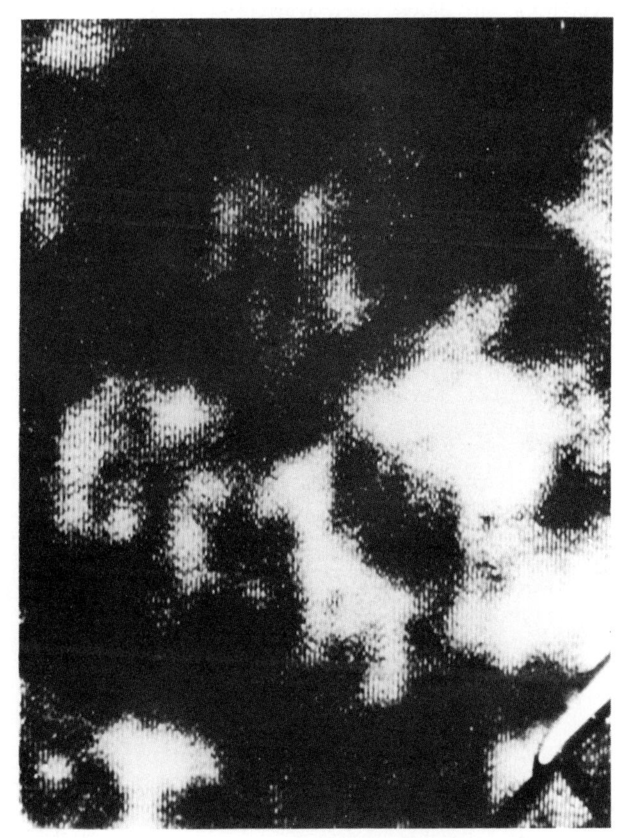

Fig. 15 Intensity pattern in the peak region of Figure 14 visualised using a pyroelectric vidicon tube.

5. CONCLUDING REMARKS

It is evident that optical frequency scattering experiments are playing an important role in advancing our understanding of non-Gaussian fluctuations in scattered waves. In particular the feed-back of information between theory and experiments in which reliable and accurate statistical data can be acquired on laboratory systems operating in a controlled environment has led not only to the development of a widely applicable non-Gaussian noise model but also to a more general appreciation of the significance of the multiscale nature of many scattering systems. At a qualitative level, the visual appearance of a scattered intensity pattern can now be used to distinguish between smoothly varying and fractal scatterers. Solutions of Maxwell's equations for the appropriate model can then in principle be used to make quantitative comparisons with experimental data and thus obtain estimates of significant parameters characterising the scatterer. In practice a number of anomalies remain - for example the observed strong geometrical optics features generated when waves are scattered by turbulent media which are usually considered to be describable in terms of the (fractal) Kolmogorov spectrum. Non-Gaussian optical scattering experiments on a well characterised turbulent system in which inner and outer scale effects are measured are thus a highly desirable objective for the future.

REFERENCES

Chandley, P. and Escamilla, 1979, "Speckle from a rough surface when the illuminated region contains few correlation areas" Opt. Commun. **29** 151-154.

Dainty, J.C., Ed., 1975, "Laser Speckle and Related Phenomena" Vol 9 of Topics in Applied Physics Springer-Verlag, Berlin.

Jakeman, E., 1984, "Speckle statistics with a small number of scatterers" Opt. Eng. **23** 453-461.

Jakeman, E., 1980, "On the statistics of K-distributed noise" J.Phys.A **13** 31-48.

Jakeman, E., 1982, "Fresnel scattering by a corrugated random surface with fractal slope" J. Opt. Soc. Am. **72** 1034-1041.

Jakeman, E. and Jefferson, J.H., 1984, "Scintillation in the Fresnel region behind a sub-fractal diffuser" Optica. Acta in the press.

Jakeman, E. and McWhirter, J.G., 1981, "Non-Gaussian scattering by a random phase screen" Appl. Phys. **1326** 125-131.

Jakeman, E., Parry, G., Pike, E.R. and Pusey, P.N., 1978, "The twinkling of stars" Contemp. Phys. **19** 127-145.

Jakeman, E. and Pusey P.N., 1976, "A model for non-Rayleigh microwave sea echo" IEEE Trans. Antennas Propag. **AP-24** 806-814.

Jakeman, E. and Pusey, P.N., 1975, "Non-Gaussian fluctuations in electromagnetic radiation scattered by a random phase screen, I:theory" J.Phys. A **8** 369-391.

Jakeman, E. and Pusey, P.N., 1978, "Significance of K-distributions in scattering experiments" Phys. Rev. Letts **40** 546-550.

Jordan, D.L., Hollins, R.C. and Jakeman, E., 1983 "Experimental measurements of non-Gaussian scattering by a fractal diffuser" Appl. Phys. **B31** 179-186.

Jordan, D.L., Hollins, R.C. and Jakeman, E., 1984 "Infrared scattering by a fractal diffuser" Optics Commun. **44** 1-5.

Levine, B.M. and Dainty, J.C., 1983, "Non-Gaussian image plane speckle: measurements from diffusers of known statistics" Opt. Commun. **45** 252-257.

Ohtsubo, J. and Asakura, T., 1978, "Measurement of surface roughness properties using speckle patterns with non-Gaussian statistics" Opt. Commun. **25** 315-319.

Parry, G., Pusey, P.N., Jakeman, E. and McWhirter, J.G., 1977, "Focussing by a random phase screen" Opt. Commun. **22** 195-201.

Parry, G., Pusey, P.N., Jakeman, E. and McWhirter, J.G., 1978, "The statistical and correlation properties of light scattered by a random phase screen" Coherence and Quantum Optics IV. Eds L. Mandel and E. Wolf, Plenum, New York.

Pusey, P.N., 1977, "Statistical properties of scattered radiation" in Photon Correlation spectroscopy and velocimetry ed by H.Z. Cummins and E.R. Pike, Plenum, New York.

Pusey, P.N. and Jakeman, E., 1975, "Non-Gaussian fluctuations in electromagnetic radiation scattered by a random phase screen, II: Applications to dynamic light scattering in a liquid crystal" J.Phys.A **8** 392-410.

Walker, J.G. and Jakeman, E., 1984, "Observation of sub-fractal behaviour in a light scattering system", Optica Acta **31** 1185-1196.

Walker, J.G. and Jakeman, E., 1982, "Non-Gaussian light scattering by a ruffled water surface" Optica Acta **29** 313-324.

SCATTERING OF RADIO WAVES BY POLAR ICE SHEETS

M.E.R. Walford
(H.H. Wills Physics Laboratory, Bristol University)

ABSTRACT

Radio-glaciology presents many problems of considerable interest because ice is an excellent medium for the propagation of radio-waves over a wide range of frequencies. We discuss the wave physics of the formation of stratified echoes in polar ice sheets. The problem of relating the stratified echo record to borehole profiles of acidity may be solved by using a phase-sensitive echo-sounding instrument. Spatial fading and the problem of inferring the geometry of internal reflecting surfaces are considered in terms of wavefront dislocations in the echoing wavefield.

1. INTRODUCTION

1.1 The purpose of this paper

The vast horizons which confront the traveller in Antarctica are echoed in structures found within the ice. These we explore using radio waves. The purpose of this paper is to consider the wave physics of the beautiful radio-echo records produced by the internal stratifications of the Antarctic Ice Sheet. (Figure 1). First however we briefly review radio-wave scattering in Polar Ice.

1.2 A review

By using the appropriate radio wavelength we may explore the surface snow layers, the deep internal structure or the bedrock underneath the Antarctic Ice Sheet. For example routine satellite observations have been made at sub-metric wavelengths of thermal radiation welling up from the surface layers of the Ice Sheet and from sea-ice. Zwally and others (1983) present maps which illustrate the value of such data. The data are plotted at intervals of 5 Kelvins in 30 km x 30 km cells. Monthly fluctuations of sea-ice extent are recorded clearly and one also sees annual variations of brightness temperature superimposed on the mean profile across the whole continent.

Active microwave studies have been made at 2.2 cm using the Seasat altimeter. They were motivated partly by the hope that Antarctica would provide the oceanographer with a convenient natural benchmark for satellite altimetry in the Southern Oceans. Unfortunately such benchmarks are hard to establish: significant long-and short-term variations probably occur in the height of many candidate snow surfaces. Fortunately one could monitor such changes using a ground-based glacier-sounding radio-echo system! It should be pointed out that there will be a need for good benchmarks if, as we hope, satellite altimetry continues to be used in Antarctic regions (Report of Working Group C, in Boerner, 1985).

We use longer (approximately metric) radio wavelengths in order to study the deep inner structure and bed of the Antarctic Ice Sheet. This cannot be carried out from satellites: the equipment must be mounted on a surface vehicle or a rather low-flying aircraft in order to avoid clutter from the snow surface obscuring the echoes of interest (Robin and others, 1983). By such means we have greatly extended our knowledge of Antarctica over the past 20 years or so, as an examination of the recently published Antarctic folio shows (Drewry, 1983). This fine folio of maps, glaciological and geophysical, relies heavily on radio-echo data as prime source-material. The measured quantities include travel times, amplitudes, polarization states and spatial-fading statistics of pulsed radio-echoes. These data provide information about the depth, the geometry, the roughness and dielectric properties of both the medium and the radio-echo targets: ice, rock, seawater or subglacial lake.

In the Bristol group, our approach to radio-glaciology has been from the viewpoint of physical optics and signal processing. We stress the value that the phase of pulsed-echoes can have in certain circumstances and have consequently introduced some new radar techniques. These have been tested using computer and physical modelling techniques, as well as field trials.

For example we have shown that one can sometimes establish a a three-dimensional frame of reference, in effect fixed to the bedrock, by carefully measuring the spatial fading pattern of radio-echo amplitude and phase. We have measured the horizontal movement of a glacier with respect to this frame and are now using it to study the vertical strain of an ice cap in Devon

Island, Canada (Walford, 1972; Walford and others, 1977). The echo-sounder is thus being used as a length gauge with a precision of about 10 cm. We hope to use it in this way to study melting from the bottom of Ice Shelves and also to measure the deformation produced in a thick glacier by tidal forces acting near a grounding line.

The amplitude and phase of pulsed radio-echoes can also be employed in imaging the glacier bed. We have used scalar wave diffraction theory to produce an image from data collected over a synthetic aperture, (Walford and Harper, 1981). We hope next year to extend this technique at a suitable site to include polarization information.

The great flexibility of radio-echo techniques available to the glaciologist will no doubt receive lively attention now that a first survey of the Antarctic Ice Sheet is rather well established.

2. ECHOES FROM STATIFIED ICE SHEETS

2.1 Echo characteristics

Figure 1 displays a range of radio-echo records from Antarctica produced using an airborne pulsed echo-sounder. The frequency was 60 MHz, the pulse-length 250 m, the aerial had a polar diagram in air 20° wide laterally and 120° along the line of flight. The aircraft flew at about 1000 feet above the snow surface.

In many inland regions layered echoes of remarkable continuity are found (Figure 1a). More commonly layering is still clearly pronounced but spatial fading decorates some of the echo layers (Figure 1b). Finally, when flying near mountains, the bed echo is interrupted by clusters of hyperbolic echo-trajectories and the distorted layers of internal stratifications give rise to echo-focussing effects: the near vertical lines on the Figure 1c have been interpreted in terms of caustics produced by these nested concave reflections, (Harrison, 1972).

Millar (1981) discusses radio-echo layering in Polar Ice Sheets in considerable detail. In the present paper we confine ourselves to the wave physics which underlies Figures 1a and 1b and we briefly discuss some problems of interpretation.

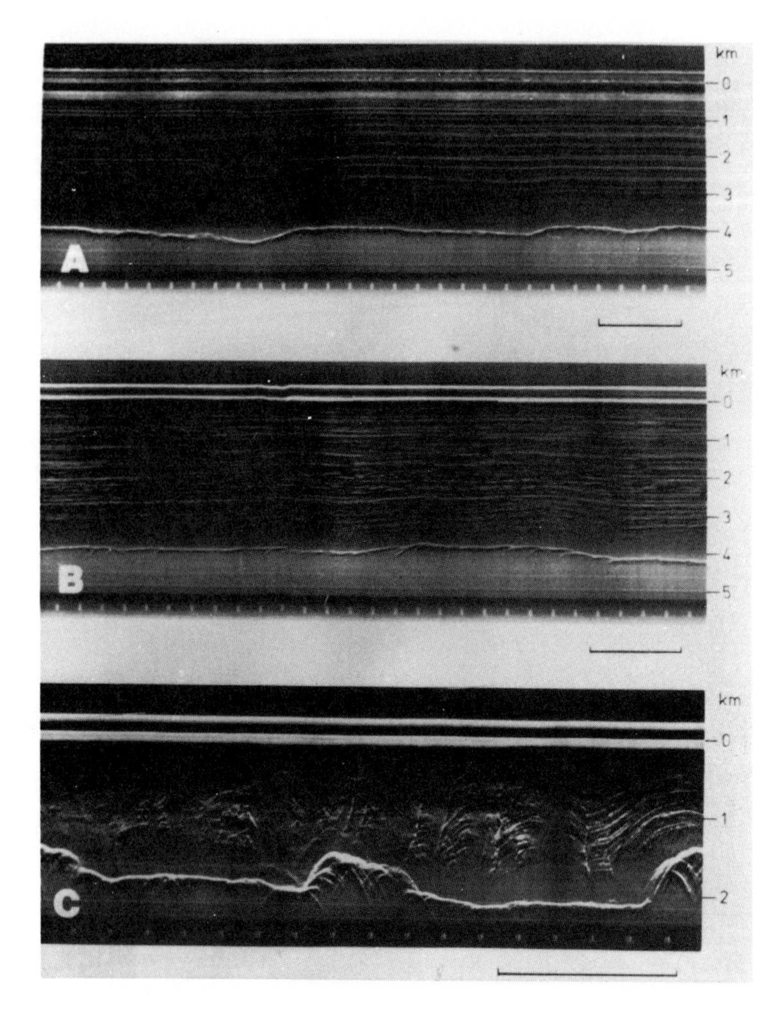

Fig. 1. Examples of the different appearances of internal layer echoes in the Antarctic obtained using a 60 MHz radio-echo sounder with a 250 ns pulse length. Scale bar = 10 km. [Photograph provided by D.H.M. Millar.]

2.2 Wave theory of layered echoes

The theory of diffraction of pulses of scalar waves by rough reflecting surfaces has been formulated by Berry (Berry, 1973) with radio-glaciology in mind. He shows that when a pulse F(t) is transmitted from an isotropic source, the pulsed echo ψ as a function of t can be written

$$\psi(t) = -\frac{1}{2\pi v}\int_0^\infty g(r)\ F'(t-2r/v).dr \qquad (1)$$

g here is a geometrical function of range r from the source-receiver point to an r-contour line in the reflecting surface; g takes account of the inverse square law and of the inclination factor and can itself be written as an integral taken around the r-contour(s) in the surface. v is the phase velocity in the ice which Berry supposes to be a uniform isotropic medium. The expression for $\psi(t)$ is a paraxial, far field approximation.

Equation 1 may easily be integrated for the case of a single plane reflecting surface at depth z with an amplitude reflection coefficient A: as we might have expected, we find the expression for the echo is

$$\psi(t) = \frac{A}{2z} F(t-2z/v) \tag{2}$$

We wish now to adapt this expression to the case of a horizontally stratified polar ice sheet. This problem is greatly simplified by the observation that radio echoes from internal layers are extremely weak: Millar remarks that the reflected wave peak power is in general about 80dB down on the peak power in the incident wave. We suppose therefore that, at least below a few hundred metres depth, fluctuations in the dielectric constant of ice are extremely small. To a good approximation, v is constant.

Furthermore we may anticipate that multiple reflections between internal layers do not contribute significantly to the echoing wavefield and the forward wave is not significantly affected by the layering.

The dielectric constant of ice does not vary significantly with frequency within the bandwidth of a pulsed-echo so we write the complex dielectric constant:

$$\varepsilon^* = \varepsilon' - j\varepsilon'' \tag{3}$$

Paren and Robin (1975) argue strongly that, at least below a few hundred metres depth, it is fluctuations in the imaginary part ε'' of the dielectric constant which are responsible for internal reflections. Consider therefore the reflection from a step-like discontinuity where the dielectric constant changes from ε^* to $\varepsilon^* - j\Delta\varepsilon''$. The amplitude reflection coefficient at this interface is

$$A = \frac{\sqrt{\varepsilon^*} - \sqrt{\varepsilon^* - j\Delta\varepsilon''}}{\sqrt{\varepsilon^*} + \sqrt{\varepsilon^* - j\Delta\varepsilon''}} \approx \frac{j\Delta\varepsilon''}{4\varepsilon'} \tag{4}$$

Because of the factor j, the pulsed echo is a different function of time from the incident pulse: for example if F(t) consists of a sinusoidal carrier wave within a symmetrical envelope, $\psi(t)$ is a cosinusoidal carrier wave within the same evelope.

We show later that there are physical reasons for representing $\varepsilon''(z)$ as a smooth differentiable function of depth rather than a fractal function. In this case the change $\Delta\varepsilon''$ associated with a small increment of depth Δz may be written to a good approximation:

$$\Delta\varepsilon'' = \frac{d\varepsilon''(z)}{dz} \Delta z \tag{5}$$

The ice sheet may be modelled as a sequence of layers of thickness Δz and we may obtain the reflected wavefield simply by superimposing the weak reflections from each discontinuity. In the limit as $\Delta z \Rightarrow 0$ therefore:

$$\psi(t) = \frac{j}{8\varepsilon'} \int_0^\infty \frac{1}{z} \frac{d\varepsilon''}{dz}, F(t - 2z/v)\, dz \tag{6}$$

which follows from equations 2, 4 and 5.

Equation 6 is formally a convolution integral and can in principle be deconvoluted in the frequency domain to provide a useful estimate of $\frac{d\varepsilon''(z)}{dz}$. Because of the echoing geometry, it will be the case that a frequency component w in the echo carries information about the spatial frequency $k = 2w/v$ in A(z). In practice the problem we anticipate is that the signals-to-noise ratio for stratification echoes is low and therefore any deconvolution will be severely bandwidth-limited. Time-domain procedures which work well on model data (Corones et al, 1984) will be subject to similar limitations. To overcome these difficulties sufficiently well to extract a useful record of $\frac{d\varepsilon''(z)}{az}$ from the layered echoes, it may be necessary to use a high-powered, phase sensitive, echo-sounder with a wide band-width and a long integration time.

The records shown in Figure 1 are typical of those available at present and they do not in principle permit us to perform the deconvolution at all! The reason is that these records are derived from $\psi(t)$ by non-linear processes that destroy all phase information. $\psi(t)$ is square-law detected and smoothed to give a relatively slowly varying power function P(t). This is differentiated with respect to time and applied as intensity modulation to an oscilloscope trace, which is photographed on a slowly transported film. The resulting records are beautiful and informative in some respects but cannot be used for deconvolution. Neither is there necessarily a simple relationship between observed lines on the photograph and the physical layers in the ice. A simple relationship could hold only if significantly reflecting layers were all separated by at least $\frac{1}{2}$ $v\Delta t$ where Δt is the duration of the transmitted pulse.

Paren and Robin show that the reflecting layers are probably manifestations of fluctuations in the acidity of ice as a function of depth. Hammer and others (1977) show from a study of Greenland ice cores that there is a background level of acidity against which events, often caused by major volcanic eruptions, occur. An event has a characteristic rise and fall in acidity which lasts for a few years in the stratification record. Broadly speaking, events which reach twice the background level occur in the Greenland record approximately every ten or twenty years, events five times the background acidity occur perhaps once in a thousand years, larger events are even less frequent.

Consider the history of a particular acid layer. Vertical compressive strain takes place within the ice sheet. In the upper few hundred metres, densification of snow into ice increases the acidity (per unit volume) by a factor of order 3 and decreases the layer thickness by the same factor. As the acid layer descends further it becomes progressively thinner owing to plastic flow but the peak acid concentration remains roughly constant until near the bed of the ice sheet. Here diffusion and an irregular geometry of flow destroy the layering.

Consider a model layer which has a gaussian acidity profile, $c(z)$ initially:

$$c(z) = B \exp\left[-\left\{\frac{z-z_0}{D}\right\}^2\right], \tag{7}$$

where B, D and z_0 are constants. After densification by a factor α, the layer has mean depth z_1 and:

$$c(z) = \alpha B \exp\left[-\left\{\frac{\alpha(z-z_1)}{D}\right\}^2\right] \tag{8}$$

A further, linear compressive strain β is produced by plastic flow and the mean depth becomes z_2:

$$c(z) = \alpha B \exp\left[-\left\{\frac{\alpha(1+\beta)\ (z-z_2)}{D}\right\}^2\right] \tag{9}$$

In a real ice sheet the strain rate $\dot{\beta}$ increases rapidly near the bed and the layer thickness $D/\alpha(1+\beta)$ decreases without limit.

In Figure 2 are sketched the acidity profiles for a single layer with $D = 0.5$m and $\alpha = 3$, typical values for central Antarctica.

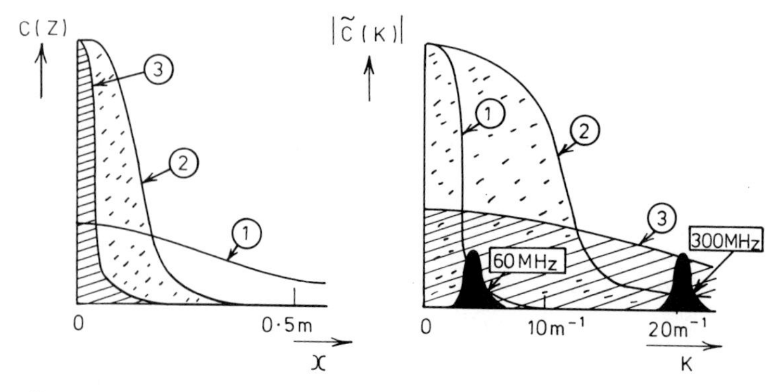

Fig. 2. Real space and k-space representations of a gaussian model acid layer. Curve 1: at the snow surface with halfwidth $D = 0.5$m. Curve 2: after compaction from snow to ice with $\alpha = 3$. Curve 3: after plastic flow with vertical strain, $\beta \approx 3$. The bands explored by a 60 MHz and a 300 MHz pulsed-echo sounder each of video bandwidth 10 MHz are shown.

The Fourier transforms of these profiles are also sketched together with the spectral windows sampled by a 60 MHz and a 300 MHz pulsed echo sounder, each of 10 HMz bandwidth. The diagram illustrates how the energy sampled by the echo sounder is rather sensitive to the strain in the ice column. This effect may contribute to the apparent lack of internal stratifications in shallow ice and at depths below about 3000 m. The echo from deep ice is also reduced because of the inverse square law and those effects near the bed which destroy the layering.

In addition there arises a statistical effect because, in deep ice, many layers contribute to the echo received at a given time. Consider a single layer whose thickness $\Delta z = \{D/\alpha(1+\beta)\}$is much less than the wavelength of the radio waves in ice, λ_i. It reflects a radio pulse, approximately, as two plane surfaces Δz apart, which have reflection coefficients A and -A respectively (equation 2-5). The reflection coefficient of this pair of surfaces is:

$$A_D = \left(\frac{1-e^{j\delta}}{1-Ae^{j\delta}}\right) A$$

where $\delta = 4\pi\Delta z/\lambda_i$. Since $A<<1$ we have:

$$A_D = (1-e^{j\delta})A.$$

For a thin layer $\delta<<1$ and so:

$$A_D = -j\delta A = \{\pi D\Delta\varepsilon''/\varepsilon'\lambda_i \alpha(1+\beta)\}.$$

(The quantity $\Delta\varepsilon''$ is, we suppose, proportional to the peak acidity, αB, in the layer.)

Many thin layers contribute to the echo at time t. Now, suppose that the width of the transmitted pulse in ice is σ_i, and that initially the mean separation between acid layers is d. After densification and compressive flow, the mean number of layers contributing to the echo at time t is $N = \frac{\sigma_i}{2d}\alpha(1+\beta)$.

The mean echo power is therefore:

$$\langle P(t)\rangle \propto N \; A_D^2/z^2 \;, \; z = \tfrac{1}{2}vt.$$

$$\propto 1/\alpha(1+\beta)t^2.$$

Since β increases without limit near the glacier bed, $\langle P(H)\rangle$ must tend to zero. More realistically we know that the acidity B varies from layer to layer with larger peaks occurring more rarely: $\langle P(t)\rangle$ must depend upon this distribution, which however is not well known at present.

Figure 3 shows records from the same region of Antarctica taken with two different echo sounders with similar pulse widths σ_i but with centre frequencies 300 MHz and 60 MHz respectively. Records at these two different frequencies are not strongly correlated although, at 'Dome C', between 1 and 2 km depth, one can pick out some layer echoes in common. This may be because here major acid layers separated by at least $\sigma_i/2$ are the dominant reflectors. But we cannot be sure that the coincidence of the layers observed at the two different frequencies is not a statistical accident. In general records at 300 MHz show heavier spatial fading than those at 60 MHz. This is not surprising: a criterion for a surface to reflect in a specular fashion is that $S \lesssim \lambda_i/10$ where S is the r.m.s height of the surface roughness. This criterion is more easily met at 60 MHz where $\lambda_i/10 = 0.3$ m than at 300 MHz where $\lambda_i/10 = 0.06$ m.

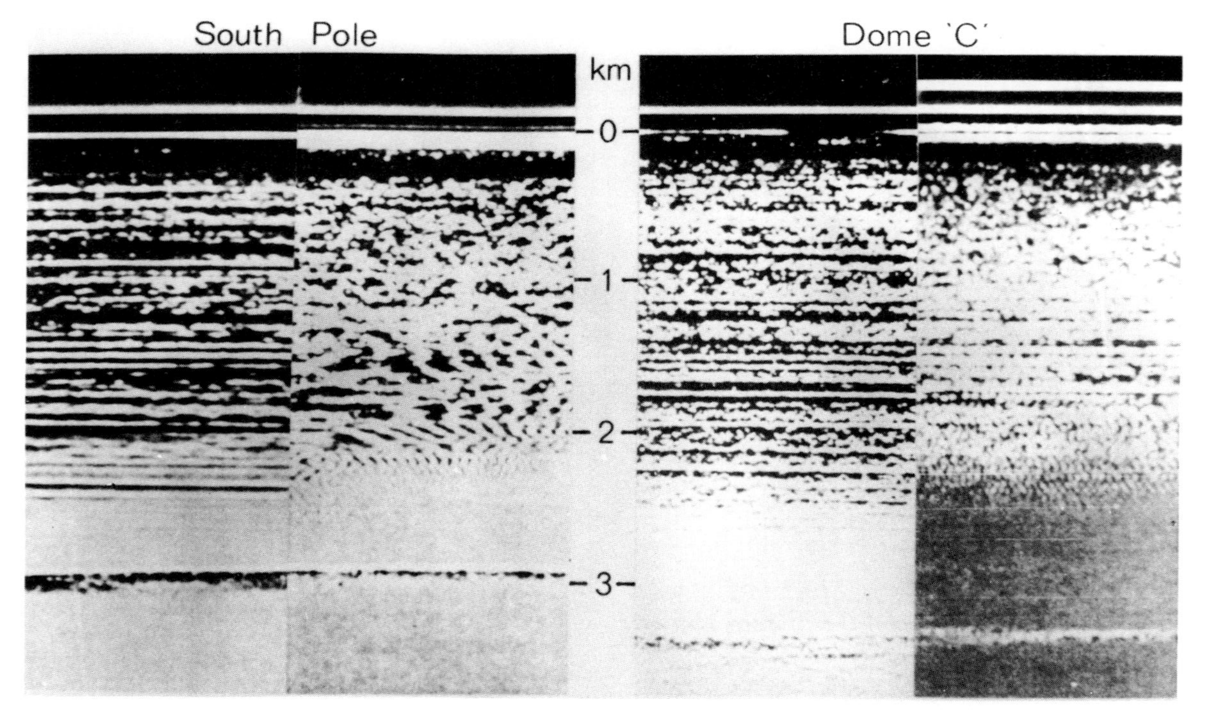

Fig. 3. 60 MHz (left) and 300 MHz (right): z-sccpe records from South Pole and Dome C, Antarctica. The spatial fading patterns seen between 2 and 2.5 km depth on the 300 MHz records are rather similar to spatial fading patterns produced by rough glacier beds and are well understood (Walford, 1972). (Photographs by D.H.M. Millar).

It is interesting to notice that the spatial fading pattern near the bottom of the 300 MHz record has quite a different character from the rest of the internal record. It shows an extended vertical structure very similar to the familiar, and well-understood, spatial fading pattern produced by a rough glacier bed (Berry, 1973). Perhaps here too the dominant reflecting surfaces are separated by $\sigma_i/2$ and appear rough only at 300 MHz. By contrast, most of the internal echo records exhibit fading in the form of a quasi-periodic modulation of the intensity of a given stratification, apparently independent of the behaviour of the neighbouring stratifications. We give a possible explanation of this curious behaviour in the next section.

Records at 60 MHz, with 250ns and 60ns pulses respectively, are shown in Figure 4. The two records do seem to show some significant correlation: the broader layered echoes can be found at nearly the same depth in each record. The weaker echoes however show significant spatial fading and do not seem correlated in the two records. Probably, where correlations occur, the record is dominated by reflections from major acid layers separated by at least $\sigma_i/2$ (where σ_i is the width of the 250ns pulse so $\sigma_i/2 \approx 20$m). Elsewhere more than one layer contributes significantly to the echo $\psi(t)$ received at a given instant, and the precise form of $\psi(t)$ depends upon the interference of more than one superposed pulse, and so, consequently, upon the widths of those pulses.

Some profiles of the acidity in deep ice cones taken at Byrd Station, Antarctica are to be published shortly (Hammer, personal communication). It will be interesting to compare the most prominent layers with existing radio-echo records but it would probably be much more revealing if a complete amplitude and phase record of the echo $\psi(t)$ were available. The most we can do at present is to solve the forward problem by calculating the expected radio-echo record, (of the type shown in Figure 1,) from the acidity profile and the known transmitter pulse. But it would ultimately be more valuable to deconvolute $\psi(t)$ in order to estimate the acidity profile down through the ice sheet.

3. SPATIAL FADING OF ECHOES FROM STRATIFIED ICE SHEETS

The most impressive feature of Figure 1a is the almost complete lack of any spatial fading in the record at all: each reflecting layer must be a very close approximation to a specular reflector at the radio wavelength $\lambda_i = 3$m. This implies constraints on the possible horizontal variations of

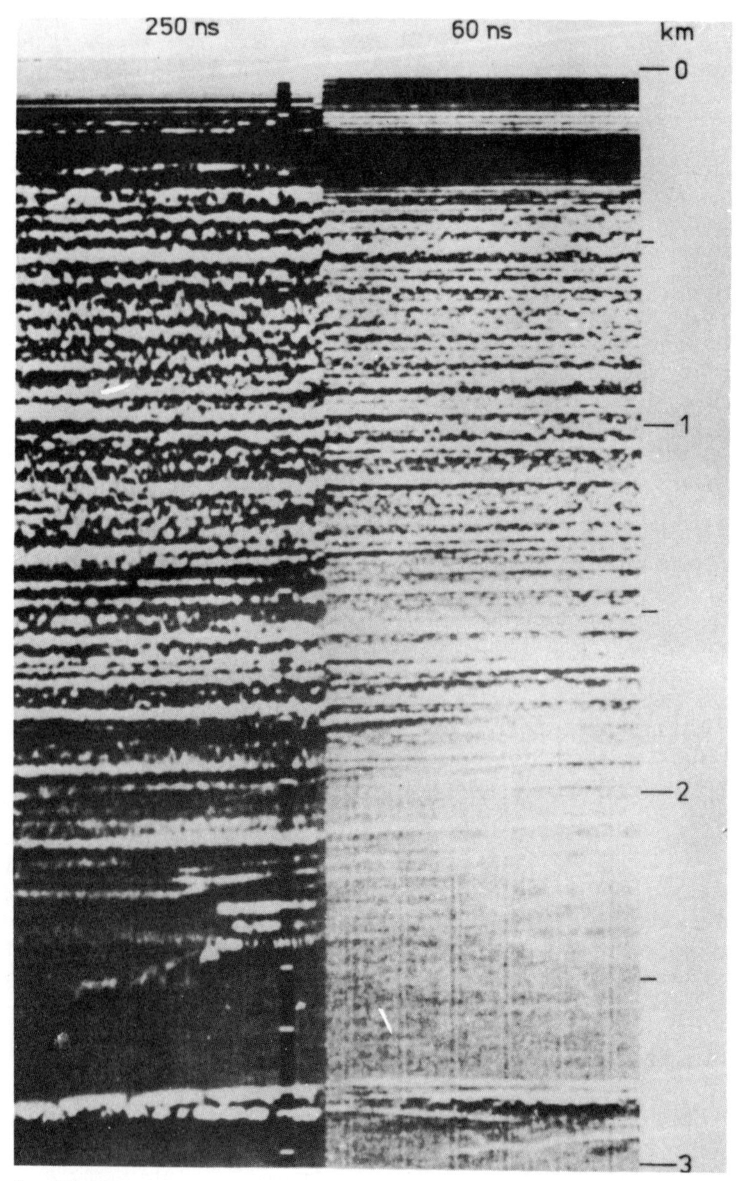

Fig. 4. 60 MHz z-scope records with 250ns (left) and 60ns (right) pulse lengths, recorded near South Pole. [Photograph by D.H.M. Millar.]

acid concentration in each layer and on the vertical scales of geometrical roughness of each layer. Millar estimates that the r.m.s amplitude of roughness cannot exceed approximately 10 cm. This is consistent with the enhanced spatial fading observed at 300 MHz which we discussed in section 2.2 and with the observed roughness of the snow surface in Antarctica, if we allow for compaction.

Spatial fading increases away from selected regions of Central Antarctica (Millar, 1981). The gradual onset of this fading is physically interesting and potentially quite informative. Consider the first effects we should expect to see in a radio-echo record as stratified ice flows over a gently wavy landscape. If, locally, the ice is mechanically uniform and isotropic we expect the finest scales of bedrock roughness to produce effects confined to layers near the bed. The longest scales on the otherhand will produce effects which penetrate upward further into the ice sheet. At distances of order 1km above the bed only very long wave effects occur. Their horizontal scale exceeds the horizontal dimensions of the radio-echo footprint and so in effect they may be regarded as introducing only a differential strain into the ice sheet which varies rather slowly with the position of the echo sounder. It is this extending and compressing of the vertical ice column which introduces the first signs of spatial fading into the radio-echo record. It appears that for certain delay times, the echo $\psi(t)$ is, by accident, particularly sensitive to the effects of such slight vertical strain. This may be partly the result of the system of photographic data recording: there is a highly non-linear relationship between density on the film and voltage applied to the oscilloscope and much detail is inevitably lost in the photographic shadows and highlights. However there is a more fundamental reason why some stratifications in the record are very sensitive to vertical strain in the ice and neighbouring stratifications are not.

An elegant explanation of this effect is afforded by the theory of wavefront dislocations (Nye and Berry, 1974). Dislocations are spatial singularities produced by the interference of waves. When pulses of waves interfere, the dislocations are localized, linear singularities which in general move as functions of time to sweep out surfaces in space. By definition, at a dislocation the wave amplitude is strictly zero and the phase is indeterminate. If a wavefront dislocation sweeps past a receiver placed in the field, then the envelope of the received signal is instantaneously reduced to zero, at a rate limited by the receiver bandwidth. It is important to realize that, although the detailed geometry of a dislocation is sensitive to the pulse shape and the distribution of the sources, nevertheless the dislocations have a distinctly stable existence: like vortices in frictionless fluids or dislocations in crystals they are destroyed only by mutual annihilation processes.

Consider now the dislocation wavefield produced by two plane, parallel reflectors separated by not more than $\sigma_i/2$ (Figure 5). If multiple reflections are negligible, then two point-like images of the transmitter appear at I_1 and I_2. Dislocations may arise by interference in the region where the pulsed wavefields produced by I_1 and I_2 overlap. The wavefield has axial symmetry about the line through I_1 and I_2, and the dislocations are therefore ring-shaped edge dislocations. They appear at points such as P in Figure 5 and travel upwards with the echo wavefield tracing out trajectories which are hyperboloid surfaces with I_1 and I_2 as foci.

Suppose the pulsed echo receiver is placed at the point Rz in Figure 5, coincident with the transmitter. If the separation of the two plane parallel reflectors is decreased the separation of I_1 and I_2 decreases and the dislocation trajectories collapse on the the line I_1, I_2, one after another, and vanish. At the critical separation of I_1 and I_2 equal to $(n + \frac{1}{2})\lambda_i$ where n is an integer less than σ_i/λ_i, the dislocation ring has zero radius and passes through the receiver. We observe it as a true zero in the signal envelope, at the appropriate delay time. This observation is rather sensitive to the separation of I_1 and I_2: σ_i/λ_i is approximately 15 and so the sensitivity is comparable with that of a reflecting monochromatic etalon, set to the n-th order interference minimum where $n < 15$.

In fact the polar ice sheet contains a distribution of specular reflectors and the echoing wavefield is a continuous function much longer than the transmitter pulse. It contains many dislocation rings which sweep upwards like a stream of assorted smoke rings. Some of these rings will be of a small radius where they pass through the snow surface. When this occurs the echo signal at the receiver may react extremely sensitively: the slightest vertical strain in the corresponding region of the ice sheet may cause the dislocation trajectory to collapse rapidly through the line I,Rz, and produce a true zero in $\psi(t)$. This sensitivity of $\psi(t)$ may cause a corresponding sensitivity in the photographic records, despite the non-linear processing involved.

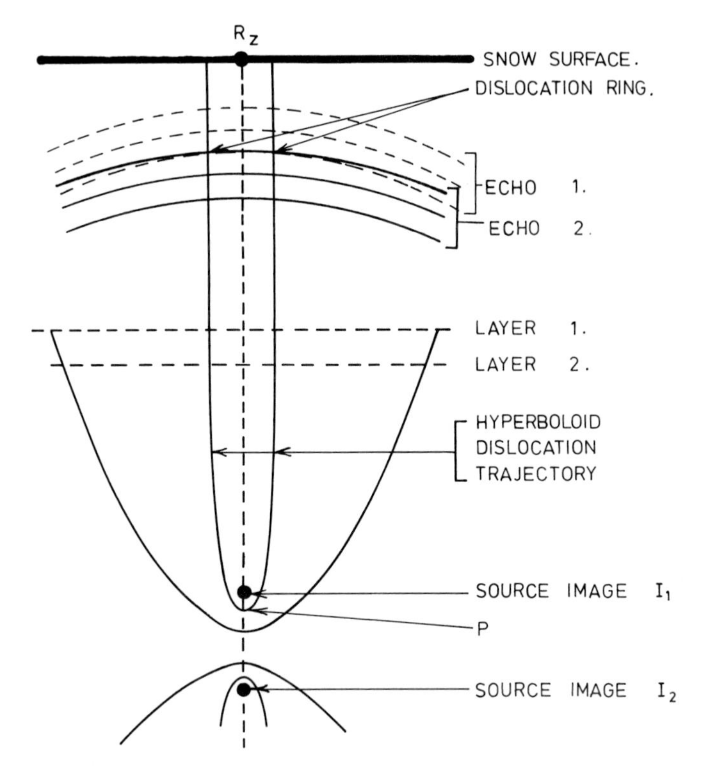

Fig. 5. Wavefront dislocations in the pulsed echo wavefield reflected by two plane parallel weak specular surfaces.

We suggest that such a condition can account for a feature of the echo record seen for example in Figure 1a at 25 km from the left-hand edge. Here the ice appears to be flowing over a broad valley which must produce some vertical strain in the ice sheet. The effect on the radio-echo record can be seen at depth 2 km but above and below this sensitive stratification continuous stratifications are found which seem to be quite unaffected by the event!

A similar explanation may account for the curious pattern of spatial fading which occurs in, for example, Figure 1b. As we pointed out in section 2.2, spatial fading of internal echoes is usually confined to one stratification: adjacent stratifications appear continuous or exhibit spatial fading quite independently. The reason may be that a sensitive stratification arises at a certain delay time when a dislocation ring passes very close to the receiver. Suppose now that the stratifications are slightly rough, either geometrically or in their lateral distribution of acidity. This puckers the dislocation ring. As we traverse over the snow surface, the shape of the puckered ring will vary quasi-periodically with our position. (This is because of the periodicity of the transmitted wave pulse; such quasi-periodicity of a spatial fading-pattern is normal and well-understood (Berry, 1973)). As a result, true zeros of the envelope of $\psi(t)$ appear, periodically, at the corresponding delay time and produce therefore precisely that type of spatial fading pattern which is confined to one particular sensitive stratification in the record.

Finally we can appreciate the importance of dislocations from a slightly different point of view. The echo signal may be written:

$$\psi(t) = R\{a^*(t)\} \cos \omega_o t$$

where ω_o is the fixed carrier frequency and $a^*(t)$ is a complex wave amplitude whose real part $R\{a^*(t)\}$ determines the amplitude of the echo and its phase $a^*(t)$ varies relatively slowly as a function of time at a rate determined by the bandwidth of the system. If we plot $a^*(t)$ as a time-dependent phasor on an Argand diagram it shows on the average neither right-handed nor left-handed behaviour. Statistically this even-handedness must be maintained whatever vertical strains may occur in the ice sheet. It may be shown that this is possible only because dislocations are created or destroyed, thereby accomodating the wavefield to extensive or compressive vertical strain.

4. RADIO-ECHO STRATIFICATIONS AND ICE SHEET STRATIGRAPHY

Although radio-echo stratifications are clearly related to the stratigraphy of an ice sheet, it would be an over-simplification to assume a one-to-one relationship between them, except in the simplest cases perhaps. If dislocations are present in the echoing wavefield, gross errors of interpretation may occur if due consideration to them is not given. To illustrate this consider the following experiment carried out with a phase-sensitive radio-echo sounder at the Ice cap, Mer de Glace Agassiz, in Ellesmere Island, Canada, (Harper, 1981).

An echo-sounding traverse was carried out between two boreholes 1.2 km apart one of which was 210 m deeper than the other. As expected the pulsed echo range increased by 210 m $\pm$ 5 m during the traverse. The amplitude and phase structure of the radio-echoes were also recorded along the traverse. The records showed a clear tendency for the echo-phase to increase with increasing depth along the traverse, although there were places where the echo signal amplitude was small and the phase could not be clearly determined. Nevertheless we attempted to measure the total phase change observed during the traverse in the expectation that, on the average, no serious error would arise from regions where the signal was bad. We obtained a value of 440 radians equivalent to 110 m increase in depth. The "error" of 100 m was surprisingly large and we attribute it to the presence of wavefront dislocations where, by definitions the phase of the echo signal is indeterminal. Our assumption of negligible net phase change near the dislocations was systematically in error and the net effect was gross error in our attempt to infer the geometry of the reflecting glacier bed from observations of echo-phase along the line of traverse. One should either be content with pulsed-echo travel-time observations or carry out a fully-detailed analysis using information from an adequate two-dimensional aperture in the snow surface as described by Walford and Harper (1981).

Our conclusion is that gross errors may possibly occur if one attempts to relate radio-echo stratifications to the layering in an ice sheet in too simplistic a way. One should seek closure tests or borehole comparisons with the radio-echo record wherever possible.

5. CONCLUSIONS

Acid layering in the Antarctic ice sheet gives rise to stratified radio-echo records. A record of the amplitude and phase of the pulsed echoes would permit one to infer the distribution of acid layering with depth, in principle. In practice an excellent signals-to-noise ratio over as wide a bandwidth as possible would be necessary. The concept of wavefront dislocations provides a useful insight into the curious spatial fading found in the presently available radio-echo records. However these dislocations can be a source of error in attempting to infer the geometry of layering from radio echo records.

6. ACKNOWLEDGEMENTS

My thanks are due to Dr. D.H.M. Millar for providing the recordings reproduced in figures 1, 4 and 5, and to my wife Alison for not giving birth to Imogen Morwenna until after the conference. Prof. J.F. Nye commented helpfully on the manuscript.

7. REFERENCES

Berry, M.V., 1972, On deducing the form of surfaces from their diffracted echoes, *J. Phys. A: GEN-PHYS.*, Vol.**5**, p.272-291.

Berry, M.V., 1973, The statistical properties of echoes diffracted from rough surfaces, *Phil. Trans. Roy. Soc. Lond.*, A., Vol. **273**, No. 1237, p.611-658.

Boerner, W.-M, 1985, Inverse methods in electromagnetic imaging, Proc. NATO-ARW, Bad Windsheim, 1983. NATO ASI Series. Series C: Mathematical and Physical Scienes. vol. 143. D. Reidel, Dordrecht, Holland.

Corones, J.P., Davison, M.E. and Krueger, R.J., 1984, Dissipative inverse problems in the time-domain, in Boerner, 1985.

Drewry, D.J., 1983, Antarctica: Glaciological and Geophysical Folio, Scott Polar Research Institute, University of Cambridge, U.K.

Hammer, C.U., Clausen, H.B., Dansgaard, W., Gundestrup, N., Johnsen, S.J. and Reeh, N., 1978, Dating of Greenland ice cores by flow models, isotopes volcanic debris and continental dust, *J. Glacial*, Vol. **20**, No. 82, p. 3-26.

Harper, M.F.L., 1981, Detailed observations of glacier beds using phase-sensitive radio-echo techniques, Ph.D. Thesis, University of Bristol.

Harrison, C.R., 1972, Radio propagation effects in glaciers, Ph.D. Thesis, University of Cambridge, U.K.

Millar, D.H.M., 1981, Radio-echo layering in polar ice sheets, Ph.D. Thesis, University of Cambridge, U.K.

Nye, J.F. and Berry, M.V., 1974, Dislocations in wave trains, *Proc. Roy. Soc. Lond.*, A. **336**, p.165-259.

Paren, J.G. and Robin, G.de Q, 1975, Internal reflections in polar ice sheets, *J. Glaciol*, Vol. **14**, p.251-259.

Robin, G. de Q., Drewry, D.J. and Squire, V.A., 1983, Satellite observations of polar ice fields, *Phil. Trans. Roy. Soc. Lond.*, A **309**, p.447-461.

Walford, M.E.R., 1972, Glacier movement measured with a radio echo technique, *Nature*, Vol. **239**, No. 5367, p.95-96.

Walford, M.E.R., Holdorf, P.C. and Oakberg, R.G., 1977, Phase-sensitive radio-echo sounding at the Devon Island Ice Cap, Canada, *J. Glaciol,* Vol. **18**, p.217-229.

Walford, M.E.R., and Harper, M.F.L., 1981, The detailed study of glacier beds using radio-echo techniques, *Geophys. J.R. astr. Soc.* No. **67**, p.487-514.

Zwally, H.J., Comiso, J.C., Parkinson, C.L., Campbell, W.J., Carsey, F.D. and Gloerson, P., 1983, Antarctica sea ice, 1973-1976: Satellite passive microwave observations, *NASA SP; 459,* U.S. Government Printing Office, Washington, D.C.

SPECKLE FROM PSEUDO-RANDOM STRUCTURES

J.C. Dainty and D. Newman
(Blackett Laboratory, Imperial College, London)

ABSTRACT

Photon correlation techniques are used to study the spatio-temporal properties of dynamic speckle from compound objects consisting of a stationary phase grating and a moving phase diffuser. Measurements of either the space-only or time-only correlation can be used to detect the presence of the grating under circumstances where ordinary mean intensity measurements in the far-field would not reveal it. The work has potential application to information coding.

1. INTRODUCTION

Speckle is the name given to the random intensity structure that appears in the diffraction or image plane of an optically random medium, such as a rough surface, when it is illuminated by light with some finite degree of coherence. The statistics of speckle patterns have been discussed extensively in the literature [Goodman (1975), Dainty (1976)] and several applications have been found [Dainty (1975), Erf (1978), Francon (1979)].

There are, in fact, many different types of speckle patterns whose statistical properties differ depending upon the coherence of the illumination, the nature of the scattering medium and the diffraction or imaging geometry. In this paper we describe some aspects of the speckle pattern formed in the far-field of a coherently illuminated composite diffuser that consists of a thin phase grating placed in contact with a phase diffuser. In this case the overall speckle pattern consists of the interference of a number of identical speckle patterns each centred at a diffraction order of the grating. This pattern has interesting spatial correlation properties which facilitate the detection of the presence of the grating under conditions in which detection by conventional means is difficult or impossible.

In a series of papers, the group of Baltes has suggested that this phenomenon could be used as a means of information encoding [Baltes et al (1981a,b,c,d), Glass et al (1982a,b), Glass (1982), Baltes and Huiser (1984)]. They have also demonstrated the feasibility of detecting hidden structures using amplitude interferometry [Jauch et al (1981, 1982a,b)]. We present the results of experiments based on the technique of photon correlation.

If the diffuser moves with respect to the grating, then the temporal and spatio-temporal statistics of the composite speckle pattern reveal the presence of the grating through a sinusoidal modulation of the correlation function. This may provide a simple method of detecting hidden gratings or other structures.

The main results of our theory and experiments have recently appeared in the literature [Dainty and Newman (1983), Newman and Dainty (1984)] and here we shall give a more tutorial description of this phenomenon. Glass et al (1984) have also used photon correlation techniques to investigate this problem.

2. FAR-FIELD INTENSITY CORRELATION

The basic terminology is illustrated in Fig. 1. A Gaussian laser beam of amplitude distribution $P(\xi,\eta)$ and radius a illuminates a phase grating of complex amplitude transmittance $T(\xi,\eta)$ followed by a random phase diffuser $D(\xi,\eta,t)$ which is allowed to move in a straight line at velocity $\underline{v}$. The complex amplitude $U(\xi,\eta,t)$ immediately after the diffuser is assumed to be given by the product

$$U(\xi,\eta,t) = P(\xi,\eta)\ T(\xi,\eta)\ D(\xi,\eta,t) \quad , \tag{1}$$

where, in our experiments,

$$P(\xi,\eta) = I_0^{1/2} \exp\left[-(\xi^2+\eta^2)/4a^2\right] \tag{2}$$

$$T(\xi,\eta) = \exp\left[i\,\alpha\,\sin(2\pi\xi/b)\right]$$

$$= \sum_{n=-\infty}^{\infty} g_n \exp\left[2\pi i\xi/b\right] \tag{3}$$

and

$$D(\xi,\eta,t) = \exp [\ i\ \phi(\xi - v_\xi t,\ \eta - v_\eta t)]\ . \qquad (4)$$

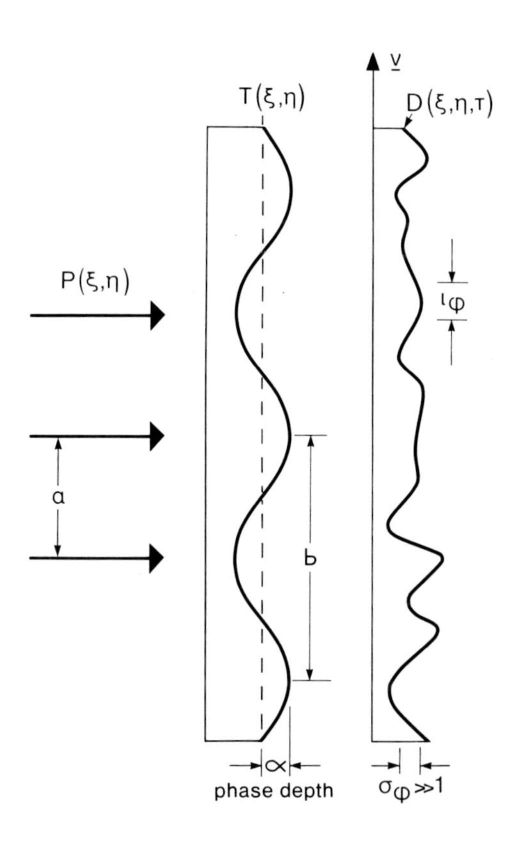

Fig. 1. Notation for interaction of laser beam with phase grating and phase diffuser.

In Eq. (3), α is the optical depth of the sinusoidal phase grating, b is the period of the grating and g_n are the Fourier coefficients. In our experiments, $\alpha \sim .85$, b = 5.1 and 9.2 μm and only the first four Fourier coefficients had significant values. In Eq. (4), ϕ is a statistically stationary Gaussian random process with a Gaussian spatial correlation of correlation length l_ϕ and variance $\sigma_\phi^2 > \pi^2$. It follows that the spatial correlation function of $D(\xi,\eta,t)$ is also Gaussian of width $L \sim l_\phi/\sigma_\phi$.

The basic quantity of interest is the spatio-temporal correlation function of complex amplitude in the far-field, $\Gamma\ (x_1, y_1, x_2, y_2, \tau)$. This is related to the spatio-temporal correlation of the object wave by,

$$\Gamma\ (x_1, y_1, x_2, y_2, \tau) = \iint_{-\infty}^{\infty} \iint_{-\infty}^{\infty} \langle U(\xi_1, \eta_1, t)\ U^*(\xi_2, \eta_2, t+\tau) \rangle$$

$$\exp\left[\frac{-2\pi i}{\lambda R} (x_1\xi_1 - x_2\xi_2 + y_1\eta_1 - y_2\eta_2) \right]$$

$$d\xi_1\ d\xi_2\ d\eta_1\ d\eta_2\ , \qquad (5)$$

where R is the distance from the object plane (ξ, η) to the far-field plane (x,y) .

Evaluation of Eq. (5) requires that we first evaluate the correlation of complex amplitude due to the Gaussian phase diffuser, namely

$$\langle \exp\left[i \{ \phi(\xi_1, \eta_1, t) - \phi(\xi_2, \eta_2, t+\tau) \} \right] \rangle .$$

If the diffuser is optically rough, i.e. $\sigma_\phi > \pi$, then it can be shown [Goodman (1975)] that the correlation of the complex amplitude is also Gaussian of width equal to $L = l_\phi / \sigma_\phi$. We shall only be concerned with measurements made along the x-axis and it is convenient to use the following sum and difference coordinates:

$$s = (x_1 + x_2) / 2R = (\sin\theta_1 + \sin\theta_2) / 2$$

$$\delta = (x_1 - x_2) / R = \sin\theta_1 - \sin\theta_2 .$$

A straightforward but tedious calculation [Newman and Dainty (1984)] leads to the result:

$$\Gamma\ (s, \delta, 0, 0, \tau) = \exp\left[- |\underline{v}|^2\ \tau^2 / 8a^2 \right]$$

$$\sum_{\text{all } n,m} g_n g_m^* \exp\left[\frac{-2\pi i v \xi \tau}{b} \left(\frac{s}{\sin\theta'} - \frac{n+m}{2} \right) \right]$$

$$\exp\left[-k^2 a^2 / 2 \left(\delta - (n-m)\frac{\lambda}{b} \right)^2 \right]$$

$$\exp\left[-k^2 L^2 / 2 \left(s - (n+m)\frac{\lambda}{2b} \right)^2 \right] . \qquad (6)$$

Equation (6) describes the spatio-temporal correlation of the complex amplitude in the far-field for points separated by δ whose mid-point is s (both are angular measures along the x-axis) and separated by τ in time. The first diffraction order is at an angle $\sin\theta' = \lambda/b$.

Photon correlation experiments do not measure Γ directly. They provide an (unbiased) estimate of the correlation of intensity fluctuation which in one dimensional notation is defined by

$$C_I(x_1,x_2,\tau) = \frac{< I(x_1,t)\ I(x_2,t+\tau) >}{|<I(x_1,0)><I(x_2,0)>|} - 1 . \qquad (7)$$

If L/a >> 1, the far-field speckle has Gaussian statistics so that the intensity correlation is simply related to the amplitude correlation:

$$C_I(x_1,x_2,\tau) = \frac{|\Gamma(x_1,x_2,\tau)|^2}{|\Gamma(x_1,x_1,0)|\,|\Gamma(x_2,x_2,0|} = |\gamma(x_1,x_2,\tau)|^2 , \qquad (8)$$

where $\gamma(x_1,x_2,\tau)$ is called the complex degree of coherence and satisfies $0 \le |\gamma| \le 1$ by virtue of the Schwarz inequality.

3. SPATIAL CORRELATION AT ZERO TIME DELAY

For zero time delay, $\tau = 0$, the complex degree of coherence $\gamma(s,\delta,0)$ and the normalised intensity fluctuation correlation $|\gamma(s,\delta,0)|^2$ are special cases of Eq. (6). The average intensity in the far-field exhibits broad peaks centred on the grating diffraction orders $\sin\theta_n = n\lambda/b$. The angular width of these peaks is approximately equal to $(kL)^{-1}$. The complex degree of coherence in the so-called anti-symmetric scan, $\gamma(s=0,\delta=2\sin\theta,\tau=0)$ is sharply peaked whenever $\sin\theta = \pm n\lambda/b$, i.e. whenever one correlates pairs of diffraction orders. The angular width of the correlation peaks is approximately equal to $(ka)^{-1}$ and is narrow compared with $(kL)^{-1}$ in cases of practical interest.

The experimental setup is shown in Fig. 2. The detectors consist of optical fibres of 50 μm core diameter coupled to Hamamatstu R928 photomultipliers. The signals are passed through a preamplifier/discriminator and the photon counts analysed in a Langley-Ford DC128 correlator with 128 channels each with time increment $\delta\tau$. Typical time increments used in our experiment were $\delta\tau$ = 50 - 200 μs and photon rates of approximately one per $\delta\tau$ or $5 \times 10^3 - 2 \times 10^4\ s^{-1}$ were used with a dark count of approximately 100 s^{-1}. Typical measurement times were 10-20 s or about 10^5 samples. Measurements of the spatial cross-correlation in the anti-symmetric scan are estimated from the temporal cross-correlation at zero time delay.

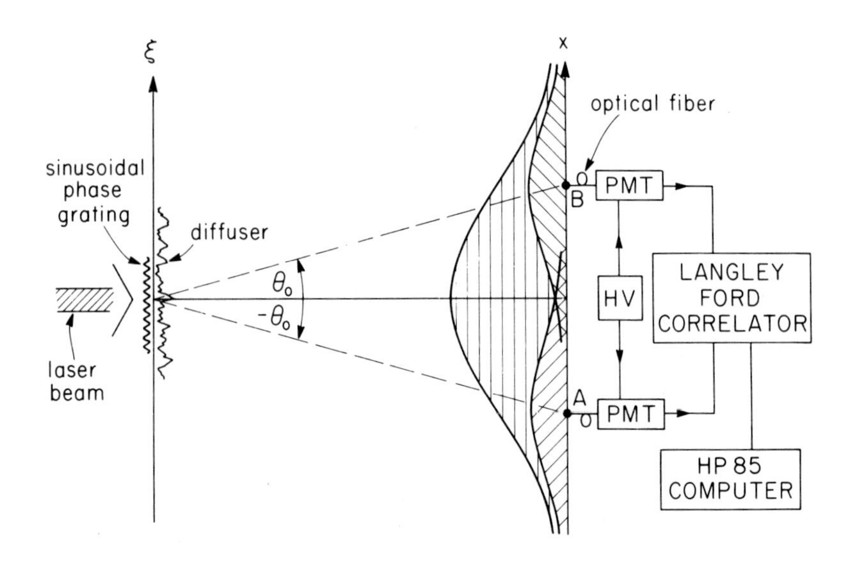

Fig. 2. Experimental arrangement.

A series of diffusers were produced in photoresist. Multiple exposure of these plates to statistically independent speckle patterns and subsequent development yields an optically rough surface whose surface height probability density function is approximately Gaussian [Gray (1978), Levine and Dainty (1983)]. The standard deviation σ_h of this distribution and the correlation length l_h of each diffuser are measured using a Dektac profilometer (mechanical stylus device). The resulting value of the correlation length L of the complex amplitude transmittance is then determined using the relation $L = l_h/\sigma_\phi$ and a value of 1.67 for the refractive index of photoresist at a wavelength of 633 nm. The diffusers used in this experiment had correlation lengths L of 0.9, 1.4, 2.3, 2.7, 3.3, 6.4, 7.1 and 10.3 μm.

Two thin sinusoidal phase gratings of spatial periods b = 5.1 and 9.2 μm were produced holographically by using Kodak 131-02 holographic film. The optical depth of these gratings was estimated from standard intensity measurements and both have a depth of $\alpha \sim .85$. The values of the first four Fourier coefficients are then

$$g_0 \sim 0.826$$

$$g_1 \sim 0.385$$

$$g_2 \sim 0.080$$

$$g_3 \sim 0.012$$

The experimental error associated with α is less than 5% and the error in the measurement of the correlation length L is 5-10%.

A precise estimate of the spatial cross-correlation peak is difficult because of the sharpness in this peak. A typical measurement across the peak is shown in Fig. 3. The sharpness of this peak may be appreciated by considering our experimental arrangement. For a laser beam of width $2a \sim 0.66$ mm and $\lambda \sim 633$ nm, the angular width of the peak is $1/ka \sim 3 \times 10^{-4}$ rad corresponding to a width of approximately 0.6 mm at the far-field distance of 2 m.

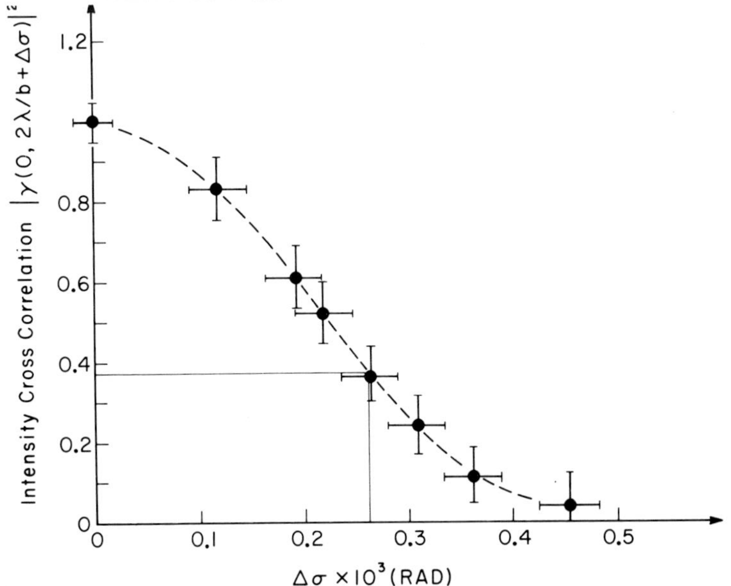

Fig. 3. Measurement of the angular width of the spatial correlation peak in the antisymmetric scan ($\Delta\sigma$ in Figure is same as δ in text).

The diffuser translation was effected by a continuous rotation in the (ξ,η) plane. The diffuser's relative component of velocity parallel to the far-field x-axis, v_ξ, is equal to $2\pi\omega r_\eta$, where r_η is the vertical displacement of the laser beam with respect to the centre of rotation and ω is the angular speed of the diffuser. The relative velocity v_η is then equal to $2\pi\omega r_\xi$, where r_ξ is the horizontal displacement. For measurements involving only one dimensional translations, a variable speed linear translator was used that allowed for accurate measurements of v_ξ and hence the modulation period as a function of b/v_ξ.

Figure 4 compares experimental results and theory for a measurement of the normalised intensity correlation in the anti-symmetric (first-order) scan, $|\gamma(0,2\lambda/b,0)|^2$, against the parameter L/b. The agreement between theory and experiment is excellent. The intensity correlation measurements reveal the presence of the grating for all $L/b \gtrsim 0.15$. On the other hand, the far-field average intensity is diffuse for $L/b \lesssim 0.33$, so that there is a small but important range of values $0.15 \lesssim L/b \lesssim 0.33$ for which average intensity measurements would fail. The exact range of values of L/b for which the hidden periodicity is revealed depends on the form of the grating (via the Fourier coefficients in Eq. (6)).

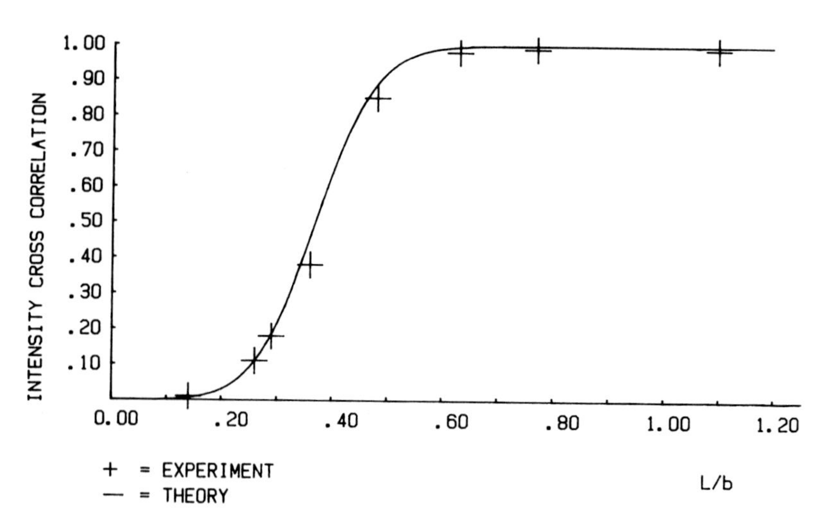

Fig. 4. Peak values of the spatial intensity correlation measured at angles θ' and -θ' as a function of L/b (sin θ' = λ/b).

4. FAR-FIELD TEMPORAL CORRELATION

Equation (6) gives the general expression for the spatio-temporal correlation function. It simplifies greatly if we consider only the temporal correlation at a single point, i.e. $s = (\sin\theta_1 + \sin\theta_2)/2$ and $\delta = 0$. Restricting the observation point to lie along the x-axis between the +1 and -1 grating orders, i.e. $-\lambda/b \leq s \leq \lambda/b$, and values of $L/b \geq 0.05$, it can be shown [Newman and Dainty (1984)] that the normalised intensity correlation is given by

$$C_I(s,0,\tau) \sim \frac{\exp(-|\underline{v}|^2\tau^2/4a^2)\,[A(q,K) + B(q,K)\cos(\omega_0\tau)]}{A(q,K) + B(q,K)} \tag{9}$$

where $q^2 = 2\pi^2L^2/b^2$,

$\omega_0 = 2\pi v_\xi/b$,

$K = s/\sin\theta' = \sin\theta / \sin\theta'$,

$A(q,K)$ and $B(q,K)$ are functions of the grating and diffuser and $-1 \leq K \leq 1$.

Thus the temporal autocorrelation at any point along $-\lambda/b \leq \sin\theta \leq \lambda/b$ consists of a Gaussian envelope of temporal width $t_0 = 2a/|\underline{v}|$ modulated by a cosine of period $t_M = b/v_\xi$, where v_ξ is the diffuser's component of velocity parallel to the x-axis in the far-field. Whenever $t_M < t_0$ and $B(q,K) > 0$, this modulation is clearly present and Eq. (9) shows that its period is independent of the detector location K. The modulation $S(q,K) = B(q,K)/A(q,K)$ depends on the detector location K, the overlap factor q and the Fourier coefficients g_n of the grating. Figs. 5-7 illustrate these dependencies.

Figure 5 shows the general behaviour of the temporal intensity correlation. The overall envelope has a width equal to $2a/|\underline{v}|$ and the modulation has a period equal to b/v_ξ as predicted by theory. Figure 6 shows experimental measurements of the intensity correlation for a linearly translating diffuser, $v_\xi = |\underline{v}|$ and $v_\eta = 0$, for two gratings of different period b. The period of the cosinusoidal variation and its modulation are both smaller for the finer grating in accordance with theory. Figure 7 compares theoretical and experimental values of modulation strength $S(q,K)$ as a function of the intensity overlap L/b at two measurement positions K = 0 and 1 (i.e. on-axis and at the first diffraction order).

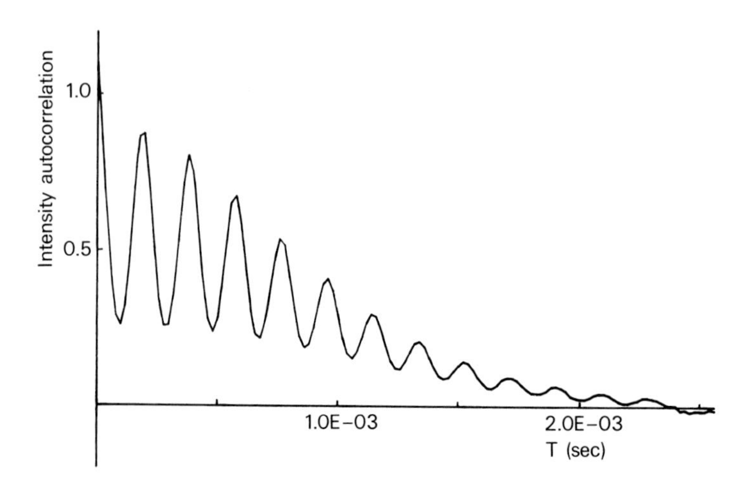

Fig. 5. Typical experimental measurement of the temporal correlation function.

The Gaussian envelope of width $2a/|\underline{v}|$ is a well-known property of dynamic speckle [Asakura and Takai (1981)]. It may be understood in a rather elementary way as the time taken for a speckle in the far-field to evolve to an essentially new speckle with no memory of its previous configuration. The speckle field evolves because of the motion of the diffuser with respect to the laser beam. After the diffuser has translated the laser beam of width 2a, the diffraction field is formed by an entirely new set of scatterers and hence the decorrelation time of each speckle in the far-field is approximately $2a/|\underline{v}|$.

Figures 7(a) and (b) show how the modulation of the cosinusoidal variation increases dramatically for $L/b \lesssim 0.5$ and essentially disappears above this value. This may be understood by considering a diffuser correlation cell of linear dimension L traversing the phase grating of spatial period b and noting that the grating is uniform in the η direction and periodic in the ξ direction. Whenever $L/b \ll 1$, the instantaneous random phase transmittance associated with a particular point on the diffuser surface has added to it a time-dependent periodic component due to the grating. This gives rise in the far-field to a strong cosinusoidal modulation of period b/v_ξ. As L increases, the random phase transmittance still has a periodic component added to it. However, the strength of the component is now smaller since it is the value of the phase of the grating averaged over a distance L and hence the modulation disappears as $L/b \to \infty$.

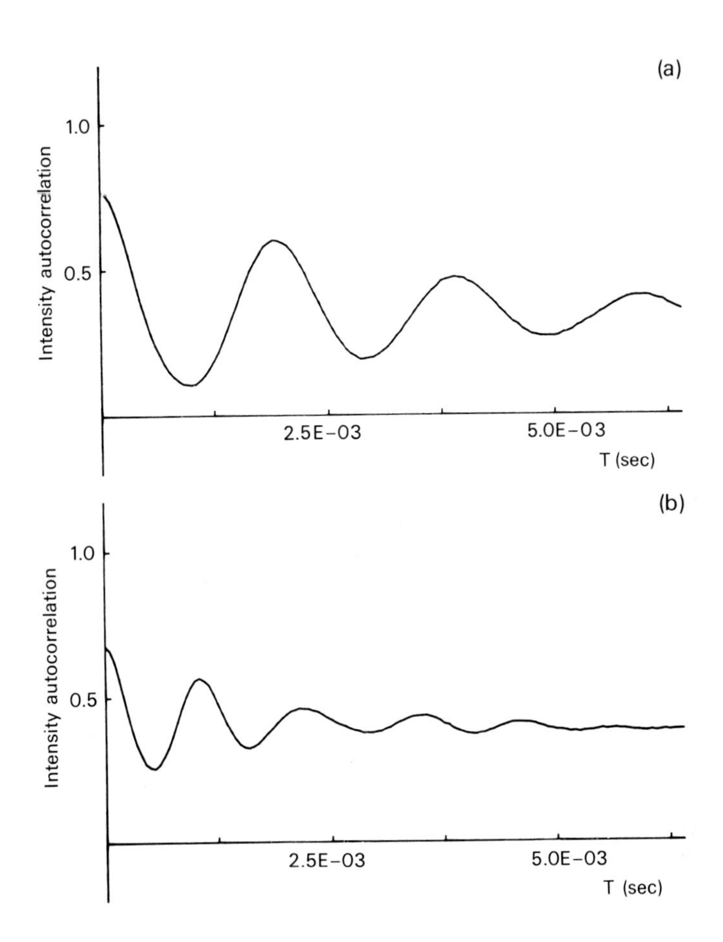

Fig. 6. Measurements of the temporal intensity correlation at the centre of the far-field ($s=0$, $\delta=0$) for $v_\xi = 0.5$ cm s^{-1}, $v_\eta = 0$, $L = 0.9$ μm and (a) $b = 9.2$ μm, (b) $= 5.1$ μm.

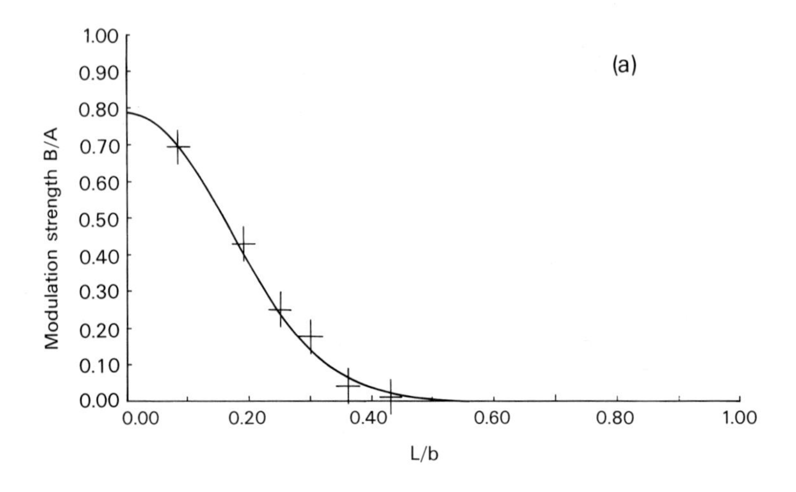

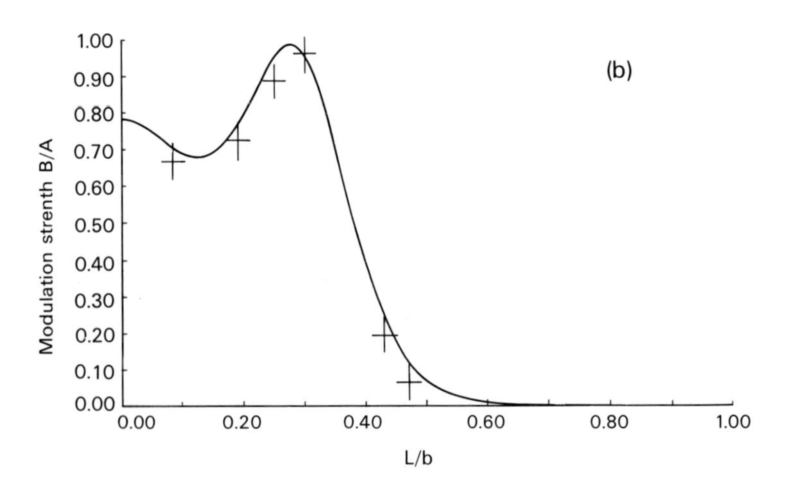

Fig. 7. Modulation strength as a function of L/b at a fixed measurement position: (a) K = 0, (b) K = 1. The solid curve is the theoretical plot.

5. CONCLUSIONS

We have verified experimentally that the presence of a phase grating of period b hidden behind a diffuser of correlation length L can be revealed by measurements of the spatial and/or temporal intensity correlation of the scattered field. In the case of the detection of the grating by measurement of the spatial correlation of intensity, we require that $L/b \gtrsim 0.15$ and for temporal correlation that $L/b \lesssim 0.5$.

In our experiments, the diffuser was allowed to move with respect to the grating and this was essential for the detection of the grating via temporal autocorrelation measurements. In practical information coding schemes, it may be desirable that the two are fixed together (e.g. bonded in a 'credit' card). In that case one can still use the spatial correlation method but the ensemble average required in the correlation is replaced in the experiment by a spatial average. For example, one could correlate the outputs of two charge-coupled-devices (CCDs). Experiments involving such spatial averages have been reported by Glass et al (1984).

6. ACKNOWLEDGEMENTS

This research was carried out at The Institute of Optics, University of Rochester with support from the US Army Research Office (DAAG-29-80-K-0048) and US Air Force Office of Scientific Research (AFOSR-81-0003).

7. REFERENCES

Asakura, T. and Takai, N., 1981 Dynamic Laser Speckles and their Application to Velocity Measurements of a Diffuse Object, Appl. Phys., **25**, 179-194.

Baltes, H.P., Ferwerda, H.A., Glass, A.S., and Steinle, B., 1981a, Retrieval of Structural Information from the Far-Zone Intensity and Coherence of Scattered Radiation, Opt. Acta, **28**, 11-28.

Baltes, H.P. and Ferwerda, H.A., 1981b, Inverse Problems and Coherence, IEEE Trans., **AP-29**, 405-406.

Baltes, H.P., Glass, A.S., and Jauch, K.M., 1981c, Multiplexing of Coherence by Beamsplitters, Opt. Acta, **28**, 873-876.

Baltes, H.P. and Jauch, K.M., 1981d, Multiplex Version of the Van Cittert-Zernike Theorem, J. Opt. Soc. Am., **71**, 1434-1439.

Baltes, H.P. and Huiser, A.M.J., 1984, Inverse Scattering and Grating Reconstruction, in Optics in Modern Science and Technology (Proceedings of ICO-13, Sapporo, Japan).

Dainty, J.C., 1976, The Statistics of Speckle Patterns, in Progress in Optics, **14**, Ed. E Wolf (North-Holland, Amsterdam).

Dainty, J.C. and Newman, D., 1983, Detection of Gratings Hidden by Diffusers using Photon-Correlation Techniques, Opt. Lett., **8**, 608-610.

Erf, R.K., (Editor), 1978, Speckle Metrology (Academic Press, NY).

Francon, M., 1979, Laser Speckle and Applications in Optics (Academic, NY).

Glass, A.S. 1982, The Significance of Image Reversal in the Detection of Hidden Diffractors by Interferometry, Opt. Acta, **29**, 575-583.

Glass, A.S. and Baltes, H.P., 1982a, The Significance of Far-Zone Coherence for Sources or Scatterers with Hidden Periodicity, Opt. Acta, **29**, 169-185.

Glass, A.S., Baltes, H.P. and Jauch, K.M., 1982b, The Detection of Hidden Diffractors by Coherence Measurements, Proc. SPIE, **369**, 681-686.

Glass, A.S., Jauch, K.M., Pike, E.R. and Rarity, J., 1984, Unmasking Hidden Diffractors by Photon Correlation, in Optics in Modern Science and Technology (Proceedings of ICO-13, Sapporo, Japan).

Goodman, J.W., 1975, Statistical Properties of Laser Speckle Patterns, in "Laser Speckle and Related Phenomena", Ed. J.C. Dainty (Springer-Verlag, Heidelberg).

Gray, P.F., 1978, A Method of Forming Optical Diffusers of Simple Known Statistical Properties, Opt. Acta, **25**, 765-775.

Jauch, K.M. and Baltes, H.P., 1981, Coherence of Radiation Scattered by Gratings Covered by a Diffuser. Experimental Evidence, Opt. Acta, **28**, 1013-1015.

Jauch, K.M. and Baltes, H.P., 1982a, Reversing Wavefront Interferometry of Radiation from a Diffusely Illuminated Phase Grating, Opt. Lett., **7**, 127-129.

Jauch, K.M., Baltes, H.P. and Glass, A.S., 1982b, Measurements of Coherence of Radiation from Diffusely Illuminated Beamsplitters, Proc. SPIE, **369**, 687-690.

Levine, B.M. and Dainty, J.C., 1983, Non-Gaussian Image Plane Speckle: Measurements for Diffusers of Known Statistics, Opt. Commun., **45**, 252-257.

Newman, D. and Dainty, J.C., 1984, Detection of Gratings Hidden by Diffusers using Intensity Interferometry, J. Opt. Soc. Am. A, **1**, 403-411.

ON THE IMAGING OF OCEAN WAVES BY SYNTHETIC APERTURE RADARS

K. Ouchi
(Physics Department, King's College London)

ABSTRACT

The theory of synthetic aperture radar (SAR) imaging of dynamic ocean surface waves is presented. The principles are described of single- and multi-look processing of SAR data from a stationary and moving point target with special emphasis on the motion effects inherent to multi-look processing. The theory is extended to the imaging of diffusely scattering ocean waves, where image modulation by backscatter cross sections, velocity bunching and tilt or range bunching is discussed. Image degradation by defocusing and the finite lifetime of scatterers are also considered.

1. INTRODUCTION

In June 1978, an oceanographic satellite SEASAT carrying a synthetic aperture radar (SAR) was put into orbit around the Earth. The SEASAT-SAR provided the fine radar images (resolution of 25 m) of the ocean on a global basis. The success of this mission was followed by the Shuttle Imaging Radar-A (SIR-A) in November 1982 and SIR-B in October 1984. Similar satellites are also being planned by the European Space Agency (ERS-1 in 1987), Canada (RADASAT in 1990) and Japan (in late 1980).

SAR is a high resolution imaging radar. Fine range (cross-track) resolution is achieved with short pulses using pulse compression and fine azimuth (along-track) resolution is achieved by coherently synthesizing a large aperture using the radar platform motion. The principle is well understood and described in known literature (Harger 1970, Kovaly 1976, Leith 1978). In order to synthesize a large aperture, it is necessary to combine successive return signals over a period of time

(~ 2.0s for SEASAT-SAR single-look). Motions of targets during this integration time introduce considerable effects into the imaging process. Ocean surfaces are almost always in motion and it is the dynamic nature of the ocean that has caused great difficulties in interpreting the images. At present, the relationship between the ocean and the SAR images is only partially understood.

The purpose of this article is to discuss the theory of SAR imaging of dynamic ocean surface waves. The principles are first described of the single- and multi-look processing of SAR data from a stationary and moving point target. Although the effects of target's motion are well known (e.g. Raney 1971), those having particular reference to multi-look processing are not discussed in detail. Such multi-look processing techniques are generally used to reduce speckle by the incoherent addition of sub-images. However, due to the fact that the sub-apertures are synthesized at different centre times, they contain information about the target at different times. As a result, the impulse response function from a moving target differs from one sub-image to another. Such images cannot be improved by this technique as much as those of stationary targets. On the other hand, since the image structure depends on the look number and the nature of the motion, the temporal change may be estimated from the comparison of the sub-images.

In the second part, the theory is extended to the imaging of diffusely scattering ocean waves. It seems certain that backscatter cross sections play a dominant role in the image modulation of ocean waves, except those travelling in an interval of angles close the the azimuth, (Alpers et al 1981). Other image modulations include velocity bunching (Larson et al 1976, Alpers and Rufenach 1979a, Swift and Wilson 1979) and tilt or range bunching (Gower 1983, Ouchi 1984). The former modulation is a result of non-uniform azimuth image shift caused by the temporal change in wave amplitude, and the latter is based on the effect of radar foreshortening by the spatial change in wave amplitude. However, no experimental data have been reported to support these two modulations. The images of ocean waves may be degraded by the finite lifetime of scatterers (Raney 1980) and by defocusing (Jain 1978). The theories postulated in order to describe the effects are discussed.

2. SAR PRINCIPLE

2.1 *Single-look processing*

As shown in Fig. 1, SAR operates on board of a moving platform. The antenna transmits a sequence of frequency modulated (FM) or chirped pulses at an angle Θ to the vertical.

A single pulse has a form,

$$E_c(\tau)=W_r(\tau)\exp(i\omega\tau)\ \exp(-i\ \frac{\pi\alpha}{2}\ \tau^2) \tag{1}$$

where τ is the slant-range time variable, W_r is the pulse envelope, ω is the pulse centre frequency and α is the linear FM rate. Notations used in the article are listed in Table 1 and 2. The return signal from a local single scattering element at a position (x,y) on the surface can be written as

$$E_s(t,\tau;x,y)=W_a(t-x/V)\ U(x,y;t)\ E_c(\tau-2r/c) \tag{2}$$

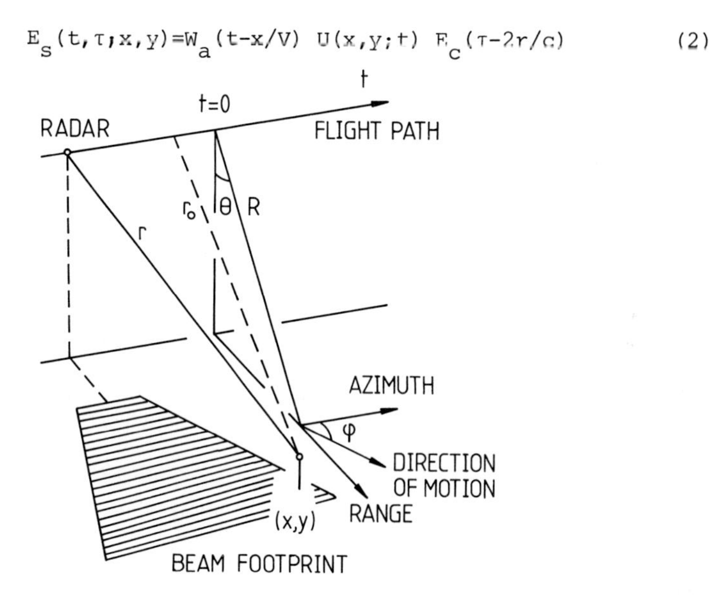

Fig. 1. SAR geometry

In the expression, c and V are the velocities of the radio wave and the radar platform respectively, W_a is the envelope of the azimuth illuminating beam, r is the slant range distance between the radar and the element, U is the scattering amplitude of the element at time t, and t is a discrete azimuth time variable. This variable can be regarded as continuous since an appropriate pulse repetition frequency is generally chosen to satisfy the sampling theorem. At this point, it is convenient to replace the time variables by the equivalent spatial variables,

$$\xi = Vt \tag{3}$$

$$\eta = c\tau/2 - R$$

with R being the distance from the radar at time t=0 to the centre of the reference ground co-ordinate system. The return signal may then be expressed as

$$E_s(\xi,\eta;\ x,y) = W_a(\xi - x) W_r(\eta + R - r)\ U(x,y;\xi)$$

$$\cdot \exp(i2k(\eta + R - r))\ \exp\left(-i \frac{2\pi\alpha}{c^2} (\eta + R - r)^2\right) \qquad (4)$$

where $k=2\pi/\lambda$ and λ is the radar wavelength.

Table 1. Nomenclature

E	total return signal
E_n	look n impulse response function
E_r	reference signal
E_s	return signal from a point target
I	image intensity
J_m	Bessel function of mth order
L_n	spatial centre of look n sub-aperture
M	fractional scattering amplitude modulation
M_c	fractional cross section modulation
M_{SAR}	image modulation function
N	total look number
T_n	centre time of look n sub-aperture
T_{SAR}	SAR transfer function
U	backscattering amplitude
$U_{o,r}$	deterministic and random part of U
$W_{a,r}$	azimuth and slant-range envelope of return signal
$W_{x,y}$	azimuth and slant-range envelope of reference signal
X,Y	azimuth and range image spatial variable

$a_{r,y}$	slant and ground range acceleration
c	velocity of radio wave
h	surface height
k	radar wavenumber
k_o	ocean wavenumber
n	look number configuration
r	slant range distance
t,τ	azimuth and slant range time
t_B	lifetime of Bragg wave
v_o	phase velocity of ocean wave
v_p	defocusing parameter of azimuth reference signal
$v_{x,r,y}$	azimuth, slant range and ground range velocity
x,y	azimuth and range surface spatial variable
ξ,η	azimuth and slant range spatial variable equivalent to t,τ
σ	backscatter cross section
ϕ	direction of target motion
λ_o	wavelength of ocean wave

Consider now that the scattering element has height h from the mean surface level and let r_o be the slant-range distance when the radar is abeam of the element. With reference to Fig. 1 a simple relation holds,

$$r^2=r_o^2 + (\xi-x)^2 \tag{5}$$

followed by a good approximation,

$$r \simeq r_o + (\xi-x)^2/2r_o \tag{6}$$

with

$$r_o = R + y \sin \Theta - h(x,y;\xi) \cos \Theta \tag{7}$$

The return signal is

$$E_s(\xi,\eta;x,y)=E_o\ W_a(\xi-x)\ W_r(\eta+R-r_o)\ U(x,y;\xi)$$

$$\cdot\ \exp(i2k(\eta - y \sin\Theta + h(x,y;\xi)\cos\Theta))\ \exp(-i\frac{k}{r_o}(\xi-x)^2)$$

$$\cdot\ \exp(-i\frac{2\pi\alpha}{c^2}(\eta - y\sin\Theta + h(x,y;\xi)\cos\Theta)^2) \quad (8)$$

In the equation, only the dominant terms have been retained; other terms including those higher than quadratic and azimuth-range cross product terms have been absorbed in the normalizing constant E_o. This return signal is then correlated with a reference signal E_r to produce the impulse response function E. The correlation process is

$$E(X,Y;x,y)=\iint_{-\infty}^{\infty} E_s(\xi+X,\eta+Y\sin\Theta;\ x,y)\ E_r(\xi,\eta)\ d\xi\ d\eta \quad (9)$$

where X and Y are the azimuth and range spatial variable in the image plane respectively and the sine term is a result of simple geometrical transformation from the slant range to ground range scale. The reference signal is a matched filter having a form;

$$E_r(\xi,\eta)=W_x(\xi)\ W_y(\eta)\ \exp(i\frac{k\ \xi^2}{R+Y\sin\Theta})\ \exp(i\frac{2\pi\alpha}{c^2}\eta^2) \quad (10)$$

with W_x and W_y being the envelopes of the reference signal in azimuth and range direction respectively. In practice, system resolution is limited by the extent of the reference signal. This means that the integrals with respect to ξ and η in Eq. (9) are determined only by W_x and W_y. Note that the azimuth reference signal is dependent on range due to the change in the azimuth illuminating beam width from the near-range to far-range as indicated in Fig. 1. This implies that the maximum azimuth correlation occurs when

$$r_o=R+y\sin\Theta \quad (11)$$

and the azimuth resolution changes from the near- to far-range if the width of W_x is not varied appropriately. In the present analysis, we assume that these corrections have already been made. Then we may put

$$R \gg Y\sin\Theta\ ,\ y\sin\Theta - h(x,y;\xi)\cos\Theta \quad (12)$$

and consider that the extent of W_x is constant.

The shapes of W_x and W_y depend on the design of correlators. Gaussian or raised-cosine function are occasionally used for side-lobe reduction (Corr 1979) but there is no loss of generality by assuming rectangular envelopes given by

$$W_x(\xi) = \begin{cases} 1 & -L/2 \leq \xi \leq L/2 \\ 0 & \text{otherwise} \end{cases} \quad (13)$$

$$W_y(\eta) = \begin{cases} 1 & -c\tau_o/2 \leq \eta \leq c\tau_o/2 \\ 0 & \text{otherwise} \end{cases} \quad (14)$$

where L and τ_o are the full synthetic aperture length and the pulse duration respectively, and also L=VT with T being the full integration time. The impulse response function may now be expressed as

$$E(X,Y;x,y) = E_o \iint_{-\infty}^{\infty} d\xi\, d\eta\, W_x(\xi)\, W_y(\eta)\, U(x,y;X+\xi)$$

$$\cdot \exp(i2k \sin\Theta\, (Y-y+h(x,y;X+\xi) \cot\Theta))$$

$$\cdot \exp(-i \frac{2k}{R} (X-x)\xi) \exp(-i \frac{4\pi\alpha}{c^2} \sin\Theta\, (Y-y+h(x,y;X+\xi)\cot\Theta)\eta) \quad (15)$$

where

$$E_o = \exp(-i \frac{k}{R}(X-x)^2) \exp(-i \frac{2\pi\alpha}{c^2} \sin^2\Theta\, (Y-y+h(x,y;X+\xi) \cot\Theta)^2) \ldots \quad (16)$$

If the point target is stationary, the impulse response function is

$$E(X,Y;x,y) = E_o\, U(x,y) \exp(i2k \sin\Theta\, (Y-y+h(x,y) \cot\Theta))$$

$$\cdot \operatorname{sinc}(\frac{kL}{R}(X-x)) \operatorname{sinc}(\frac{2\pi\alpha\tau_o}{c} \sin\Theta\, (Y-y+h(x,y) \cot\Theta)) \quad (17)$$

For a moving target, the effects are well known and good summaries are available in several papers (Raney 1971, Tomiyasu 1978).

2.2 Origin of speckle

It is well known (Goodman 1975) that speckle in images arises from the coherent addition of impulse response functions with random phase shifts and, or, random amplitude changes. In SAR images there are three main contributions to speckle formation.

The first contribution is the backscattering amplitude U in Eq. (17). When radio wave is backscattered from a rough surface, the amplitude modulation may vary randomly. The properties of speckle formed through this process depend on the statistics of random spatial fluctuations of the backscattered wave. But for general ocean surfaces, it is reasonable to assume that more than 8 scatterers exist within one resolution cell. Furthermore, it is well accepted (e.g. Wright 1978) that the predominant scatterers are small-scale (order of radio wavelength) Bragg resonant ocean waves. Under these conditions, the scattering surface may be modelled as an assembly of many independent scattering elements. The resultant speckle is known as Gaussian or classical speckle and the statistical properties do not depend on the detailed properties of the scattering surface.

The second term, $\exp(-i2ky \sin \Theta)$, in Eq. (17) contributes to the speckle formation through the addition of impulse response functions with random phase shifts. It is caused by the random distribution of scatterers in range direction. For the formation of Gaussian speckle the surface should introduce a phase change into the return signal that is much greater than unity. This implies that the separation of scatterers must be much greater than $\lambda/4\pi \sin \Theta$. For SEASAT-SAR, it is 6 cm and such a condition can easily be met on general sea surfaces. Since the term is independent of x, speckle formed through this process appears only in range direction.

The third term, $\exp(i2kh(x,y)\cos \Theta)$, depends on random surface height. The condition for Gaussian speckle is that the height change should be greater than $\lambda/4\pi \cos \Theta$. This value is 2 cm for SEASAT-SAR, which is a plausible condition for sea surfaces of moderate to high states.

Thus, apart from external noise such as rain drops and electronic noises (Tomiyasu 1978) SAR speckle can be produced through any one of or combinations of these three contributions and its statistical properties do not seem to depend on the detailed structure of scattering surfaces, since it obeys Gaussian statistics. Indeed, it has been shown theoretically and experimentally that speckle in SEASAT-SAR images of static surfaces is of a Gaussian type and does not carry much information about scatterers (Raney 1980, Ouchi 1981, Barber 1983). It should also be mentioned that if only few scatterers are present within a resolution cell, the speckle patterns do not obey Gaussian statistics. It is known (e.g. Welford 1978) that non-Gaussian speckle carries more information about scatterers but such cases have not been investigated in detail for SAR speckle. Also not well examined is the speckle pattern from moving scatterers. Although a few papers appear (Raney 1980, Ouchi 1981) on the subject, the theory is not complete and needs further investigation.

2.3 Multi-look processing

Multi-look processing of SAR data is a technique of speckle reduction that can be achieved, at the expense of system resolution, by dividing a full synthetic aperture into smaller sub-apertures or "looks", through which the images are formed and summed on an intensity basis to produce a final image. The speckle patterns formed through the non-overlapping sub-apertures are uncorrelated, while deterministic targets are correlated provided that they are stationary. The technique is well described in literature with supporting experimental results (Porcello et al 1976, Zelenka 1976, Bennett and Cumming 1978, Corr 1979). However, it has recently been noted (Ouchi 1985) that the images of moving targets cannot be improved by the method so much as those of stationary targets. This arises due to the fact that sub-apertures are synthesized at different centre times so that a time-lapse exists between looks. If targets are in motion, the information content and hence the images differ from look to look. The theory is described as follows.

Multi-look processing is achieved by applying an azimuth envelope of the reference signal of Eq. (13),

$$W_x^n(\xi) = \begin{cases} 1 & -L/2N \leq \xi - L_n \leq L/2N \\ 0 & \text{otherwise} \end{cases} \tag{18}$$

where n is the look number configuration, N is the total number of looks and L_n is the centre of the look n sub-aperture. L_n is related to the centre time T_n by

$$L_n = VT_n \tag{19}$$

and

$$T_n = \frac{T}{2N}(2n-N-1) \tag{20}$$

Now, consider a moving point target having the velocity components v_r and v_x, and the acceleration components a_r and a_x in slant-range and azimuth direction respectively. Then, in place of Eq. (5), the following relation holds.

$$r^2 = (R+v_r t+\tfrac{1}{2}a_r t^2)^2 + (Vt-v_x t-\tfrac{1}{2}a_x t^2)^2 \tag{21}$$

Let h=0 for simplicity and ignore image smearing and distortion. If only the velocity components are considered then the look n impulse response function takes a form,

$$E_n(X,Y)=E_o \operatorname{sinc}(\frac{2\pi\alpha\tau_o}{c} \sin\Theta\ (Y-v_yT_n))$$

$$\cdot \int_{-L/2N}^{L/2N} d\xi \exp(i\frac{2kv_x}{RV}\xi^2) \exp(-i\frac{2k}{R}(X-v_rR/V-2v_xT_n)\xi), \qquad (22)$$

where v_y is the ground range component of v_r.
Note that the azimuth resolution is reduced by a factor N. As can be seen from Eq. (22) the range impulse response function is centred at

$$Y=v_yT_n \qquad (23)$$

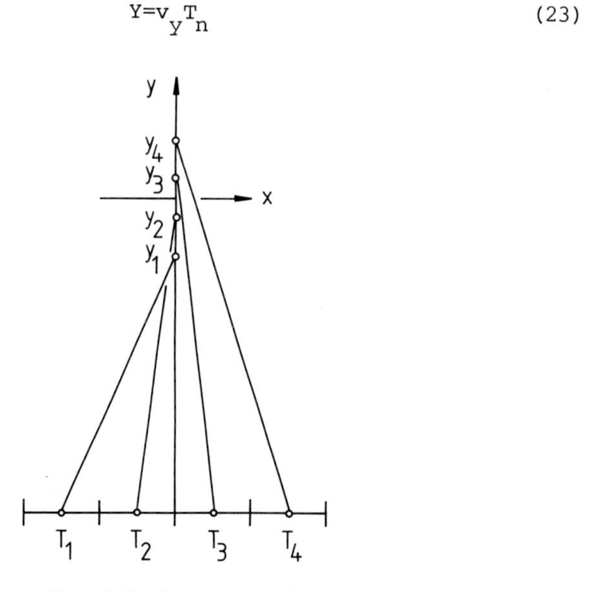

Fig. 2. Illustrating the 4-look processing of SAR data from a point target moving in range direction.

This position corresponds to that of the target at time $t=T_n$ on the surface. The effect is illustrated in Fig. 2. At time $t=T_1$ the radar is situated at the centre of the look 1 sub-aperture and the target is at $y=y_1$. Hence the look 1 impulse response function is formed at $Y=y_1$. As the radar travels to the centre of the look 2 sub-aperture at time $t=T_2$ the target also moves to $y=y_2$ and the look 2 impulse response function is centred at $Y=y_2$. From this viewpoint, the SAR multi-look imagery in range direction may be regarded as a time-lapse imaging system.

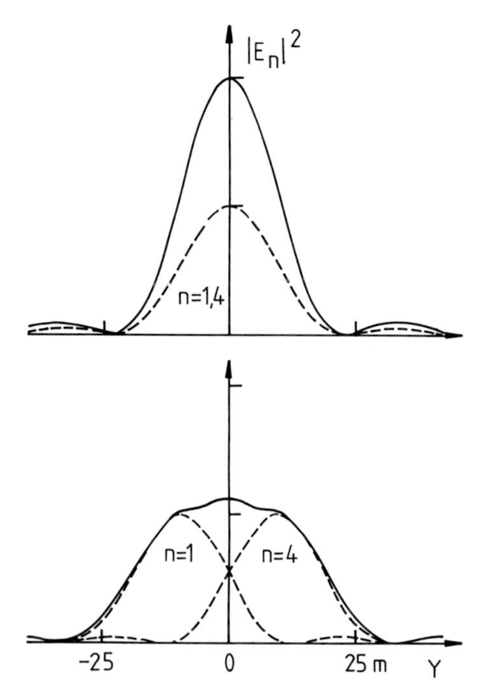

Fig. 3. The range component of a look n intensity impulse response function $|E_n|^2$ for n=1 and 4. A point target has the slant-range velocity v_y=10 m/s (lower graph) and it is stationary (upper graph). The solid lines represent the incoherent addition of the two looks. The SAR parameters are listed in Table 2.

Fig. 3 shows an example where the target is assumed to have v_y=10 m/s for the SAR parameters listed in Table 2. Since the image position of moving targets differs from look to look, the incoherent addition of the images cannot enhance the image quality compared with cases of stationary targets as shown in Fig. 3.

The quadratic phase term inside the integral of Eq. (22) causes defocusing. A tolerable amount of defocusing may be defined such that the phase does not vary more than $\pi/2$. The limit corresponds to the Strehl criterion for optical aberrations (e.g. Welford 1974). Thus defocusing is negligible if

$$v_x \leq \frac{\lambda RV}{2}\left(\frac{N}{L}\right)^2 \tag{24}$$

Table 2. SAR parameters

Full synthetic aperture length	L	16 km
Full integration time	T	2.5 s
Radar platform velocity	V	6.4 km/s
Reference slant-range distance	R	850 km
Linear FM rate	α	0.57 MHz/μs
Radar look angle	θ	23°
Radar wavelength	λ	0.235 m
Pulse duration	τ_o	33.4 s

For SEASAT-SAR single-look processing, $v_x \leq 2.5$ m/s and $v_x < 40$ m/s for 4-look processing. The linear phase term of Eq. (22) implies that the azimuth pulse response function is centred at

$$X = -v_r R/V + 2v_x T_n \tag{25}$$

This arises because the imaging process in azimuth direction is different from the range imaging in such a way that SAR locates the position of a target via its doppler returns over the integration time. If the target moves in range direction, it produces a linear phase shift and consequently the optimum correlation between the return and reference signal occurs when $X=v_r R/V$ as in Eq. (25). This is a well known effect of azimuth image shift (e.g. Raney 1971). The shift is 133 m for $v_r = 1$ m/s and SEASAT-SAR. A typical example is shown in Fig. 4, where ships travelling in range direction are displaced from the wakes. If the target moves in azimuth direction, it changes the relative velocity of the radar platform and causes defocusing as noted above. If a sub-aperture of a non-zero centre time is applied to the return signal, an additional linear phase shift is introduced and as a result the azimuth impulse response function is shifted by a distance $2v_x T_n$. For SEASAT-SAR and $v_x=10$ m/s, the shift is 38 m between looks 1 and 4.

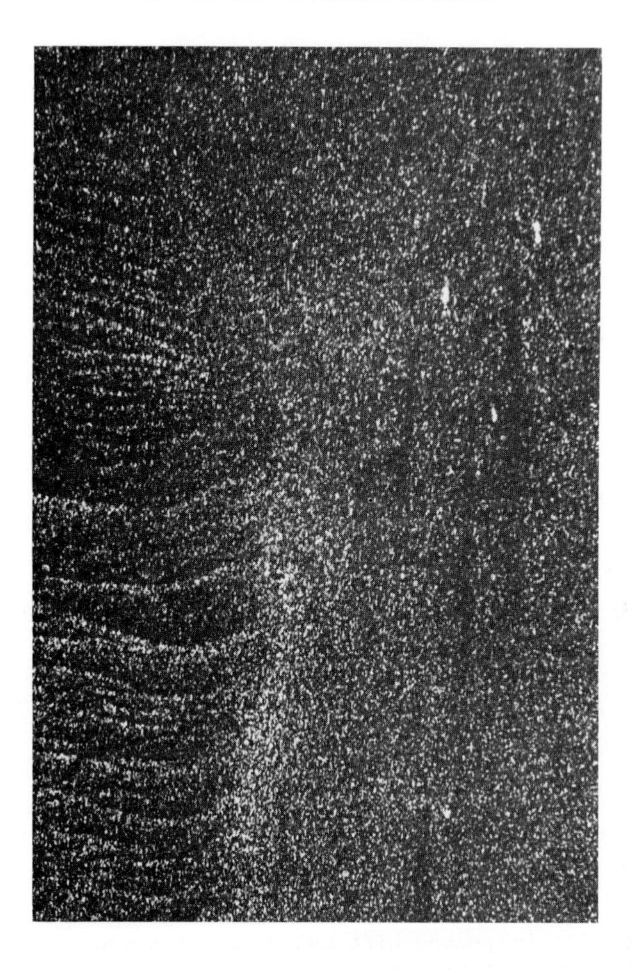

Fig. 4 A SEASAT-SAR optically processed image of the English Channel. Three ships in upper right and one in lower right were moving in range (vertical) direction and the images that appear as bright spots are shifted in azimuth (horizontal) direction from the wakes in dark vertical lines. Quasi-periodic patterns on the left are considered to be a result of tidal current interaction with sea beds. The image size is approximately 8 km in azimuth and 12 km in range. The system resolution is 16 m in azimuth and 22 m in range (courtesy of Dr. T.D. Allan of the Institute of Oceanographic Sciences).

In the presence of the slant-range acceleration, the look n impulse response function is (Ouchi 1985),

$$E_n(X,Y)=E_o \operatorname{sinc}\left(\frac{2\pi\alpha\tau_o}{c}\sin\theta\,(Y-\tfrac{1}{2}a_r T_n^2)\right)$$

$$\cdot \int_{-L/2N}^{L/2N} d\xi \exp\left(-i\frac{ka_r}{V^2}\xi^2\right)\exp\left(-i\frac{2k}{R}\left(X+\frac{a_r R}{V}T_n\right)\xi\right) \tag{26}$$

The range impulse response function is centred at

$$Y=\tfrac{1}{2}a_r T_n^2 \tag{27}$$

The shift is characteristic of time-lapse imagery. For general values of SAR and a_r the displacement is small and may be ignored. The quadratic phase term of Eq.(26) causes defocusing. If the same tolerance level as for Eq. (22) is used, defocusing is negligible if

$$a_r \leq \lambda(N/T)^2. \tag{28}$$

For SEASAT-SAR single-look $a_r \leq 0.04$ m/s^2 and $a_r \leq 0.6$ m/s^2 for 4-look processing. The linear phase term inside the integral shows the azimuth image shift of

$$X= -a_r RT_n/V. \tag{29}$$

Between the look 1 and 4 the image separation is 249 m for $a_r = 1$ m/s^2, which is a significant amount.

The azimuth component of acceleration produces defocusing and image shift but the effects are negligible for general parameters of SAR and acceleration.

3. IMAGERY OF THE OCEAN

3.1 Radio wave scattering

Scattering of radio waves from statistically rough surfaces depends mainly on the angle of incidence, wavelength and polarization of waves, dielectric constant and surface roughness (Beckmann and Spizzichino 1963, Ishimaru 1978, Bass and Fuks 1979). Valenzuela (1978) and Wright (1978) have reviewed the subject with particular reference to ocean surface scattering.

In the problems of radio wave backscattering from the ocean, two major approaches seem to exist, namely a Kirchoff or physical optics model (Barrick 1968) and composite-surface model (Wright 1968). The former model applies to backscattering with small incidence angles and rough surfaces where the radii of curvature of surface irregularities are large compared with the radio wavelength. This model considers a rough surface as an assembly of tilted plane facets and only those tilted at an angle normal to the incidence direction provide specular reflection back to the radar. The second model applies to scattering with incidence angles greater than ~ 20°. The model assumes that the sea surface is composed of a large number of slightly rough areas whose irregularities are small enough to satisfy the Rayleigh criterion (Rice 1951, Wright 1968) so that perturbation theory may be applied. To first order, radio waves are strongly backscattered from sea waves propagating radially from the radar and satisfying the Bragg resonance condition,

$$k_B = 2k \sin \theta_i \tag{30}$$

where $k_B=2\pi/\lambda_B$, λ_B is the Bragg wavelength and θ_i is the local incidence angle. For SEASAT-SAR λ_B =30 cm. The backscatter cross section σ from such sea surfaces is well known and given by (Wright 1968),

$$\sigma\ (\theta_i)=4\pi\ k^4 \cos^4\theta_i\ |g(\theta_i)|^2\ \Phi(2k \sin \theta_i,0) \tag{31}$$

where $\sigma = |U|^2$, g is the first-order scattering coefficient and Φ is the two-dimensional Fourier spectrum of surface height fluctuations. σ in terms of θ_i is evaluated in any published papers (e.g. Valenzuela 1978). In the presence of long ocean waves, the local tilt and incidence angle changes, giving rise to the spatial variation in the cross section modulation.

A problem arises when the incidence angle is in the transitional region between the two models. Indeed this is the case for SEASAT-SAR (θ_i ~ 23°). The relative importance of the two models may then have to be taken into consideration. Such cases have recently been treated by Bahar et al (1983). The backscatter cross section is also modulated by other factors including wind stresses, currents and bathymetric conditions (Evans and Shemdin 1980, Phillips 1981) but discussion on the subject is beyond the scope of this paper.

For the ocean surfaces described above, it is reasonable to assume that the scattering amplitude is a product of a deterministic term U_o representing large scale changes and a probablistic term U_r due to small scale Bragg waves that are distributed randomly, i.e.,

$$U(x,y;t)=U_o(x,y;t)\ U_r(x,y;t) \tag{32}$$

We further assume that the spatial and temporal component of U_r are uncorrelated, so that

$$U_r(x,y;t)= u_s(x,y)\ u_t(t|\ x,y) \tag{33}$$

where u_s is the spatial change of U_r and u_t is the temporal change of u_s at a position (x,y) on the surface. As noted in the previous section, the SAR images are immersed in the random speckle noise. Such image structures may best be described in terms of the local mean (expected) image intensity. Then it is necessary to take the spatial and temporal correlations of U. If there are more than ~ 8 Bragg waves within one resolution cell and the random processes are ergodic, we may replace the spatial and temporal averages by ensemble averages denoted by $< >$. The correlation function of U_r is

$$<U_r(x_1,y_1;t_1)\ U_r(x_2,y_2;t_2)> = <u_s(x_1,y_1;t_1)\ u_s(x_2,y_2;t_2)> \cdot <u_t(t_1|x_1,y_1)\ u_t(t_2|x_2,y_2)> \tag{34}$$

Since the average size of Bragg waves is comparable with the radar wavelength, very little correlation exists between neighbouring waves and the spatial correlation function can be approximated by the Dirac delta function δ. Denoting the temporal correlation function by Γ Eq.(34) becomes

$$<U_r(x_1,y_1;t_1)\ U_r(x_2,y_2;t_2)> = \delta(x_1-x_2,y_1-y_2)\ \Gamma(t_1-t_2) \tag{35}$$

For the temporal correlation function, we assume a Gaussian form;

$$\Gamma(t_1-t_2) = \exp\left(-\left(\frac{t_1-t_2}{t_B}\right)^2\right) \tag{36}$$

where t_B is the lifetime of a Bragg wave at a position $(x_1=x_2,y_1=y_2)$ on the surface.

Thus, the ocean surface is modelled in such a way that primary scatterers are small scale Bragg resonant waves that are spatially uncorrelated and distributed randomly, having a finite lifetime. These waves are modulated by the local surface slope of large scale undulations, giving rise to the backscatter cross section.

3.2 Ocean surface waves

Fig. 5. A SEASAT-SAR optically processed image of ocean surface waves around the North Rona, Scotland. The quasi-periodic waves propagating in range (vertical) direction are disturbed by the island and possibly by sea bed topography. The image size is 16 km in azimuth and 24 km in range. Rest as Fig. 4. (courtesy of Dr. T.D. Allan of the Institute of Oceanographic Sciences).

Fig. 5 shows a SAR image of quasi-periodic ocean waves around the North Rona, Scotland. Ocean surface waves similar to these have been frequently observed by SEASAT-SAR. Comparison of surface data with the images shows that measurements of ocean wavelength and direction are in good agreement (Gower 1981, Vesecky and Stewart 1982). However, the imaging mechanisms of ocean waves are only partially understood. To describe the theories of imaging processes, we first derive an expression for the image complex amplitude and the local mean intensity of a diffusely scattering extended surface. The image complex amplitude is given by simply integrating Eq. (15) over the surface,

$$A(X,Y) = \iint_{-\infty}^{\infty} E(X,Y;x,y)\, dx\, dy \tag{37}$$

At this point, we assume a defocused azimuth reference signal. This is because the images of moving sea waves are often improved by applying the reference signal with an appropriate amount of defocusing (Jain 1978, Shuchman and Zelenka 1978, Ivanov 1982). We put, in place of Eq. (10),

$$E_r(\xi,\eta) = W_x(\xi)\, W_y(\eta)\, \exp(i\frac{k}{R}\,(1+v_p/V)^2\xi^2)\, \exp(i\,\frac{2\pi\alpha}{c^2}\,\eta^2) \tag{38}$$

where v_p is the defocusing parameter. Using the surface model in the section 3.1., i.e., Eqs. (32) -(37), it can be shown that the local mean image intensity is

$$\langle I(X,Y)\rangle = \langle I_o\rangle \iiiint\!\!\iint_{-\infty}^{\infty} dx\, dy\, d\xi_1\, d\xi_2\, d\eta_1\, d\eta_2\, W_{xy}(\xi_1,\xi_2;\eta_1,\eta_2)$$

$$\cdot\, U_o(x,y;X+\xi_1)\, U_o(x,y;X+\xi_2)\, \exp(-(\frac{\xi_1-\xi_2}{Vt_B})^2)\, \exp(i\frac{2kv_p}{RV}(\xi_1^2-\xi_2^2))$$

$$\cdot\, \exp(i2k\cos\theta\,(h_1-h_2))\, \exp(-i\,\frac{2k}{R}\,(X-x)(\xi_1-\xi_2))$$

$$\cdot\, \exp(-i\frac{4\pi\alpha}{c^2}\sin\theta\,[\,(Y-y+h_1\cot\theta)\eta_1-(Y-y+h_2\cot\theta)\eta_2])\tag{39}$$

where

$$W_{xy} = W_x(\xi_1)W_x(\xi_2)W_y(\eta_1)W_y(\eta_2)$$

$$h_{1,2} = h(x,y;X+\xi_{1,2})$$

and $\langle I_o\rangle$ is a normalizing constant. For multi-look processing W_x should be replaced by W_x^n given by Eq. (18).

3.2.1. Backscatter cross section

The image modulation by backscatter cross sections is regarded to be dominant for most ocean waves except those propagating in azimuth and near-azimuth directions (Keller and

Wright 1975, Alpers and Jones 1978, Alpers et al 1981). For simplicity, we assume a sinusoidal wave of waveheight,

$$h = h_o \cos(k_x x + k_y y - k_o v_o t) \tag{40}$$

where h_o is the wave amplitude, $k_o = 2\pi/\lambda_o$ is the ocean wavenumber, λ_o is the wavelength, v_o is the phase velocity and k_x and k_y are the azimuth and range component of k_o respectively, i.e.,

$$k_x = k_o \cos\phi \tag{41}$$

$$k_y = k_o \sin\phi$$

with ϕ being the wave direction as in Fig. 1. The back-scattering amplitude may be expressed as

$$U_o = 1 + M \cos(k_x x + k_y y - k_o v_o t + \delta\phi) \tag{42}$$

and the backscatter cross section is

$$\sigma = |U_o|^2 = 1 + M^2/2 + 2M \cos(k_x x + k_y y - k_o v_o t + \delta\phi)$$
$$+ (M^2/2) \cos(2(k_x x + k_y y - k_o v_o t + \delta\psi)) \tag{43}$$

where $\delta\phi$ is the phase shift between the wavefield of Eq. (40) and the scattering amplitude of Eq. (42). It can then be shown that the local mean image intensity is

$$\langle I(X,Y)\rangle = 1 + M^2/2 + 2MD_1G_1Q_1 \cos(k_x X + k_y Y + \delta\phi)$$
$$+ (M^2/2) D_2G_2Q_2 \cos(2(k_x X + k_y Y + \delta\phi)) \tag{44}$$

where

$$D_m = \mathrm{sinc}\left[\frac{m\pi T}{T}\left(1 - m\left|\frac{\delta x}{\lambda_x}\right|\right)\left(1 + \frac{2v_p}{v_o}\cos\phi\right)\right] \tag{45}$$

$$G_m = \exp(-(m\delta_B/\lambda_x)^2) \tag{46}$$

$$Q_m = (1 - m|\delta_x/\lambda_x|)(1 - m|\delta_y/\lambda_y|) \tag{47}$$

Q_m has the condition,

$$m|\delta_x/\lambda_x| \geq 1 \tag{48}$$
$$m|\delta_y/\lambda_y| \geq 1$$

m is an integer, λ_x and λ_y are the azimuth and range component of λ_o respectively and T_o is the ocean wave period given by $T_o = \lambda_o/v_o$. δ_x and δ_y are the azimuth and range resolution cells respectively, where

$$\delta_x = \frac{\pi R}{kL}$$

$$\delta_y = \frac{c}{2\alpha\tau_o \sin\theta} . \tag{49}$$

δ_B is the effective azimuth resolution cell arising from the finite lifetime of Bragg waves and is given by

$$\delta_B = \frac{\pi R}{kVt_B} . \tag{50}$$

At this point, it is useful to define the SAR transfer function in a same manner as the optical transfer function (e.g. Born and Wolf 1978, Welford 1974). The linear SAR transfer function is defined such that given the backscatter cross section of the form,

$$\sigma = 1 + M_c \cos(k_x x + k_y y) \tag{51}$$

the local mean intensity of the SAR image is

$$\langle I(X,Y)\rangle = 1 + M_c |T_{SAR}| \cos(k_x X + k_y Y + \Delta) \tag{52}$$

where

$$\Delta = \arg(T_{SAR}) \tag{53}$$

and M_c is the fractional modulation of the cross section. The modulus of the SAR transfer function is a measure of the reduction in contrast from backscatter cross section to image over the wave spectrum, and the phase angle Δ represents the commensurate relative phase shift.

From the comparison between Eq. (43) and (44), we may consider the modulus of the linear SAR transfer function as

$$|T_{SAR}| = D_1 G_1 Q_1 \tag{54}$$

with zero relative phase shift. It can be seen that Q_1 given by Eq. (46) is the modulus of the SAR transfer function for time-invariant surfaces. It is equivalent to the transfer function of a real aperture radar with a beam footprint corresponding to the SAR resolution patch. Wave motions have no effect on the range component of T_{SAR} but the azimuth component is degraded as in Eq. (54). The degradation by G_1 is caused by the finite lifetime t_B of Bragg waves. If t_B is much greater than integration times, it has no effect on T_{SAR}. This is illustrated in Fig. 6 for different values of t_B, where the dispersion relation (e.g. Phillips 1977, Lighthill 1978) is used for the parameters of surface gravity waves, i.e.,

$$v_o=(g_o k_o)^{\frac{1}{2}} \tag{55}$$

where g_o=9.82 m/s^2 is the Earth's gravity. At present, the magnitude of t_B is a subject of active debate (Raney and Shuchman 1978, Raney 1980, Rufenach and Alpers 1981, Ouchi 1982b).

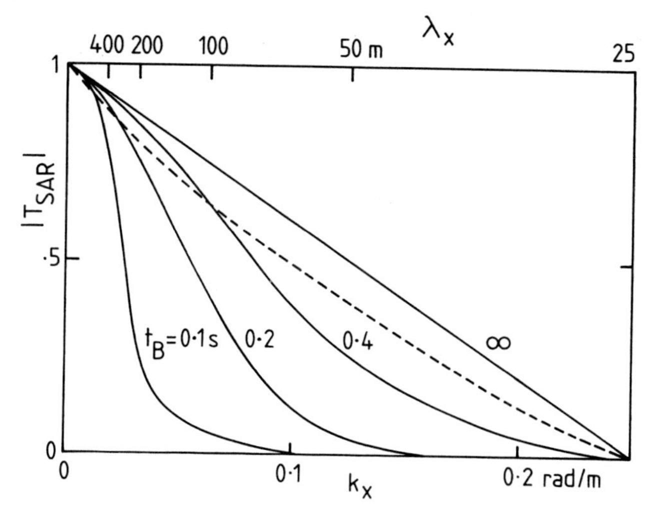

Fig. 6. The modulus of SAR transfer function $|T_{SAR}|$ for azimuth waves with different values of the lifetime t_B of Bragg waves. The broken line represents the transfer function degraded by defocusing.

The term D_1 of Eq. (44) corresponds to defocusing. It has been known (Jain 1978, Shuchman and Zelenka 1978) that the images of ocean waves are defocused if the reference signal is matched to stationary surface. They can be enhanced by applying a defocused azimuth reference signal. The effect of defocusing is shown also in Fig. 6 where the broken line represents the degraded SAR transfer function. The optimum image enhancement is achieved by putting

$$v_p = -\frac{v_o}{2\cos\phi}. \quad (56)$$

This result has been obtained by Ivanov (1982, 1983a,b), Ouchi (1982b,1983) and Rotheram (1983). However, there are two more postulated theories by Jain(1978) and Shuchman and Zelenka (1978) and by Alpers and Rufenach (1979b). The former authors claim that the azimuth component of the wave phase velocity changes the relative velocity of the radar platform and hence defocusing. According to the theory, the optimum focal setting is

$$v_p = -\frac{v_o}{\cos\phi} \quad (57)$$

Eq. (57) differs by a factor of 2 from Eq. (56). The latter authors consider that defocusing is due to the slant-range acceleration associated with the wave orbital motion, and the optimum focus lies in the range,

$$-\left|\frac{R}{2V}a_r^{max}\right| \leq v_p \leq \left|\frac{R}{2V}a_r^{max}\right| \quad (58)$$

where a_r^{max} is the maximum slant-range acceleration of a Bragg wave. They suggest that the value of v_p happens to be of the same order of magnitude as v_o. At present , there is not enough experimental data to compare with the theories.

3.2.2. Velocity bunching

The image modulation by velocity bunching is considered (Larson et al 1976, Alpers and Rufenach 1979a). It is based on the effect of azimuth image shift, where the image of a range moving target is shifted in azimuth direction. For the monochromatic ocean wave of Eq. (40), a Bragg scatterer on the wave has the slant range velocity component,

$$v_r = -\cos\theta\,\frac{\partial h}{\partial t} = -k_o h_o v_o \cos\theta\,\sin(k_x x + k_y y - k_o v_o t). \quad (59)$$

From Eq.(2.3.8) the impulse response function is formed at

$$X = x + \frac{R}{V}k_o h_o v_o \cos\theta\,\sin(k_x x + k_y y) \quad (60)$$

where the acceleration is ignored. According to Eq. (60) the impulse response functions are shifted and bunched periodically, producing the image corresponding to the wave field. The process is graphically illustrated in Fig. 7.

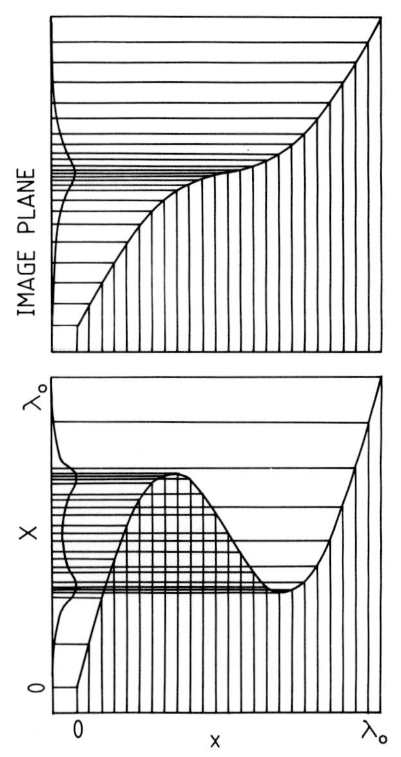

Fig. 7. Graphical illustration of velocity bunching for weak (upper graph) and moderate (lower graph) modulation. The horizontal and vertical axes are the surface and image plane in azimuth direction respectively.

If the wave travels with the radar ($\phi = 0^\circ$) and if the amplitude of Eq. (60) is small, the impulse response functions are shifted by small amounts and bunched in the regions corresponding to the troughs of the wave. If the amplitude is sufficiently large such that the concentration spreads into two regions, two intensity peaks appear within one ocean wavelength as in the lower diagram of Fig. 7. We can predict that further increase in the amplitude leads to large image shifts and the intensity maxima appear near the regions corresponding to the wave crests. For waves propagating against the radar, the process occurs in the reverse manner.

A rigorous solution can be derived by setting Eq. (40) into (39). Assuming that U_o is uniform and the height dependence on range direction is neglected, it can be shown that the local image intensity formed by velocity bunching alone is

$$\langle I(X,Y)\rangle = 1+2\sum_{m=1}^{\infty}(-1)^m S_m \cos(m(k_x X+k_y Y)), \tag{61}$$

where

$$S_m = D_m G_m Q_m J_m(H_m) \tag{62}$$

$$H_m = 4kh_o \cos\theta \sin\left(\frac{m\pi T\delta_x}{T_o\lambda_x}\right),$$

and J_m is the Bessel function of order m of the first kind. Eq. (61) is evaluated in Fig. 8, where the predicted changes in image structure can be observed.

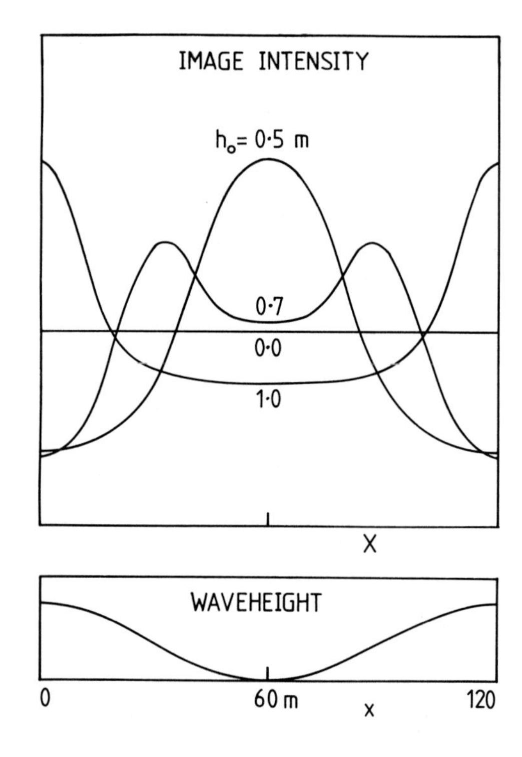

Fig. 8. The local mean image intensity in azimuth direction by velocity bunching. The ocean wavelength is taken as λ_o=120 m, the wave amplitude as h_o= 0, 0.5, 0.7 and 1 m and the dispersion relation for deep-water gravity waves are assumed for other parameters. The lower graph is the wave surface height profile.

It is difficult to describe the process of modulation in terms of the linear SAR transfer function, since the transfer is highly non-linear except when

$$|S_1| \gg |S_{m>1}|. \tag{63}$$

This condition is not satisfied for ocean waves in general sea states and it may probably be better to describe the process graphically as in Fig. 8.

Thus it seem that velocity bunching dominates the image modulation of ocean waves if the variation in backscatter cross section is small such as azimuth travelling waves. The effect disappears for range waves. For those propagating in any other directions, the two contributions may well be coupled, but the relative importance has not been discussed in detail.

3.2.3. Tilt (range) bunching

A significant but not dominant image modulation exists for ocean waves travelling in range and near-range directions. The effect is termed as "tilt or range bunching" (Gower 1983, Ouchi 1984). It is based on the well known effect of radar layover or foreshortening (Sabins 1978, MacDonald 1980), where the image of a scatterer of height h at a position (x,y) is shifted to

$$Y=y-h(x,y)\cot\theta. \tag{64}$$

The equation can easily be deduced from Eq. (17). The effect is often visible in the images of large undulations as shown in Fig. 9. No simple interpretation can be made for such images and this is the main reason why SAR is suitable to imaging surfaces of low to moderate relief. For the sinusoidal waveheight of Eq. (40), the impulse response functions from scatterers on the wave are displaced according to

$$Y=y-h_o\cot\theta\cos(k_x x+k_y y). \tag{65}$$

The process is analogous to velocity bunching. It is not easy to derive a simple expression for the image of ocean waves travelling in arbitrary directions except in range direction.

For a range wave we put $\phi=90^\circ$ in Eq. (40) and assume a sinusoidal backscatter cross section

$$\sigma=1+M_c\cos(k_o(y-v_o t)+\pi/2) \tag{66}$$

where M_c is the fractional cross section modulation and the $\pi/2$ phase shift in Eq. (66) is due to the composite surface model discussed in the section 3.1. According to the model, the backscatter cross section is maximum at surfaces of minimum local incidence angle, hence the $\pi/2$ phase shift. Then it can be shown that the local mean image intensity is

$$\langle I(X,Y)\rangle = 1+2\sum_{m=1}^{\infty}(-1)^m\left\{P_{2m}\cos(2mk_oY)+P_{2m-1}\cos((2m-1)k_oY)\right\} \tag{67}$$

where

$$\begin{aligned}
P_m &= C_m + M_c(C_m^+ + C_m^-)/2 \\
C_m &= D_m Q_m J_m(Z_m) \\
C_m^{\pm} &= D_m Q_m J_{m\pm1}(Z_m) \\
Z_m &= mk_o h_o \cot\theta .
\end{aligned} \tag{68}$$

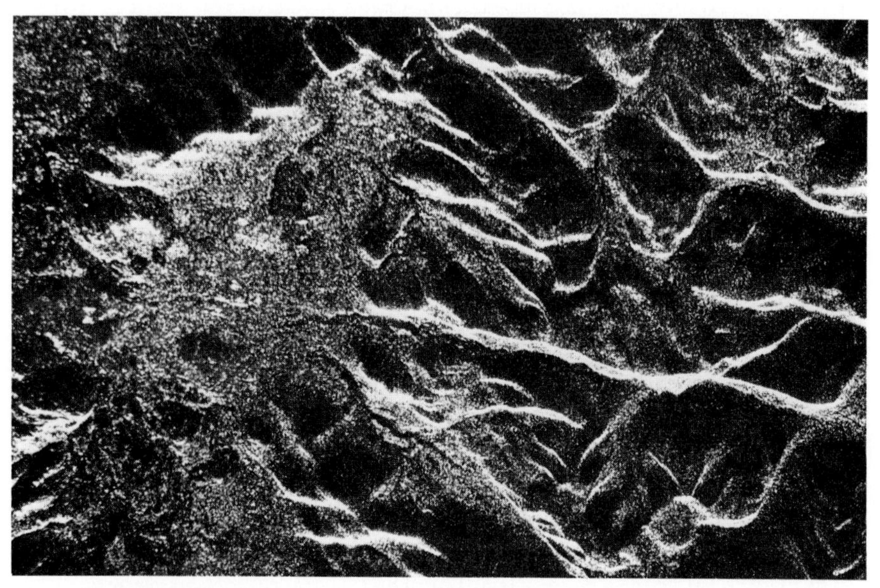

Fig. 9. A SEASAT-SAR optically processed image of the Grampian Mountains. Bright areas are due to layover and foreshortening. Rest as Fig. 5 (courtesy of Dr. T.D. Allan of the Institute of Oceanographic Sciences).

For the backscatter cross section only (h_o=0), Eq. (67) reduces to a simple form,

$$\langle I(X,Y)\rangle = 1+M_c D_1 Q_1 \cos(k_o y+\pi/2) \quad (69)$$

The modulus of the SAR transfer function is

$$|T_{SAR}| = D_1 Q_1 \quad (70)$$

with no phase shift. Note that for range waves defocusing cannot be corrected by adjusting v_p since $\phi = \pm 90°$ in Eq. (45). For tilt bunching we put M_c=0 so that P_m in Eq. (67) is replaced by C_m. The image structure is illustrated in Fig. 10.

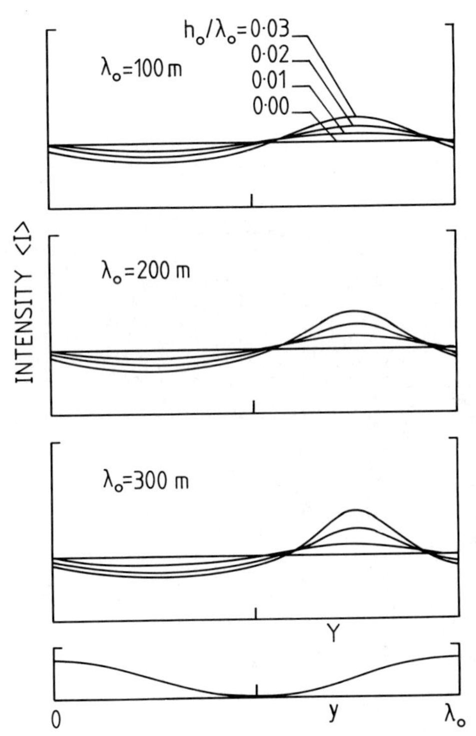

Fig. 10. The local mean image intensity in range direction by tilt bunching for different wavelength λ_o and wave slope h_o/λ_o. The bottom graph is the wave surface elevation.

It can be seen that there is no tilt bunching and, of course, no visible image if the wave slope h_o/λ_o is zero. For gentle slopes only the first term of the harmonics of Eq. (67) needs to be taken. Then

$$\langle I(X,Y)\rangle \simeq 1 + 2D_1Q_1J_1(Z_1)\cos(k_oY+\pi/2). \tag{71}$$

The process is linear and the modulus of the SAR transfer function by weak tilt bunching may be expressed as

$$|T_{SAR}| = 2D_1Q_1J_1(Z_1), \tag{72}$$

with the phase shift of $\pi/2$. The surface tends to become non-linear as the slope increases. To compare this non-linear process with the linear backscatter cross section modulation, we define the SAR image modulation function as

$$M_{SAR} = \frac{\langle I\rangle_{max} - \langle I\rangle_{min}}{\langle I\rangle_{max} + \langle I\rangle_{min}} \tag{73}$$

where $\langle I\rangle_{max}$ and $\langle I\rangle_{min}$ are the maximum and minimum local mean image intensity respectively. The function is also known as modulation index, contrast or visibility of an image. For the backscatter cross section, the image modulation is

$$M_{SAR} = M_cD_1Q_1 \tag{74}$$

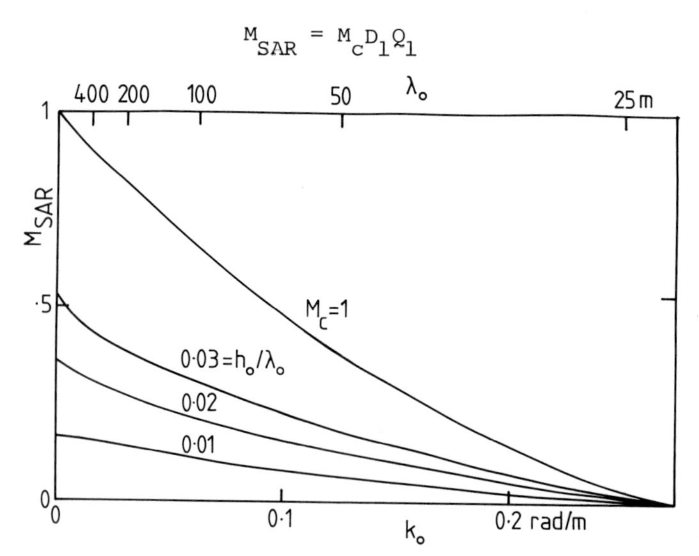

Fig. 11. The SAR image modulation function M_{SAR} in range direction for the backscatter cross section with the fractional modulation M_c=1 and tilt bunching with different wave slope h_o/λ_o.

This is the measure of the reduction in contrast for a given fractional modulation M_c. For tilt bunching,

$$M_{SAR} = 2 \sum_{m=1}^{\infty} C_{2m-1} \Big/ \left\{1+2 \sum_{m=1}^{\infty} C_{2m}\right\} \tag{75}$$

Clearly, $M_{SAR} = |T_{SAR}|$ for the cross section with M_c=1 and weak tilt bunching. The SAR image modulation function is plotted in Fig. 11 for tilt bunching with different slopes and for the backscatter cross section with M_c=1. Fig. 12 shows the image structure when both the contributions are coupled with M_c=0.5. Since the image modulations by tilt bunching and the backscatter cross section are in phase, the image is enhanced. Thus, the image modulation by tilt bunching is not dominant but cannot be ignored in the analysis.

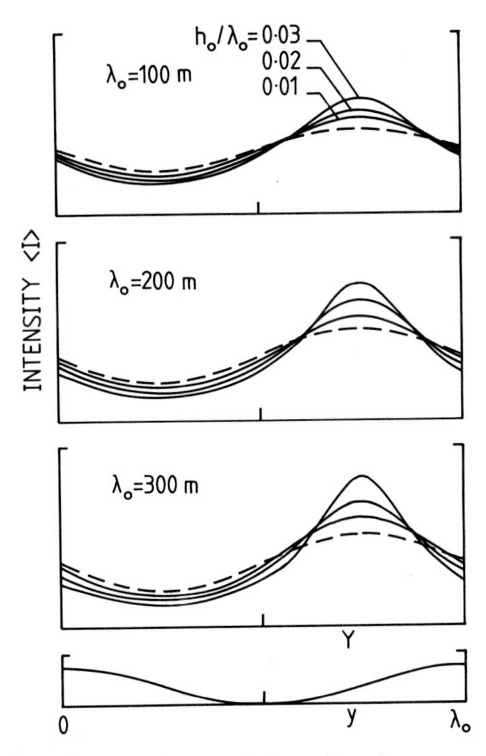

Fig. 12. The local mean image intensity in range direction when both the backscatter cross-section and tilt bunching are present. The fractional modulation is taken as M_c=0.5. Note that the two modulations are in phase so that the coupling effect enhances the image contrast.

3.2.4 Multi-look images

It has been noted in the previous section that the image of a moving point target cannot be impoved by multi-look processing so much as stationary targets due to a time-lapse between looks. The same principle applies to diffusely scattering ocean waves. For simplicity we consider the backscatter cross section only; applications to other modulations are trivial. From Eqs. (38)-(39) and (42) the local mean intensity of the look n sub-image is

$$<I_n(X,Y)> = 1+M^2/2 + 2M_1\cos(k_xX+k_yY-k_ov_o(1+\frac{2v_p}{v_o}\cos\phi)T_n)$$
$$+(M^2/2)M_2\cos(2(k_xX+k_yY-k_ov_o(1+\frac{2v_p}{v_o}\cos\phi)T_n) \qquad (76)$$

where we have put $\delta\phi=0$ and M_1 and M_2 are the appropriate modulation terms given by Eqs. (45)-(47). An important point here is that when the azimuth reference signal is set to stationary focus ($v_p=0$), the phase of the image differs by an amount $k_ov_oT_n$ between looks and that the phase difference can be corrected by defocusing the azimuth reference signal. For example, for SEASAT-SAR look 1 and 4 with $v_p=0$, the images are out of phase by a factor $(3/4)v_oT_n$ corresponding to the distance over which the ocean wave propagates during the time (~1.9 s) taken by the radar to travel from the centre of look 1 to look 4 sub-aperture. Eq. (76) is evaluated in Fig. 13.

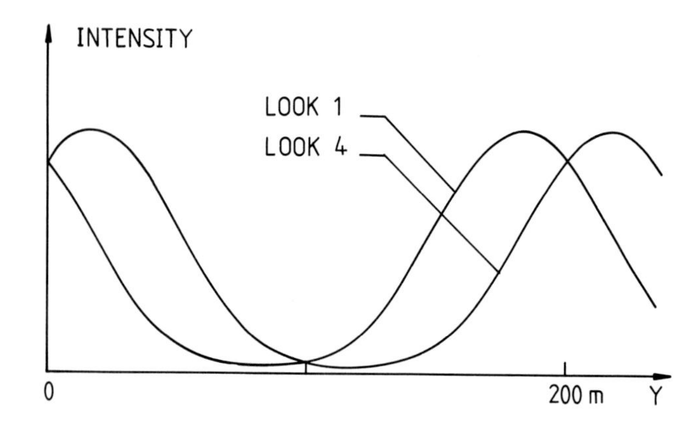

Fig. 13. The local mean intensity of the look 1 and 4 sub-image of a range wave of wavelength 200 m and the phase velocity 18 m/s. The fractional modulation is taken as $M_c=1$.

Obviously, the incoherent addition of these images cannot improve the image quality so much as those of stationary surfaces. On the other hand, since the image separation depends on the wave phase velocity and the look number, the former can be estimated from the comparison of images of different looks. The technique is not restricted to the detection of the temporal changes in ocean waves, but also applicable to any dynamic surfaces.

The phase difference between the sub-images of ocean waves can be corrected by defocusing the azimuth reference signal by an amount,

$$v_p = -\frac{v_o}{2\cos\phi} \tag{77}$$

which is the same as for the correction of motion induced defocusing given by Eq. (56). An explanation may be found in the characteristics of multi-look processing for moving targets, namely the time-lapse and image shift in azimuth direction. For simplicity we consider only the azimuth imaging. When the radar is at the centre of a look n sub-aperture, the backscattering amplitude is, from Eq. (42),

$$U_o = 1 + M\cos(k_x(x - v_o T_n/\cos\phi)) \tag{78}$$

with $\delta\phi=0$. If the azimuth reference signal is defocused, it produces image shift in azimuth direction in a similar way as the image of a target moving with a constant azimuth velocity. The shift is (c.f. Eq. (25))

$$X = x + 2v_p T_n \tag{79}$$

Substituting Eq. (79) into Eq. (78) the ocean wave field is expressed in terms of the image co-ordinate system,

$$U_o = 1 + M\cos(k_x(X - (2v_p + v_o/\cos\phi)T_n)) \tag{80}$$

Thus the change in image position can be corrected by adjusting v_p by an amount given by Eq. (77). Making the correction requires prior information about the wave phase velocity or reprocessing the SAR data for different azimuth focal settings. It can then be suggested that the images of highly dynamic surfaces, such as ocean waves, may better be presented in a single-look format since desired image enhancement cannot be achieved by multi-look processing.

4. CONCLUSIONS

The theory of SAR imagery of ocean surface waves has been discussed. In the first part, the basic SAR system has been described with particular emphasis on the multi-look processing of SAR data from a moving point target. The technique is used to enhance the speckled image by incoherent addition of sub-images. However, the images of moving targets cannot be improved as much as those of stationary targets. This is because there is a time-lapse between sub-apertures so that the images of moving targets are formed at different positions, which depend on the amount of the time-lapse and the motion. The same principle applied to diffusely scattering ocean waves. This property may be applied to estimating the temporal change of dynamic surfaces. In the second part, the image modulations by the backscatter radar cross section, velocity bunching and tilt (range) bunching have been discussed. The first modulation is a result of radio wave interaction with sea waves, and to first order strong backscatter cross sections can be obtained from small scale Bragg resonant sea waves. These waves are modified by local surface tilt of long waves, giving rise to large scale cross section modulation. This modulation seems to take a dominant role in the SAR image formation of ocean waves except those travelling in an interval of angles around the azimuth direction. The image modulation by velocity bunching arises from the temporal change in waveheight. The effect is inherent to SAR and is maximum for azimuth travelling waves. Tilt or range bunching are not dominant in image modulation compared with the former two contributions, but cannot be ignored. The effect is based on radar foreshortening and it is maximum for range travelling waves. Mention should be made that there has been no experimental evidence to support the latter two image modulations. The images of ocean waves may be degraded by the finite lifetime of Bragg waves and by defocusing induced by wave motions; these phenomena are subject to active current idscussion.

5. ACKNOWLEDGEMENTS

This work has been supported by the Science and Engineering Research Council in collaboration with Professor R.E. Burge.

6. REFERENCES

Alpers, W.R. and Jones, L., 1978, "The modulation of the radar backscattering cross section by long waves", in Proceedings 12th International Symposium on Remote Sensing of Environment, Manila, 1597-1607.

Alpers, W.R., Ross, D.B. and Rufenach, C.L., 1981, "On the detectability of ocean surface waves by real and synthetic aperture radar". *J.Geophys.Res.*, **86**, 6481-6498.

Alpers, W.R. and Rufenach, C.L., 1979a, "The effect of orbital motions on synthetic aperture radar imaging of ocean waves". *IEEE Trans. Antennas Propagat.*, AP-27, 685-690.

Alpers, W.R. and Rufenach, C.L., 1979b, "Image contrast enhancement by applying focus adjustment in synthetic aperture radar imagery of moving ocean waves" in Proceedings 2nd Seasat-SAR Processing Workshop, Frascati, Italy, 25-31.

Bahar, E., Rufenach, C.L., Barrick, D.E. and Fitzwater, M.A., 1983, "Scattering cross section modulation for arbitrary oriented composite rough surfaces: Full wave approach". *Radio Sci.*, **18**, 675-690.

Barber, B.C., 1983, "Some properties of SAR speckle" in "Satellite Microwave Remote Sensing" (T.D. Allan, ed.), 129-145 Ellis Horwood, Chichester.

Barrick, D.E., 1968, "Rough surface scattering based on the specular point theory". *IEEE Trans. Antennas Propagat.*, AP-16, 449.

Bass, F.G. and Fuks, I.M., 1979, "Wave Scattering from Statistically Rough Surfaces" Pergamon, New York.

Beckmann, P. and Spizzichino, A., 1963, "The Scattering of Electromagnetic Waves from Rough Surfaces" Pergamon, New York.

Bennett, J.R. and Cumming, I.G., 1978, "Digital technique for the multi-look processing of SAR data with application to SEASAT-A" in Proceedings 5th Canadian Symposium on Remote Sensing, Victoria, 506-516.

Born, M. and Wolf, E., 1978, "Principles of Optics" Pergamon, New York.

Corr, D.G., 1979, "Experimental SAR Processor Study and System Implementation Manual". Report C0875, Contract No. A9261502, Space Department, RAE, Farnborough, U.K.

Evans, D.D. and Shemdin, O.H., 1980, "An investigation of capillary and short gravity waves in the open ocean" *J. Geophys. Res. Lett.*, **3**, 647-650.

Goodman, J.W., 1975, "Statistical Properties of Laser Speckle Patterns" in "Laser Speckle and Related Phenomena" (J.C. Dainty, ed.) Springer-Verlag, New York.

Gower, J.F.R., (ed.) 1981, Oceanography from Space" Plenum, New York.

Gower, J.F.R. 1983, "Layover" in satellite radar images of ocean waves" *J. Geophys. Res.*, **88**, 7719-7720.

Harger, R.O., 1970, "Synthetic Aperture Radar Systems" Academic, New York.

Ishimaru, A., 1978, "Wave Propagation and Scattering in Random Media" Academic, New York.

Ivanov, A.V., 1982, "On the synthetic aperture radar imaging of ocean surface waves" *IEEE J. Ocean. Engineer.*, OE-**7**, 96-103.

Ivanov, A.V., 1983a, "On the mechanism for imaging ocean waves by synthetic aperture radar" *IEEE Trans. Antennas Propagat.*, AP-31, 538-541.

Ivanov, A.V., 1983b, "Reply to "Comments on 'On the synthetic aperture radar imaging of ocean surface waves' by William J. Plant" *IEEE J. Ocean. Engineer.*, OE-8, 300.

Jain, A., 1978, "Focusing effects in the synthetic aperture radar imaging of ocean waves". *Appl. Phys.*, **15**, 323-333.

Keller, W.C. and Wright, J.W., 1975, Microwave scattering and the straining of wind generated waves". *Radio Sci.*, **10**, 139-147.

Kovaly, J.J., 1976, "Synthetic Aperture Radar". Dedham, MA, Artech House.

Larson, T.R., Moskowitz, L.I. and Wright, J.W., 1976, "A note on SAR imagery of the ocean". *IEEE Trans. Antennas Propagat.*, AP-24, 393-394.

Leith, E.N., 1978, "Synthetic Aperture Radar" in "Optical Data Processing" (D.Casasent, ed.) Springer-Verlag, New York, 89-117.

Lighthill, J., 1978, "Waves in Fluids" Cambridge University, Cambridge.

MacDonald, H.C., 1980, "Techniques and Applications of Imaging Radars" in "Remote Sensing in Geology" (B.S. Siegal and A.D. Gillespie, eds). John Wiley, New York, 297-336.

Ouchi, K., 1981, "Statistics of speckle in synthetic aperture radar imagery from targets in random motion" *Opt. Quant. Electron.*, **13**, 165-173.

Ouchi, K., 1982a, "The effect of random motion on synthetic aperture radar imagery" *Opt. Quant. Electron.*, **14**, 263-275.

Ouchi, K., 1982b, "Imagery of ocean waves by synthetic aperture radar" *Appl. Phys.*, B-29, 1-11.

Ouchi, K., 1983, "Effect of defocusing on the images of ocean waves" in "Satellite Microwave Remote Sensing" (T.D. Allan, Ed.), 209-222, Ellis Horwood, Chichester.

Ouchi, K., 1984a, "Two-dimensional imaging mechanisms of ocean waves by synthetic aperture radars" *J. Phys. D: Appl. Phys.*, 25-42.

Ouchi, K., 1985, "On the multi-look images of moving targets by synthetic aperture radars" *IEEE Trans. Antennas Propagat.*, (to be published).

Phillips, O.M., 1977, "The Dynamics of the Upper Ocean" Cambridge University, Cambridge.

Phillips, O.M., 1981, "The structure of short gravity waves on the ocean surface" in "Spaceborne Synthetic Aperture Radar for Oceanography" (R.C. Beal, P.DeLeonibus and I.Katz, eds.) John Hopkins, Baltimore, 24-31.

Porcello, L.J., Massey, N.G., Innes, R.B. and Marks, J.M., 1976, "Speckle reduction in synthetic aperture radars" *J. Opt. Soc. Am.*, **66**, 1305-1311.

Raney, R.K., 1971, "Synthetic aperture imaging radar and moving targets" *IEEE Trans. Aerosp. Electron. Syst.*, AES-7, 499-505.

Raney, R.K., 1980, "SAR processing of partially coherent phenomena" *Int. J. Rem. Sens.*, 1, 29-51 and *IEEE Trans. Antennas Propagat.*, AP-28, 777-787.

Raney, R.K. and Schuchman, R.A., 1978, "SAR mechanisms for imaging waves" Proceedings 5th Canadian Symp. Rem. Sens., Victoria, 495-506.

Rice, S.O., 1951, "Reflection of electromagnetic waves from slightly rough surfaces" *Comm. Pure Appl. Math.*, **4**, 351-378.

Rotherham, S., 1983, "Theory of SAR ocean wave imaging" in "Satellite Microwave Remote Sensing" (T.D. Allan, ed.) 155-186, Ellis Horwood, Chichester.

Rufenach, C.L. and Alpers, W.R., 1981, "Imaging ocean waves by synthetic aperture radars with long integration times" *IEEE Trans. Antennas Propagat.*, AP-29, 422-428.

Sabins, F.F. Jr., 1978, "Remote Sensing; Principles and Interpretation" Freeman, San Francisco.

Shuchman, R.A. and Zelenka, J.S., 1978, "Processing of ocean wave data from a synthetic aperture radar" *Boundary Layer Meteorol.*, **13**, 181-191.

Swift, C.T. and Wilson, L.R., 1979, "Synthetic aperture radar imaging of ocean waves" *IEEE Trans. Antennas Propagat.*, AP-27, 725-729.

Tomiyasu, K., 1978, "Tutorial review of synthetic aperture radar with applications to imaging of the ocean surface" *Proc. IEEE*, **66**, 563-583.

Valenzuela, G.M., 1978, "Theories for the interaction of electromagnetic and ocean waves - a review" *Boundary Layer Meteorol.*, **13**, 61-85.

Vesecky, J.F. and Stewart, R.H., 1982, "The observation of ocean surface phenomena using imagery from the SEASAT synthetic aperture radar: an assessment" *J. Geophys. Res.*, **87**, 3397-3430.

Welford, W.T., 1974, "Aberrations of the Symmetrical Optical System". Academic, London.

Welford, W.T., 1978, "Optical estimation of statistics of surface roughness from light scattering measurements" *Opt. Quant. Electron.*, **9**, 269-287.

Wright, J.W., 1968, "A new model for sea clutter" *IEEE Trans. Antennas Propagat.*, AP-16, 217-223.

Wright, J.W., 1978, "Detection of ocean waves by microwave radar; the modulation of short gravity-capillary waves" *Boundary Layer Meteorol.*, **13**, 87-105.

Zelenka, J.S., 1976, "Comparison of continuous and discrete mixed integrator processor" *J. Opt. Soc. Am.*, **66**, 1295-1304.

MICROWAVE BACKSCATTER FROM THE SEA SURFACE

K.D. Ward

(Royal Signals and Radar Establishment, Malvern)

ABSTRACT

A brief survey is given of the application of microwave sea backscatter to radar, highlighting the virtues and limitations of various approaches. Interest is centred upon a compound K-distribution model for the statistics. This model is introduced through a concept of bunched discrete electromagnetic scatterers that allows the important correlation properties to be understood. Application is then demonstrated with an empirical relationship for variations of the statistics with radar and environmental parameters and a discussion of signal processing performance.

INTRODUCTION

Microwave backscatter from a disturbed sea surface is often termed sea clutter because of its effect of cluttering radar displays and reducing confidence in target detections through the presence of false alarms. Interest in sea backscatter is also generated from its use in providing remote sensing of sea condition. Whatever the case, a knowledge of the scattering mechanism is highly desirable to either invert the problem and provide information about the surface from the received radar signal; or to predict the detailed statistical properties of the received signal given the sea conditions. Unfortunately these problems are generally intractable and one must resort to making radar measurements, modelling and matching to observed conditions.

Interest here will be focussed upon the search radar application and backscatter at near grazing incidence using a high resolution radar (in this context, high resolution relates to the small area of sea resolved by the radar through the use of a short pulse length and narrow beamwidth). Under these conditions it will be shown that little is understood of the scattering mechanism but observations show that the statistics of the backscatter are very non-Gaussian and exhibit interesting correlation properties associated with the long period sea wave structures. It is worthwhile approaching a discussion of clutter modelling from the application and reviewing progress in these terms.

The detection performance of maritime reconnaissance radars is often limited by the target competing with unwanted sea echo (clutter). In the radar range equation (Skolnik, 1980), the clutter characteristics appear in two terms; σ_O, the mean clutter radar cross section per unit illuminated area, and S_{min}, the minimum detectable signal. The latter reflects the fluctuation statistics through the response of signal processing to give the necessary probabilities of false alarm and detection.

The required clutter characteristics, σ_O and the statistics, may be obtained from measurements and applied by direct simulation to the radar processing. Better and wider application may be made of the data through reduction to models for easier communication and handling. The value of a model depends upon how accurately it represents all characteristics of the data and how widely (in terms of different radar parameters and environmental conditions) it applies. Models based on a physical understanding of the processes involved provide the opportunity for extrapolation into unmeasured conditions. The test of a model is its accuracy in signal processing performance prediction.

The current understanding of sea clutter may be considered in two parts, the mean radar reflected power, σ_O, and the statistical fluctuations. Many measurements have been made of σ_O since the very early days of radar. These measurements have led to a number of empirical models being devised to assist the radar designer (Nathanson, 1969; Guinard and Daley, 1970; Sittrop, 1977; Horst, Dyer and Tuley, 1978). Although different in functional form, all the models are in broad agreement given that there is a large scatter in results taken under apparently similar conditions but difference occasions. Very briefly, the trends are that σ_O increases with grazing angle, radar frequency, and sea state; it is greater for vertical than horizontal polarization and is a maximum upwind, minimum downwind and of intermediate value for cross wind. An understanding of the physics and the generation of theoretical predictions for σ_O have also proceeded steadily. Reviews of this work are given in Long (1980) and Valenzuela (1978). It is unfortunate that the physical models developed are at their weakest for the higher radar frequencies (centimetric and above) and towards grazing incidence, for it is this area which is of most interest in search radar applications.

Investigations into the statistical fluctuations have revealed the following features. For large illuminated patch sizes and high grazing angles (>10°), it is found that sea clutter obeys the central limit theorem and has Rayleigh distributed amplitude statistics. The range correlation is commensurate with the pulse length. If the transmitter frequency is stepped by the pulse bandwidth for each pulse (frequency agility) or sufficient time elapses, the returns are independent from pulse to pulse. When the grazing angle is reduced and/or the radar resolution increased, the amplitude distribution and the spatial characteristics change and the clutter is described as "spiky". The amplitude statistics have been modelled by Lognormal (Trunk, 1972; Trunk and George, 1970), Weibull (Fay, Clarke and Peters, 1977), Contaminated-Normal (Trunk and George, 1970), Log-Weibull (Sekine, 1980) and K-distributions (Jakeman and Pusey, 1977).

As the number of parameters is increased, it is not surprising that the fit to the observed amplitude data improves, but the shortcoming of these approaches to clutter modelling is that they fail to model the temporal and spatial correlation characteristics of the clutter. This shortcoming has been recognised in the past by various workers who proposed composite distribution models for sea clutter (Trunk, 1972; Valenzuela and Laing, 1971; Blythe et al, 1969). More recently, this type of model has been developed into a closed form using a compound form of the K-distribution, which has been found not only to be a very good fit to observed amplitude measurements but which also includes the pulse-to-pulse correlation properties (Ward, 1981; Ward, 1982).

The deviations from noise-like statistics have a considerable impact on radar performance. This is to be discussed in a later section. Firstly, the compound K-distribution model will be introduced through the results from some sea echo measurements.

SEA CLUTTER MEASUREMENTS AND THE K-DISTRIBUTION

As described in the introduction, the main interest for the search radar application is sea echo at centimetric wavelengths and near grazing incidence using a high resolution radar. Over some years we have performed experiments to collect sea clutter data from both cliff-top and airborne platforms. The aim has been to provide some insight into the type of clutter seen from an airborne radar over the sea, although some of the experiments cover shipborne radar geometries. Fig. 1 shows the main operating geometry and Table 1 the radar parameters.

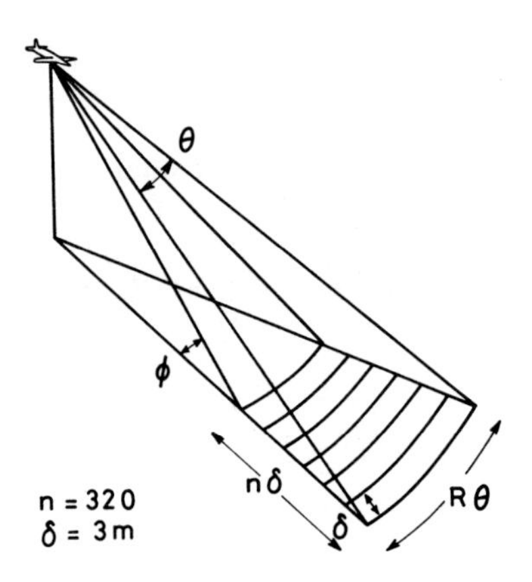

Fig. 1 Diagram to show radar geometry

When performing measurements from the aircraft, altitudes from 100m to 1000m are used. The principal clifftop site is at 80m altitude. Varying the range at which data are collected changes the across-range resolution (Rθ in Fig. 1) and also the grazing angle (φ). From the clifftop installation these are necessarily coupled due to the fixed altitude, but from the air they can be varied independently. A main requirement of the measurements is to obtain both temporal and spatial statistics for the clutter. The former is obtained using a pulse repetition frequency of 1KHz and the latter by recording a "range profile" consisting of 320 consecutive range gates matched to the pulse width (see Fig. 1). From the clifftop trials, recordings are made by setting the 960m profile gate over the area of interest and recording for the required period of time. As the aircraft is a dynamic platform, both the range to the profile gate and the aerial pointing angle are ground stabilised to the sea area of

Table 1

Summary of radar parameters

FREQUENCY BAND	I-band, 9.5-10 GHz.
TRANSMITTER POWER	1kW peak.
PULSE WIDTH	5μs compressed to 28ns (chirp).
PULSE REPETITION FREQUENCY	1kHz - 2.5kHz
AERIAL BEAMWIDTH	1.2 deg, 3dB two way.
DATA RECORDING	Range gate of 320 samples 20ns separation, digitised to 6 bits, from each echo.
MODES	Coherent or non-coherent, various polarization combinations.

interest and corrected for aircraft motion. This allows comparable records from both platforms although from the air ϕ and R will vary slightly with time.

The data presented here are non-coherent (amplitude only from an envelope detector) and contain examples of frequency agile mode where the transmitter frequency is stepped by 50MHz (pulse bandwidth) each transmission over a cycle of ten. Several other modes are available, for example, fully coherent and polarization scattering matrix. These should extend the basic modelling to be outlined here.

Data collected from the clifftop radar are presented in Figs. 2 and 3. Records of this type have been used to develop a model, a compound form of the K-distribution, which identifies two components in the clutter fluctuations. The first component is a spatially varying mean level that results from a bunching of scatterers associated with the sea swell structure. This component has a long correlation time and is unaffected by frequency agility. A second "speckle" component occurs due to the multiple scatterer nature of the clutter in any range cell. This decorrelates through internal motion of the scatterers or through the use of frequency agility.

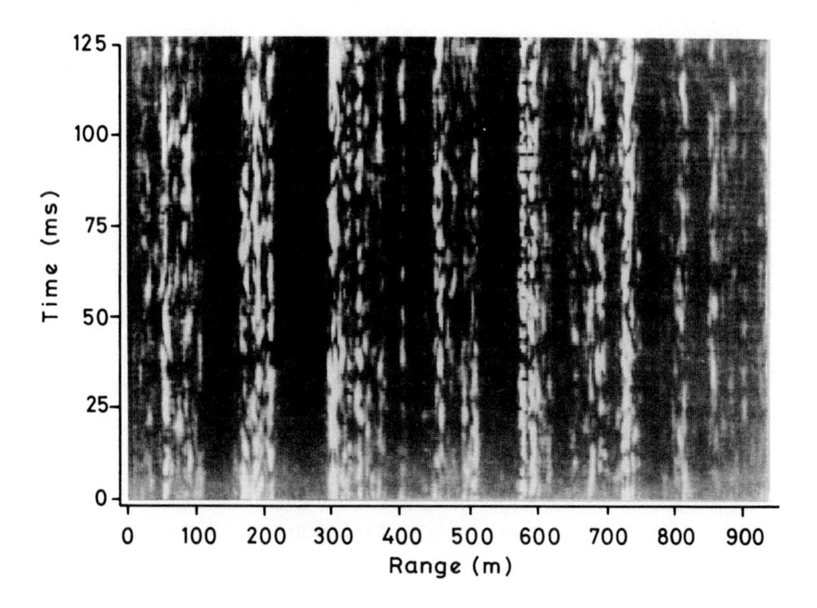

Fig. 2(a) Range-time-intensity plot of raw clutter (Fixed Frequency)

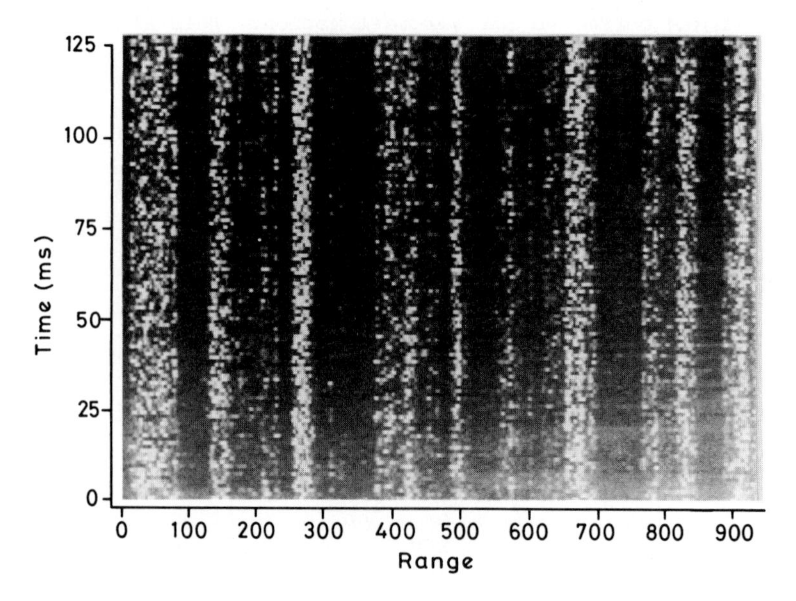

Fig. 2(b) Range-time-intensity plot of raw clutter (Frequency Agility)

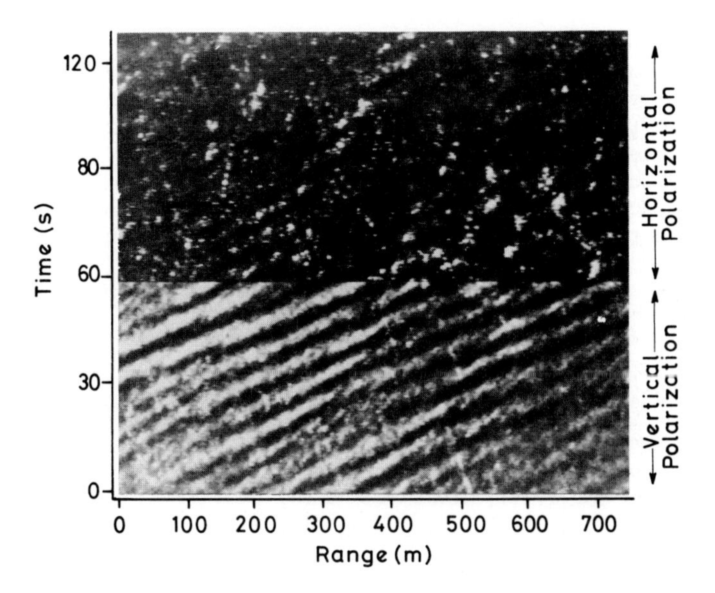

Fig. 3 Range-time-intensity plot of averaged sea clutter

These properties can be demonstrated with the aid of Figs. 2 and 3, which were collected in a medium sea state. Fig. 2a shows a Range-Time-Intensity plot of pulse-to-pulse fixed frequency clutter returns. Each line in the horizontal direction is an intensity modulation of the profile gate consisting of 320 samples. These profiles are stacked vertically to produce a time history. It is evident from the figure that there is a "bunching" in range, producing areas of locally high or low returns. At each range there is a fluctuation in the returns with a time constant of some 10ms due to the changing phase relationship of the scatterers within a patch, from their relative motion. Fig. 2b is a similar plot using frequency agility and shows the same bunching in range but the speckle is decorrelated from pulse to pulse by the change of illuminating wavelength.

The total time (125 ms) of Figs. 2a and 2b is not enough to show the temporal properties of the modulating component. In Fig. 3 averaging has been used to remove the speckle component and the plot therefore shows the "bunching" term over a longer period. Striking features are the wavelike nature of the plot and the difference between the vertical and horizontal polarization sections.

At a range of 5km where these recordings were made, the across range resolution is 100m compared to 4.2m along range. Sea waves propagating towards or away from the radar are therefore resolved much better than those in other directions and also tend to affect the radar returns more, due to the greater change in local grazing angle from the angle projection of the slopes. Figs. 2 and 3 are taken with the radar looking into the predominant wave structure and therefore show the strong periodic effect moving inward in range. Similar plots looking in other directions tend to show less inhomogeneity and much less ordered behaviour.

The difference between the radar polarizations in Fig. 3 is characteristic of most conditions. Generally σ_O, the mean echoing area, is greater for vertical (V) than horizontal (H) polarization, but H is more "spiky" than V. The difference in σ_O is generally explained assuming that the predominant echo is from wavelets that fit the Bragg resonance condition. Direct solution of Maxwell's Equations under these conditions leads to σ_O(V) greater than σ_O(H) (Wright, 1966). The difference in spikiness is suggested to be due to effects such as -

(i) Increased multipath interference in H polarization due to the lack of forward reflection of V polarization at the Brewster angle and associated phase changes (Long, 1980).

(ii) The polarization dependence of edge discontinuities or "wedges" on the sea surface structure (Kalmykov and Pustovoytenko, 1976; Lewis and Olin, 1980).

These mechanisms only, however, partly explain the observed phenonema.

Many records of the type shown in Figs. 2 and 3 have been collected for different sea conditions and aspect angles. Statistical analysis of these data shows that the speckle component fits a Rayleigh distribution (i.e. the Central Limit Theorem applies within the cell) and the modulations fit a Chi distribution (generalised for non-integer degrees of freedom). Derivation of the overall amplitude distribution yields the K-distribution,

$$p(x) = \frac{2b}{\Gamma(\nu)} \left(\frac{bx}{2}\right)^{\nu} K_{\nu-1}(bx) \quad \ldots\ldots\ldots \quad (1)$$

where x is the amplitude,

$K_{\nu}(z)$ is the modified Bessel function,

b is a scale parameter,

ν is a shape parameter.

This result is particularly significant because:

(i) Within the limit of our measurements (accuracy and coverage of parameters), the same statistical model applies. This suggests that a limit theorem is applying so that the statistics of the individual scatterers cease to contribute and,

(ii) The separation into the two components allows some of the correlation properties to be incorporated, for example, the effect of frequency agility.

The K distribution has arisen in many other scattering phenomena (Parry et al., 1977; Parry and Pusey, 1979; Jakeman et al., 1978; Jakeman, 1980) and is discussed extensively in other chapters of this book, from which a number of parallels can be drawn. To continue with our emphasis here on application, for this model to be of use, the dependence of the model parameters upon radar and environmental parameters must be established.

PARAMETER VARIATIONS

In the K distribution (1) there are two parameters, b and ν. "b" is a scale parameter and therefore only relates to the mean of the clutter or σ_0. "ν", the shape parameter, specifies the higher moments in relation to the mean and, through the compound model, the amount of inhomogeneity in the local mean (Fig. 3).

The variation of ν with external and radar parameters has been investigated in some detail using a data base of 311 airborne measurements (Ward, 1982). Limitations in the coverage with Horizontal polarization limited the conclusions that could be drawn from this mode. For Vertical polarization an empirical model for ν was developed as follows:

$$\log \nu = 2/3 \log \phi + 5/8 \log \ell + c - 1 \qquad (2)$$

where ϕ is the grazing angle (Fig. 1),

ℓ is the across range resolution ($R\theta$ in Fig. 1), and

c relates as follows to the radar pointing direction with respect to sea swell

$c = -1/3$ for up and down

$+1/3$ for across swell aspect

0 for intermediate directions or if no swell exists.

The range resolution is fixed at 4.2m.

Bearing in mind that a high ν corresponds to a noiselike distribution ($\nu = \infty$ is a Rayleigh distribution) and a low ν corresponds to spiky clutter, the main implications of (2) are:

(i) A small grazing angle ϕ implies small ν

(ii) There is no strong trend for ν against sea state as seen in σ_O.

(iii) Increasing the across range dimension of the patch (ℓ) increases ν.

(iv) Aspect angle variations depend upon swell and long wavelength sea wave content of the sea spectrum. Considerable swell content implies an aspect dependence as given above for "c". Since a locally driven sea will have more long wavelength components for higher sea states, the aspect dependence often appears to relate to sea state and is mainly due to the illuminated patch shape as discussed in the previous section.

Testing all the V polarization records against the empirical model gave an r.m.s. error of 0.24 in log ν. Comparing these with the H polarization records showed that the relative trends are similar but H generally has smaller ν than V for similar conditions, since it is more spiky.

The main limitation in this parameter modelling is the fixed range resolution of 4.2m. Also, in order to obtain sufficient samples to extract trends and significance in the variations with ℓ, results corresponding to all sea conditions had to be grouped together. With the help of the understanding obtained from the compound distribution, it is possible to simulate range resolutions changes from individual records. This will now be demonstrated. The variations of ν against r deduced from this analysis are probably a more exact model for the variations against ℓ than (2) above. Theory developed in Oliver (1984) has considerable bearing on this work.

Simulating the radar echo from a patch double the size to that used for measurements requires the coherent addition of pairs of samples in range. It is normally only possible, therefore, to do this when both amplitude and phase are recorded. However, the modelling of amplitude clutter recordings has shown that a function of the type shown in Fig. 4a (which has a Chi distribution) modulating Rayleigh noise will fully represent the clutter. The Rayleigh component is found to be not only independent from pulse to pulse if frequency agility is used, but also found to have a range correlation commensurate with the pulse length, implying independence from illuminated patch to patch. As the Rayleigh amplitude is derived from a two dimensional Gaussian process, independent samples from the Rayleigh implies a random phase. To derive the effect of increasing the patch size, it is only therefore necessary to average the power in the modulating function over the new patch and remodulate with a Rayleigh distributed function. An example of the modulating function, increasing from 28ns to 200ns pulse length, is shown in Fig. 4b.

Evaluation of the moments of the modulating function, after the patch syntheses, show that it still fits a Chi distribution, implying that the amplitude is K-distributed. Variation of the ν parameter as a function of r is plotted in Fig. 5 for a selection of records. Two extremes are expected for the gradient, G, of this log-log plot,

(i) G = 0; This is where ν is a constant and all the length scales in the clutter are much larger than r. Thus samples of the modulation function to be integrated are highly correlated.

(ii) G = 1; $\nu \propto r$, all the length scales are shorter than r and independent samples of the K-distribution are being summed.

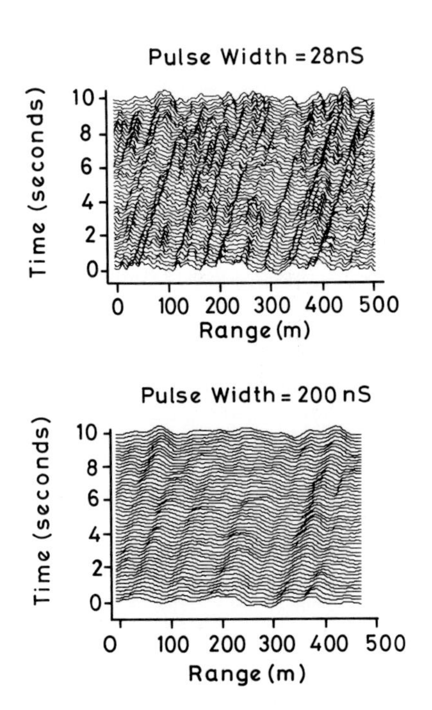

Fig. 4 Range time plots of averaged sea clutter showing the effect of different synthesised pulse widths

Intermediate situations are expected to fall between these extremes. Suprisingly the plot shows areas where the gradient is greater than unity. This is found to correspond to situations where there is a dominant wave period in the clutter return and r is of the same order. Under these conditions samples to be integrated to synthesise the larger patch are anticorrelated i.e. peaks are adding to troughs.

An autocorrelation function (A.C.F.), Fig. 6 demonstrates this effect. Unfortunately range A.C.F.'s for these data are of limited use because of the length of the profile gate used in measurements and so Fig. 6 shows a time A.C.F. of the modulating component from a single range gate. (Observation shows that this is representative of a range A.C.F. due to the propagating nature of the sea structure.) From Fig. 6a the periodic nature of the time series is evident whilst Fig. 6b shows the pronounced anticorrelation as the A.C.F. dips below $\langle x\rangle^2$. This corresponds to ν increasing faster than r, for r is the order of 120m.

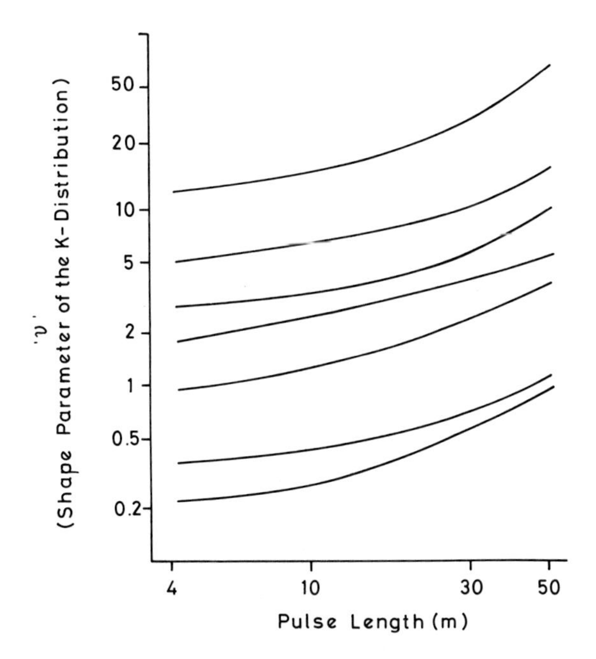

Fig. 5 Graph showing the variation of 'ν' against range resolution for a number of different records

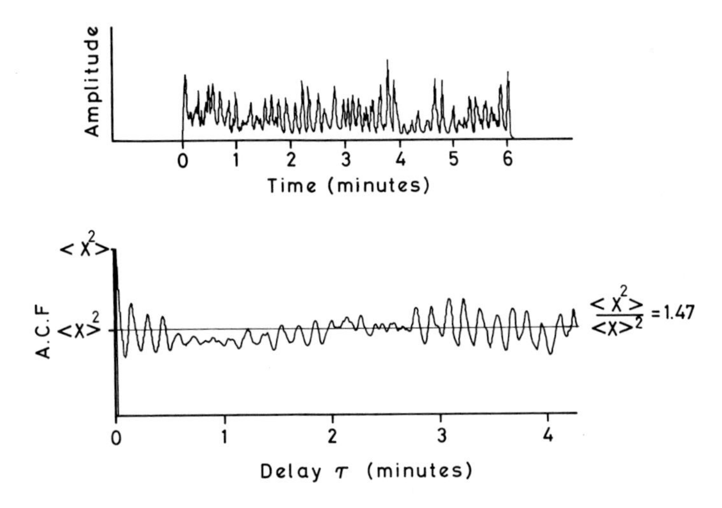

Fig. 6 A time series and autocorrelation function of the modulating function of sea clutter

In summary, equation (2) gives the approximate variation of ν with environmental and radar parameters. This can be improved upon through the type of investigation described here for range resolution changes.

APPLICATION OF THE MODEL

It was stated in the introduction that clutter deviations from noise-like statistics have considerable impact on radar performance. This is because the higher probabilities of high amplitude returns from the non-Rayleigh clutter require a higher threshold to maintain the false alarm rate, and target detection capability is reduced compared to standard detection curves in noise (Meyer and Meyer, 1973). Where a single echo and fixed threshold are used for detection, the curves calculated for Weibull and Lognormal clutter (Schleher, 1976; Schleher, 1975; Ekstrom, 1973; Chen and Morchin, 1977; Trunk and George, 1970), provide a reasonable prediction. The deviations between the clutter models are small compared to the unpredictability of the shape parameters. However, where adaptive thresholding or pulse-to-pulse integration is used, the clutter correlation properties have a marked effect upon performance. The errors incurred from assumptions of independent samples in range or time, associated with Lognormal or Weibull modelling, produce misleading results and comparative trends from one type of processing to another.

In these areas the compound formulation of the K distribution allows many of the correlation effects to be included in the calculations. This is demonstrated for within beam pulse-to-pulse integration on a scanning radar in Ward (1982), Ward and Watts (1984). This shows discrepancies, due to correlation, of up to 11dB in performance.

SUMMARY

In this chapter a brief review has been given of radar application of sea surface microwave backscatter. The application of the K distribution in this area has been used to show how many of the ideas described in the more theoretical sections of this book are contributing to the understanding of measurements and performance of systems.

REFERENCES

Blythe, J.M., et al. 1969, The CFE clutter model, with application to automatic detection, *The Marconi Review*, Vol. **32**, 185-206.

Chen, P.W. and Morchin, W.C., 1977, The detection of targets in noise and Weibull clutter background, IEEE NAECON 77, 929-933.

Ekstrom, J.L., 1973, The detection of steady targets in Weibull clutter, IEE Conf. Radar - Present and Future.

Fay, F.S., Clarke, J. and Peters, R.S., 1977, Weibull distribution applied to sea clutter, IEE Radar 77, 101-104.

Guinard, N.W. and Daley, J.C., 1970, An experimental study of a sea clutter model, *IEEE Proc.*, **58**, 543-550.

Horst, M.M., Dyer, F.B. and Tuley, M.T., 1978, Radar sea clutter model, IEE Conf. on Antennas and Propagation,

Jakeman, E., 1980, On the statistics of K-distributed noise, *J. Phys. A.*, **13**, 31-48.

Jakeman, E. and Pusey, P .N., 1977, Statistics of non-Rayleigh microwave sea echo, IEE Radar 77, 105-109.

Jakeman, E., et al., 1978, The twinkling of stars, *Contemp. Physics*, **19**, 127-145.

Kalmykov, A.I. and Pustovoytenko, V.V., 1976, On polarisation features of radar signals scattered from the sea surface at small grazing angles, *J. Geophys. Res.*, **81**, 1960-1964.

Lewis, B.L. and Olin, I.D., 1980, Experimental study and theoretical model of high-resolution radar backscatter from the sea, *Radio Science*, **15**, 815-828.

Long, M.W., 1980, Radar reflectivity of the land and sea, Lexington Books, 77-89.

Meyer, D.P. and Meyer, H.A., 1973, Radar target detection, Academic Press.

Nathanson, F.E., 1969, Radar design principles, McGraw Hill.

Oliver, C.J., 1984, A model for non-Rayleigh scattering statistics, (Optica Acta to be published).

Parry, G. and Pusey, P.N., 1979, K-distributions in atmospheric propagation of laser light, *J. Opt. Soc. Am.*, **69**, 796-798.

Parry, G., et al., 1977, Focussing by a random phase screen, *Opt. Commun.*, **22**, 195-200.

Schleher, D.C., 1975, Radar detection in log-normal clutter, IEEE Radar Conf., 262-267.

Schleher, D.C., 1976, Radar detection in Weibull clutter, *IEEE AES*, **12**, No. 6, 736-743.

Sekine, M., 1980, Log-Weibull distributed sea clutter, *Proc. IEE*, Vol. **127**, pt. F, No. 3, 225-228.

Sittrop, H., 1977, On the sea clutter dependency on windspeed, Radar 77, IEE Conf. Proc. No. 155, 110-114.

Skolnik, M.I., 1980, Introduction to radar systems, McGraw Hill.

Trunk, G.V., 1972, Radar properties of non-Rayleigh sea clutter, *IEEE Trans. AES*, **8**, 196-204.

Trunk, G.V. and George, S.F., 1970, Detection of targets in non-Gaussian sea clutter, *IEEE Trans. AES*, **6**, 620-628.

Valenzuela, G.R., 1978, Theories for the interaction of electromagnetic and ocean waves - A review, *Boundary-Layer Meteorology*, **13**, 61-85.

Valenzuela, G.R. and Laing, M.B., 1971, On the statistics of sea clutter, NRL Report 7349.

Ward, K.D., 1982, A radar sea clutter model and its application to performance assessment, Radar 82, 203-207.

Ward, K.D., 1981, Compound representation of high resolution sea clutter, *Electronics Letters*, **17**, No. 16, 561-563.

Ward, K.D. and Watts, S., 1984, Radar clutter in airborne maritime reconnaissance systems, Military Microwaves.

Wright, J.W., 1966, Backscatter from capillary waves with applications to sea clutter, IEEE AP14, 749-54.

DIRECT AND INVERSE METHODS FOR OCEAN WAVE IMAGING BY SAR

S. Rotheram and J.T. Macklin

(Marconi Research Centre, Chelmsford, Essex)

ABSTRACT

A Synthetic Aperture Radar (SAR) can form high resolution images of the sea surface. The direct problem for such an imaging system is now reasonably well understood and is summarised here for both the SAR image and its power spectrum. The inverse problem has only recently been explored and some simple approaches based on linear demodulation and speckle removal are described. These are applied to a number of SEASAT images.

1. INTRODUCTION

A Synthetic Aperture Radar (SAR) is able to form high-resolution two-dimensional images of the sea surface. Fine range resolution is achieved with short pulses using pulse compression. Fine azimuth resolution is achieved using the motion of the platform to coherently synthesise a long antenna. Typical resolutions of a few metres from aircraft and a few tens of metres from spacecraft are achieved. The SAR geometry is shown in Fig. 1.

Following SEASAT in 1978 and SIR-A in 1982, a number of spaceborne SARs are being planned including SIR-B in 1984 and ERS-1 in 1989. Like SEASAT, ERS-1 is primarily an oceanographic satellite. The main instrument on ERS-1 will be a SAR which can operate in an imaging mode for obtaining large images of the sea surface, and a wave scatterometer mode for obtaining image spectra of 5 x 5 km patches of the sea surface.

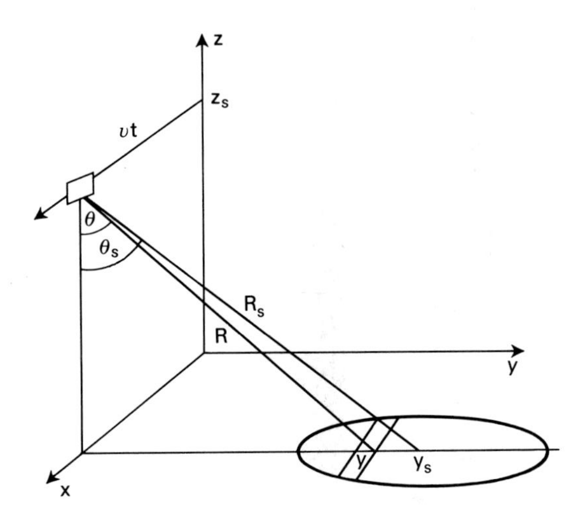

Fig. 1 SAR geometry

The main object of forming images and image spectra of the sea surface is to obtain quantitative information about the sea surface such as wave height and direction, wavenumber spectrum, storm evolution, surface winds etc. To obtain this quantitative information, the relationship between the sea surface and SAR images and spectra needs to be defined as fully as possible. There are two aspects to this relationship. These are the direct and inverse sea imaging problems.

When an imaging radar operating at a wavelength of a few centimetres illuminates the sea surface, most of the back-scattered energy is Bragg scattered from water waves of similar wavelength. The resolution of the radar is typically a few metres to tens of metres and so the Bragg waves cannot be resolved. Waves of wavelength longer than the resolution are imaged through a variety of secondary interactions between long waves and Bragg waves. As there is a large separation in wave-number space between the imaged long waves and the short Bragg waves, this leads to the idea of a two-scale theory of sea surface scattering which was originally proposed by Wright (1968). These ideas are illustrated in Fig. 2 in which the wavenumber plane of ocean waves is divided arbitrarily into long and short waves. The Bragg waves lie in the short wave region whilst the imaged waves lie in the long wave region.

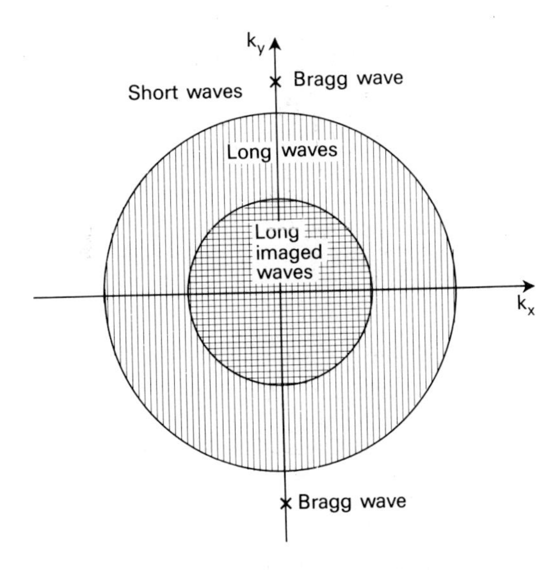

Fig. 2 Wavenumber diagram

The main secondary interactions producing imaging mechanisms have come to be known as tilting, straining, velocity bunching and Doppler splitting (Alpers et al., 1981; Vesecky and Stewart, 1982; Rotherham, 1983). Tilting occurs because the long waves tilt the short waves whilst straining occurs because the short waves propagate in the slowly varying long wave field. These can be described by a perturbation theory (Rotherham, 1983; Keller and Wright, 1975; Alpers and Hasselman, 1978). Velocity bunching, which has been studied by a number of approaches (Larson and Moskowitz, 1977; Alpers and Rufenach, 1979; Swift and Wilson, 1979; Valenzuela, 1980; Rufenach and Alpers, 1981; Raney, 1981; Ivanov, 1982), occurs because the orbital motions of the long waves Doppler shift the Bragg scattered field. A SAR determines azimuthal position by the Doppler shift, broadside being at zero Doppler, and so these periodic Doppler shifts lead to periodic scatterer displacements and thus periodic image modulations. The motion of the short waves also produces an effect known as Doppler splitting (Alpers, 1981; Rotheram, 1983). This is because there are two Bragg waves travelling in opposite directions.

The direct problem for the SAR image is to describe the image I that corresponds to a given sea S. This can be expressed by a functional relation $I = N[S]$. The functional N is nonlinear due to the effects of velocity bunching although the tilting and straining contributions are essentially linear. It is also stochastic because the Bragg scattering short wave field is random and this is the origin of speckle in SAR imagery. The direct problem for the image is described in section 2. The linear part of the relation is described in detail using a linear modulation transfer function whilst the nonlinear aspects are discussed using a computed example.

The direct problem for the SAR image power spectrum is to describe the image power spectrum P that corresponds to a given sea surface power spectrum E. This can be expressed by a functional relation P = M[E]. As well as many aspects in common with the image I, this functional contains additional aspects attributable to the stochastic nature of both the short and long wave fields. P is a quadratic functional of the image and a fourth order functional of the complex amplitude image before detection. Ensemble averaging over the short waves introduces an additional speckle contribution to P in the form of a background proportional to the static system transfer function of the SAR. Ensemble averaging over the long waves leads to an azimuthal banding due to nonlinear velocity bunching. These aspects are summarised in section 3.

Some tentative inverse methods are given in section 4 and applied to SEASAT imagery. For the image power spectrum the static system transfer function is determined from a featureless speckled image (Beal, 1982). This is then divided out of spectra to leave the wave information on a flat, exponentially distributed, background component. This is then subtracted using a thresholding technique at various confidence levels. The resultant is demodulated using a regularised modulation transfer function to give an estimate of the sea surface power spectrum. Analogous, though less soundly based, techniques are applied to the image. These techniques are exploratory in nature and are the subject of current research. It is hoped that optimum methods will be developed from these based on Bacchus-Gilbert theory.

2. THE SAR IMAGE

To first order in the rms waveheight σ_ℓ, the long waves can be represented by

$$z = \sigma_\ell \int \frac{d^2\underline{k}d\omega}{(2\pi)^3} \tilde{\zeta}_\ell(\underline{k},\omega) \; e^{-i\underline{k}\cdot\rho+i\omega t} \tag{1}$$

$$\tilde{\zeta}_\ell(\underline{k},\omega) = 2\pi\delta(\omega-g^{\frac{1}{2}}k^{\frac{1}{2}})\tilde{\zeta}_\ell(\underline{k})+2\pi\delta(\omega+g^{\frac{1}{2}}k^{\frac{1}{2}})\tilde{\zeta}_\ell{}^*(-\underline{k}) \tag{2}$$

in which z is height, $\underline{\rho} = (x,y)$ is horizontal position, x is azimuth, y is range, t is time, $\underline{k} = (k_x,k_y)$ is the wavevector of the long waves, ω is the angular frequency, $\tilde{\zeta}(\underline{k},\omega)$ is the wavenumber-frequency amplitude spectrum, $\tilde{\zeta}(\underline{k})$ is the wavenumber amplitude spectrum, g is the acceleration of gravity and (2) is the deep water gravity wave dispersion relation. If the long waves form a stationary and homogeneous random field

$$<\zeta_\ell(\underline{\rho},t)> = 0 \ , \ <\zeta_\ell(\underline{\rho},t)^2> = 1 \tag{3}$$

$$<\tilde{\zeta}_\ell(\underline{k},\omega)\,\tilde{\zeta}_\ell{}^*(\underline{k}',\omega')> = (2\pi)^3\ \delta(\underline{k}-\underline{k}',\omega-\omega')\,\tilde{E}_\ell(\underline{k},\omega) \tag{4}$$

$$\tilde{E}_\ell(\underline{k},\omega) = 2\pi\delta(\omega-g^{\frac{1}{2}}k^{\frac{1}{2}})\,\tilde{E}_\ell(\underline{k}) + 2\pi\delta(\omega+g^{\frac{1}{2}}k^{\frac{1}{2}})\,\tilde{E}_\ell(-\underline{k}) \tag{5}$$

in which $\tilde{E}_\ell(\underline{k},\omega)$ is the wavenumber-frequency energy spectrum and $\tilde{E}_\ell(k)$ is the wavenumber energy spectrum. The short waves obey a set of similar relations with ℓ changed to s. These are just the linear parts of the wave field.

The long waves amplitude modulate and Doppler shift the short waves. The amplitude modulations are known as straining. In backscattering from this surface, the radio wave is Bragg scattered from the short waves with amplitude and phase modulations from the long waves. In addition to straining, the amplitude modulations are caused by the tilting of the short waves by the long waves. The phase modulations amount to a Doppler shift by the orbital velocity component of the long waves in the look direction of the radar. Details of the short wave representations and the backscattered field are given by Rotheram (1983).

After SAR processing the complex amplitude image intensity is $W(\underline{\rho})$. The SAR image intensity is $I(\underline{\rho}) = |W(\underline{\rho})|^2$ which is the quantity displayed in a SAR image. It can be expanded in a long wave perturbation expansion

$$I(\underline{\rho}) = \sum_{n=0}^{\infty} \sigma_\ell^n\ I_n(\underline{\rho}) \tag{6}$$

In the following <•> denotes an assembly average over the short waves. The zeroth order term in (6) is the pure Bragg scattered component given by

$$\langle I_o \rangle = C[\tilde{E}_s(\underline{k}_b) + \tilde{E}_s(-\underline{k}_b)] \tag{7}$$

$$\underline{k}_b = (0, -2\, k_s \sin\theta) \;,\; \omega_b = g^{\frac{1}{2}} k_b^{\frac{1}{2}} \tag{8}$$

in which C is a constant, $\underline{k}_b$ is the Bragg wavevector, ω_b is the Bragg angular frequency, θ is the angle of incidence to the vertical (see Fig. 1) and $k_s = 2\pi/\lambda_s$ is the radar wavenumber.

The first order term $I_1(\underline{\rho})$ represents linear modulations and can be written

$$\langle I_1(\underline{\rho})\rangle = \langle I_o\rangle \int \frac{d^2\underline{k}d\omega}{(2\pi)^3}\, \tilde{\zeta}_\ell\, (\underline{k},\omega)\, \tilde{T}(k,\omega)\, e^{-i\underline{k}.\underline{\rho}+i\omega t} \tag{9}$$

in terms of a linear modulation transfer function $\tilde{T}(\underline{k},\omega)$. For simplicity the dispersion relation in (2) has been left understood in (9), as in (1) and elsewhere. It is useful to separate $\tilde{T}(\underline{k},\omega)$ into two components

$$\tilde{T}(\underline{k},\omega) = \tilde{M}(\underline{k},\omega)\, \tilde{Q}(\underline{k},\omega) \tag{10}$$

Here $\tilde{M}(\underline{k},\omega)$ is the scattering transfer function and $\tilde{Q}(\underline{k},\omega)$ is the dynamic system transfer function. The transformation from the sea surface $\tilde{\zeta}_\ell$ to the SAR image spectral density $\tilde{\zeta}_\ell\tilde{T}$ can be thought of as the two stage process $\tilde{\zeta}_\ell \rightarrow \tilde{\zeta}_\ell\tilde{M} \rightarrow \tilde{\zeta}_\ell\tilde{M}\tilde{Q}$ meaning sea surface $\rightarrow$ scatterer density $\rightarrow$ SAR image density (Vesecky and Stewart, 1982).

The scattering transfer function $\tilde{M}(\underline{k},\omega)$ is given by

$$\tilde{M}(\underline{k},\omega) = \sum_{p=\pm 1} \frac{2\tilde{E}_s(p\underline{k}_b)\, e^{-ip\omega_b t_x}}{\{\tilde{E}_s(\underline{k}_b)+\tilde{E}_s(-\underline{k}_b)\}}\, [\tilde{L}_p(\underline{k})\cos(\omega t_x/2) + i\tilde{A}(\underline{k})\sin(\omega t_x/2)] \tag{11}$$

$$t_x = \frac{TXk_x}{4\pi} \quad , \quad T = \frac{R\lambda_s}{Xv} \tag{12}$$

$$\tilde{A}(\underline{k}) = -2k_s \sin\theta . \frac{k_y}{k} + 2i\, k_s \cos\theta \tag{13}$$

$$\tilde{L}_p(\underline{k}) = \frac{k}{2} - \frac{k_x^2}{4k} - \frac{ik_y}{2}\cos\theta - 2i\, k_y \tan\theta \left[1 - \mathrm{Re}\left\{\frac{\cos\theta}{(n^2-\sin^2\theta)^{\frac{1}{2}}}\right\}\right]$$

$$- \frac{\tilde{A}(\underline{k})}{2}\, \underline{k} \cdot \frac{\partial}{\partial \underline{k}_b} \ell n\, \tilde{E}_s(p\underline{k}_b) \tag{14}$$

in which X is the length of the antenna in azimuth, v is the velocity of the SAR, T is the one look integration time and n is the complex refractive index of the sea. In brief, tilting and straining are given by the term in $\tilde{L}_p(\underline{k})$, velocity bunching by the term in $A(\underline{k})$, and Doppler splitting by the phases $e^{\mp i\omega_b t_x}$. These are discussed in detail elsewhere (Alpers et al., 1981; Vesecky and Stewart, 1982; Rotheram, 1983).

Fig. 3 shows $|\tilde{M}(\underline{k},\omega)|$ for $\omega = g^{\frac{1}{2}}k^{\frac{1}{2}}$ and SEASAT parameters ($R/v = 128s$, $\theta = 20^\circ$) in an approximation in which $\tilde{E}_s(\underline{k})$ is given by an isotropic Phillips spectrum, Doppler splitting is ignored and the moduli of $\tilde{A}$ and $\tilde{L}_p$ are formed separately. It is plotted against k for various values of the azimuth angle ψ where $\underline{k} = k(\cos\psi, \sin\psi)$ so ψ is the angle that the wave-vector makes with the azimuth or x direction. Fig. 4 shows the same data but with $|\tilde{M}|^{-1}$ plotted as a contour map in the wave-number plane.

The dynamic system transfer function $\tilde{Q}(\underline{k},\omega)$ is given by

$$\tilde{Q}(\underline{k},\omega) = \exp\left[-\frac{1}{32\pi}\left\{k_x^2 N^2 X^2 + \frac{k_y c^2 \tau^2}{\sin^2\theta} + \frac{4\omega^2\, T^2}{N^2}\right\}\right] \tag{15}$$

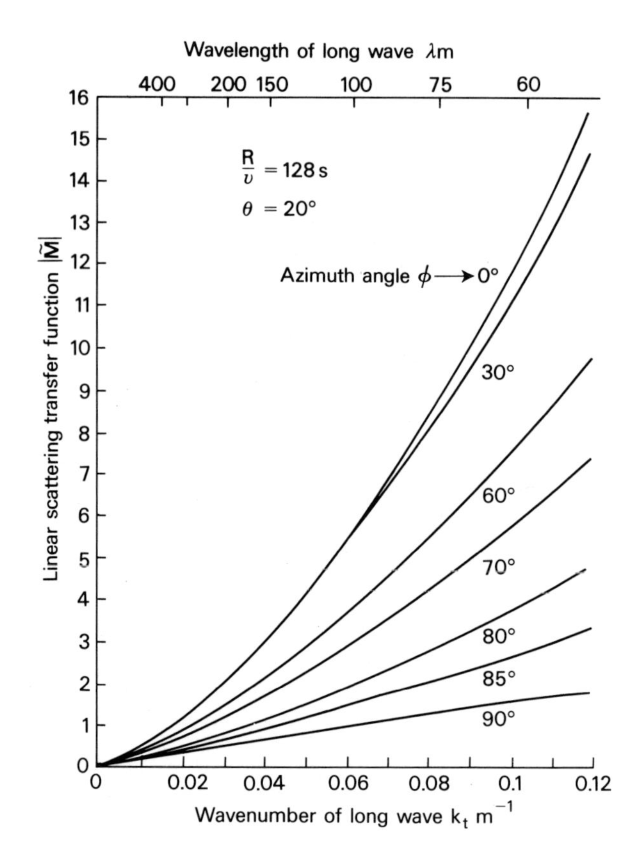

Fig. 3 Modulus of scattering transfer function $|\tilde{M}|$. SEASAT parameters

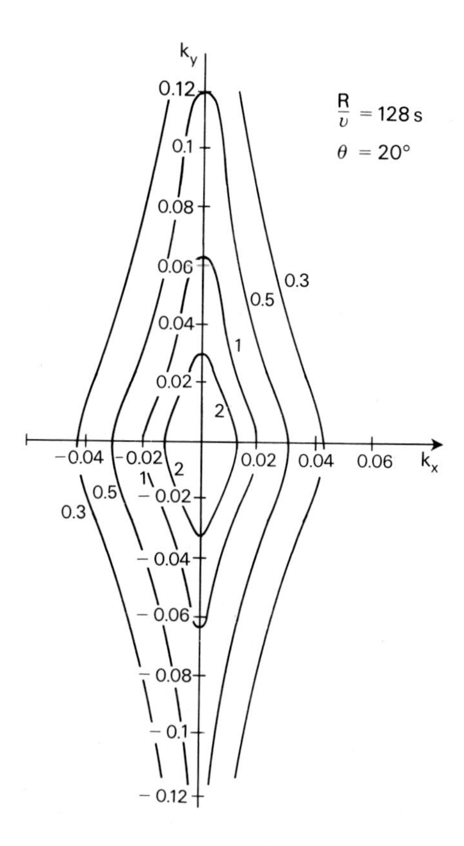

Fig. 4 Contour map of $|M|^{-1}$ SEASAT parameters

in which N is the number of azimuth looks, c is the velocity of light and τ is the compressed pulse length. It expresses the resolution of the system in azimuth, range and time by cutting off waves of large wavenumber and frequency.

A serious complication in SAR imaging theory is that velocity bunching becomes strongly nonlinear as the wave amplitude increases. The effects are described by the higher order terms in (6) and details are given by Rotheram, 1983. For a single long wave of amplitude a and wavelength λ, Fig. 5 shows the linear and nonlinear imaging regimes in the a - λ plane for various azimuth angles ϕ and SEASAT parameters.

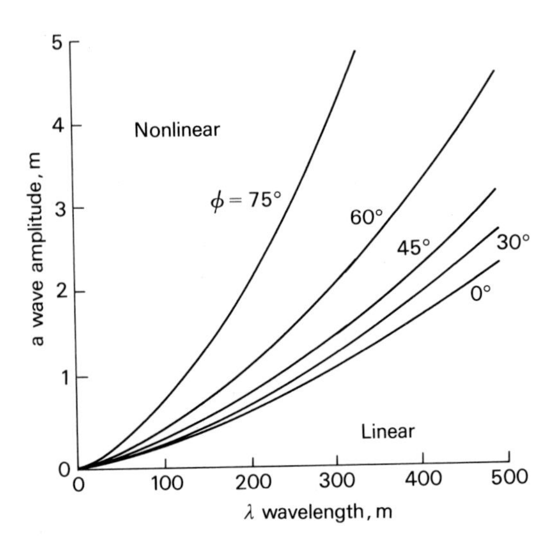

Fig. 5 Linear-nonlinear transition in velocity bunching

Fig. 6 shows a computed example for λ = 200m, ψ = 0 and increasing values of a. As the wave amplitude a increases, the image modulations increase at first but then decrease as nonlinearity becomes important.

3. THE SAR IMAGE POWER SPECTRUM

For ocean wave fields, the image power spectrum is in some respects more interesting than the image. This is reflected in the wave scatterometer mode of ERS-1. Some useful studies of the image power spectrum have been published recently (Beal, 1982; Alpers and Hasselman, 1982). The theory requires a number of significant and difficult steps beyond that for the SAR image and these are summarised here with fuller details given elsewhere (Rotheram, 1983).

Consider the effects of sampling and finite image size. Assume that $\underline{\rho}_0 = (x_0, y_0)$ are the coordinates of the bottom left hand corner of the image which is rectangular with L x J pixels each of size u x w. The sampled image is

$$I_{\ell j} = I(x_0+\ell u, y_0+jw) = \int d^2\underline{\rho} I(\underline{\rho}) \; \delta(x-x_0-\ell u, y-y_0-jw) \tag{16}$$

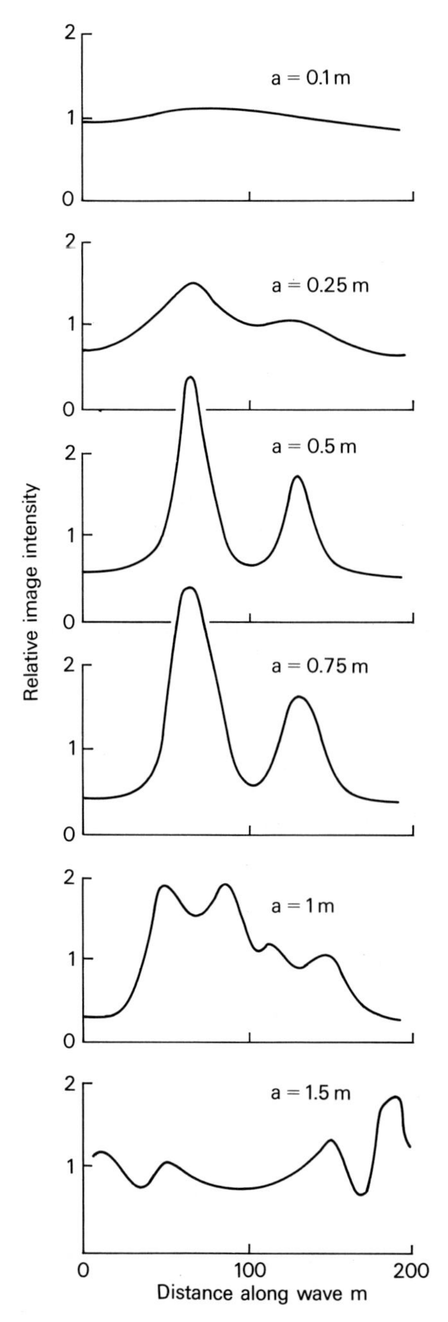

Fig. 6 Image intensity for different wave amplitudes a. 200m azimuth waves, 4-look SEASAT.

with $\ell = 0,1,\ldots,L-1$ and $j = 0,1,\ldots,J-1$. The true image spectrum $\tilde{I}(\underline{k})$ is the Fourier transform of $\tilde{I}(\underline{\rho})$ whilst the sampled image spectrum $I_s(\underline{k})$ is the DFT of $I_{\ell j}$ and these are related by

$$\tilde{I}_s(\underline{k}) = \int \frac{d^2\,\underline{k}'}{(2\pi)^2}\;\tilde{I}\;(\underline{k}')\;K(k,k') \tag{17}$$

$$\tilde{K}(\underline{k},\underline{k}') = \left[\frac{1 - e^{i(k_x-k_x')Lu}}{1 - e^{i(k_x-k_x')u}}\right]\left[\frac{1 - e^{i(k_y-k_y')Jw}}{1 - e^{i(k_y-k_y')w}}\right] \tag{18}$$

The sampled image power spectrum $\tilde{P}_s(\underline{k}) = |\tilde{I}_s(\underline{k})|^2$ and so

$$\tilde{P}_s(\underline{k}) = \int \frac{d^2\underline{k}'d^2\underline{k}''}{(2\pi)^2}\;\tilde{I}\;(\underline{k}')\;\tilde{I}^*(\underline{k}'')\tilde{K}(\underline{k},\underline{k}')\tilde{K}^*(\underline{k},\underline{k}'') \tag{19}$$

This is the quantity displayed in a SAR image power spectrum.

One is thus led to study the quantity $\tilde{I}(\underline{k}')\tilde{I}^*(\underline{k}'')$. This is related to the complex amplitude $W(\underline{\rho})$ by

$$\tilde{I}(\underline{k})\tilde{I}^*(\underline{k}') = \int d^2\underline{\rho}d^2\rho'\,|W(\underline{\rho})\;W(\rho')|^2\;e^{i\underline{k}\cdot\underline{\rho}-i\underline{k}'\cdot\underline{\rho}'} \tag{20}$$

The short wave assembly average of this requires evaluation of the fourth moment $<|W(\underline{\rho})W(\underline{\rho}')|^2>$. For Gaussian $W(\underline{\rho})$ this gives three terms, one of which is negligible. The other two terms are

$$<\tilde{I}(\underline{k})\;\tilde{I}^*(\underline{k}')>_1 = \int d^2\underline{\rho}d^2\underline{\rho}'\;<|W(\underline{\rho})|^2><|W(\underline{\rho}')|^2>\;e^{i\underline{k}\cdot\underline{\rho}-i\underline{k}'\cdot\underline{\rho}'} \tag{21}$$

$$<\tilde{I}(k)\;\tilde{I}^*(\underline{k}')>_2 = \int d^2\underline{\rho}d^2\underline{\rho}'\,|<W(\underline{\rho})W^*(\underline{\rho}')>|^2 e^{i\underline{k}\cdot\underline{\rho}-i\underline{k}'\cdot\underline{\rho}'} \tag{22}$$

The first term is $<\tilde{I}(\underline{k})><\tilde{I}^*(\underline{k}')>$ and gives a term $|<I_s(\underline{k})>|^2$ in the sampled image power spectrum. Its properties are simply the Fourier transform of those in section 2. The second term given by (22) is a new feature arising from the crossed terms in the fourth moment. It is purely a consequence of the coherent speckle in SAR imagery.

For the SAR image one is interested in a particular realisation of the sea surface. For the SAR image power spectrum one can average over an ensemble of realisations of the long wave field. A long wave assembly average will be represented by an overbar. For the term in (22) this leads simply to

$$\overline{<\tilde{I}(\underline{k})\tilde{I}^*(\underline{k}')>_2} = <I_0>^2 (2\pi)^2 \delta(\underline{k}-\underline{k}') \frac{Xc\tau}{4 \sin \theta} \tilde{Q}_0(\underline{k})^2 \quad (23)$$

$$\tilde{Q}_0(\underline{k}) = \tilde{Q}(\underline{k},0) = \exp\left[-\frac{1}{32\pi}\left\{k_x^2 N^2 X^2 + \frac{k_y^2 c^2 \tau^2}{\sin^2\theta}\right\}\right] \quad (24)$$

Here $\tilde{Q}_0(\underline{k})$ will be called the static system transfer function. It is the response to a random structureless scene. Substituting (24) in (19) gives the contribution to the sampled image power spectrum

$$<I_0>^2 \frac{Xc\tau}{4 \sin \theta} \frac{L J}{4u^2 w^2} \tilde{Q}_0(\underline{k})^2 \quad (25)$$

so this represents a background contribution that falls off like $\tilde{Q}_0(\underline{k})^2$. It contains no information about the long waves.

The contribution of the term in (21), after long-wave averaging, is much more complicated. It can be written as a long-wave perturbation expansion in which only even powers of σ_ℓ appear. An exponential series can be separated from this series with the result

$$\overline{<\tilde{I}(\underline{k})\tilde{I}^*(\underline{k}')>_1} = <I_0>^2 (2\pi) \delta(\underline{k}-\underline{k}') \exp\left[\frac{-X^2 k_x^2 \nu^2}{16}\right] \sum_{n=0}^{\infty} \sigma_\ell^{2n} \tilde{F}_n(\underline{k}) \quad (26)$$

$$\nu^2 = \frac{\sigma_\ell T^2}{\pi} \int \frac{d^2\underline{k}d\omega}{(2\pi)^3} \tilde{E}_\ell(\underline{k},\omega) \ |\tilde{A}(\underline{k})|^2\omega^2 \tag{27}$$

The exponential gives an azimuthal cut-off as described below. The zeroth order term in (26) is

$$\tilde{F}_O(\underline{k}) = (2\pi)^2\delta(\underline{k}) \tag{28}$$

and arises from the Fourier transform of the mean intensity in the SAR image. Its contribution to the sampled image power spectrum is found from (19), (26) and (28) to be

$$\langle I_O\rangle^2 \ |\tilde{K}(\underline{k},0)|^2 \tag{29}$$

This is usually sampled at $\underline{k} = 2\pi\ (\ell/Lu,\ j/J\omega)$ giving zero except at $\underline{k} = 0$ for which (29) gives

$$\langle I_O\rangle^2 \frac{L\ J}{4u^2w^2} \tag{30}$$

The first order term $\tilde{F}_1(\underline{k})$ has only been determined in an approximation in which tilting and straining have been ignored. The result is

$$\tilde{F}_1(\underline{k}) = \left| 2\tilde{A}(\underline{k})\tilde{P}(\underline{k})\sin(\frac{\omega t_x}{2})\right|^2 \ [\tilde{E}(\underline{k})\tilde{Q}(\underline{k},\omega)^2+\tilde{E}_\ell(-\underline{k})\tilde{Q}(-\underline{k},\omega)^2] \tag{31}$$

$$\tilde{P}(\underline{k}) = \sum_{p=\pm1} \frac{\tilde{E}_s(p\underline{k}_b)e^{-ip\omega_b t_x}}{\{\tilde{E}_s(\underline{k}_b)+\tilde{E}_s(-\underline{k}_b)\}} \tag{32}$$

The function $\tilde{P}(\underline{k})$ describes Doppler splitting. The contribution of the first-order term to the sampled image power spectrum is found from (19), (26) and (31) to be

$$\langle I_O \rangle^2 \frac{L\ J}{4u^2 w^2} \sigma_\ell^2 \exp\left[-\frac{X^2 k_x^2 \nu^2}{16}\right] \tilde{F}_1(\underline{k}) \tag{33}$$

As $\tilde{F}_1(\underline{k})$ is a linear functional of the wavenumber energy spectrum $\tilde{E}_\ell(\underline{k})$, it is the term containing the information in a SAR image power spectrum.

The expression in (33) is incomplete because of the neglect of tilting and straining. One consequence is that $\tilde{F}_1(\underline{k})$ in (31) is incomplete. The expression between the modulus signs should be similar to $\tilde{M}(\underline{k},\omega)$ in (11) but the precise details have not been worked out.

Higher-order terms in (26) are a consequence of nonlinearity in velocity bunching. The nth term is an nth order functional of $\tilde{E}_\ell(\underline{k})$. For a narrow band swell wave system $\tilde{E}_\ell(\underline{k})$ is sharply peaked and the higher order terms give contributions that are harmonics of the main peak.

However the principal contribution of nonlinearity is the azimuthal cut-off in (26) and (33) with ν given by (27). For a static scene the azimuthal resolution γ_x of a SAR is

$$\gamma_x = \frac{XN}{2} \tag{34}$$

The azimuthal cut-off modifies this to

$$\gamma_x' = \frac{X}{2}(N^2 + \nu^2)^{\frac{1}{2}} \tag{35}$$

For a wind-wave system one finds approximately (4)

$$\frac{X}{2}\nu \sim 2.7 \frac{R}{V} \sigma_\ell^{\frac{1}{2}} \tag{36}$$

in MKS units. This agrees well with the empirical value $2(R/v)\sigma_\ell^{\frac{1}{2}}$ given by Beal (1982). For SEASAT (35) and (36) give

$$\gamma_x' = 6(N^2 + 3400\ \sigma_\ell)^{\frac{1}{2}} \tag{37}$$

Fig. 7 shows γ_x' for N = 4 plotted against σ_ℓ. The azimuthal resolution is drastically reduced which is the reason that azimuthal waves appear less often on SAR imagery than one would expect.

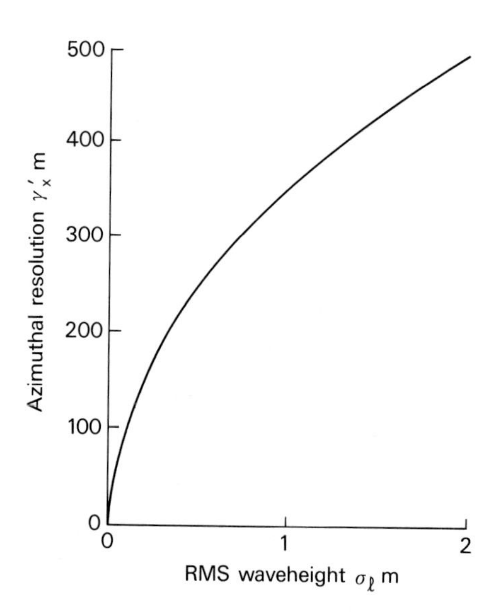

Fig. 7 Azimuthal resolution against r.m.s. wave-height. 4-look SEASAT.

4. INVERSE METHODS

4.1 SEASAT images

Before describing some simple inverse methods, it is helpful to describe the three images and spectra to which these methods are to be applied. These images and spectra are shown in Figs. 8, 9 and 10.

The images are in the top left of the figures; they are SEASAT images with the orbits and geographical locations given in Table 1.

Table 1

Properties of SEASAT Images

Image	Figure	Orbit No.	Date	Latitude	Longitude	Description
300	8	762	19/8/78	58° 26'N	5° 50'W	No waves
002	9	1087	11/9/78	60° 32'N	12° 58'W	Azimuthal waves
298	10	762	19/8/78	60° 11'N	6° 41'W	Range waves

Each image consists of 256 x 256 pixels with 12.5 m pixel spacing and 25m resolution. Although this is the resolution associated with 4 looks in azimuth, only one look is displayed. SAR processing was carried out at the Royal Aircraft Establishment, U.K.

The image power spectra are in the middle of the top row of figures. All of the spectra shown here have been smoothed with a 4 x 4 running average, and colour coded as shown in the adjacent strips, for display purposes. In all the images and spectra, the x (azimuth) and k_x co-ordinates are horizontal, and the y (range) and k_y co-ordinates are vertical. The images are 3 km square, whilst the boundaries of the spectra are at $k_x, k_y = \pm 2\pi/25\ m^{-1}$ (where wavenumber k is defined as $2\pi/\lambda$, λ being wavelength). The rings marked on the spectra show the loci of 50 m and 100 m wavelengths.

The image in Fig. 8 contains virtually no waves; its associated power spectrum consists almost entirely of the background speckle contribution, with radial fall-off due to the static system transfer function (25). (The large component at zero wavenumber, which is due to the mean intensity in the image, has been set to zero in all the spectra displayed here.) There are two small peaks in this spectrum, indicating some very weak waves.

The image in Fig. 9 shows near-azimuth waves. These are the two large peaks in the image power spectrum, at a wavelength of about 300 m with azimuth $\phi = 28°$. The two small peaks near the origin correspond to larger-scale structure with a wavelength of about 1 km. Note that the power spectra have rotational symmetry about the origin, and so there is a 180° ambiguity in the direction of the waves.

The image in Fig. 10 contains near-range waves. The double peaks in the image power spectrum show that there are two wave systems; the wavelength is about 200 m with azimuths $\phi = 80°$ and 100°. Another feature in this spectrum is the azimuthal banding due to nonlinear velocity bunching, as described by the exponential factor in (33). This feature is clearest in the spectrum on the top right, where the system transfer function has been corrected by filtering. This spectrum also shows lesser peaks at wavenumbers approximately twice those of the main peaks. The lesser peaks are probably the harmonics of the main peaks, corresponding to the n = 2 term in (26).

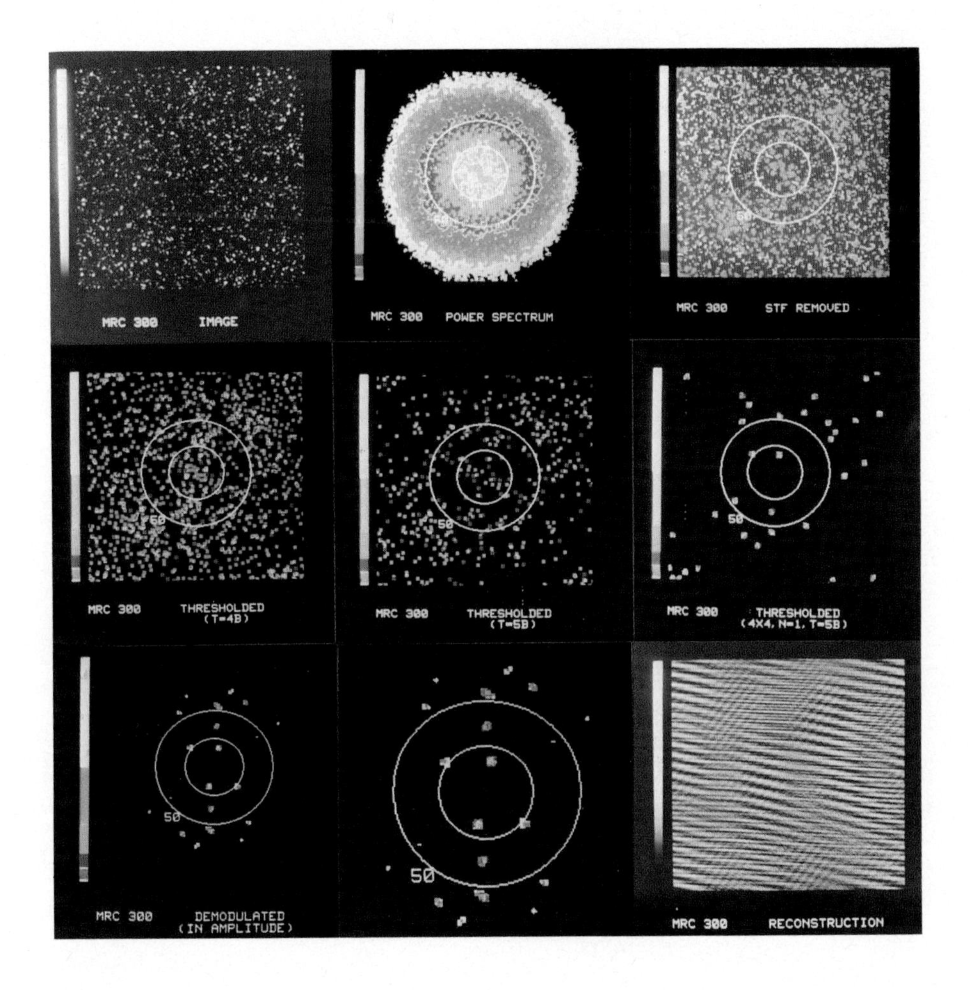

Fig. 8 SEASAT image and spectra. Orbit 762, latitude 58° 26'N, longitude 5° 50'W

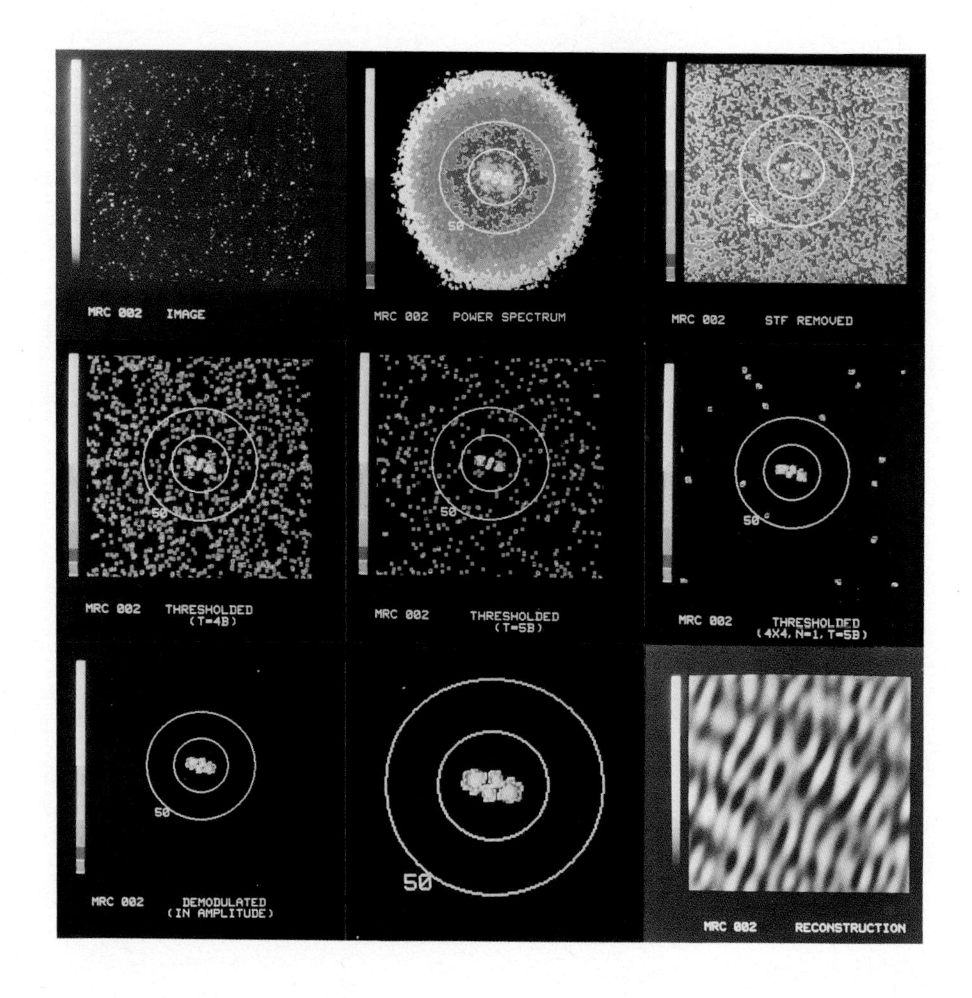

Fig. 9 SEASAT image and spectra. Orbit 1087, latitude 60° 32'N, longitude 12° 58'W

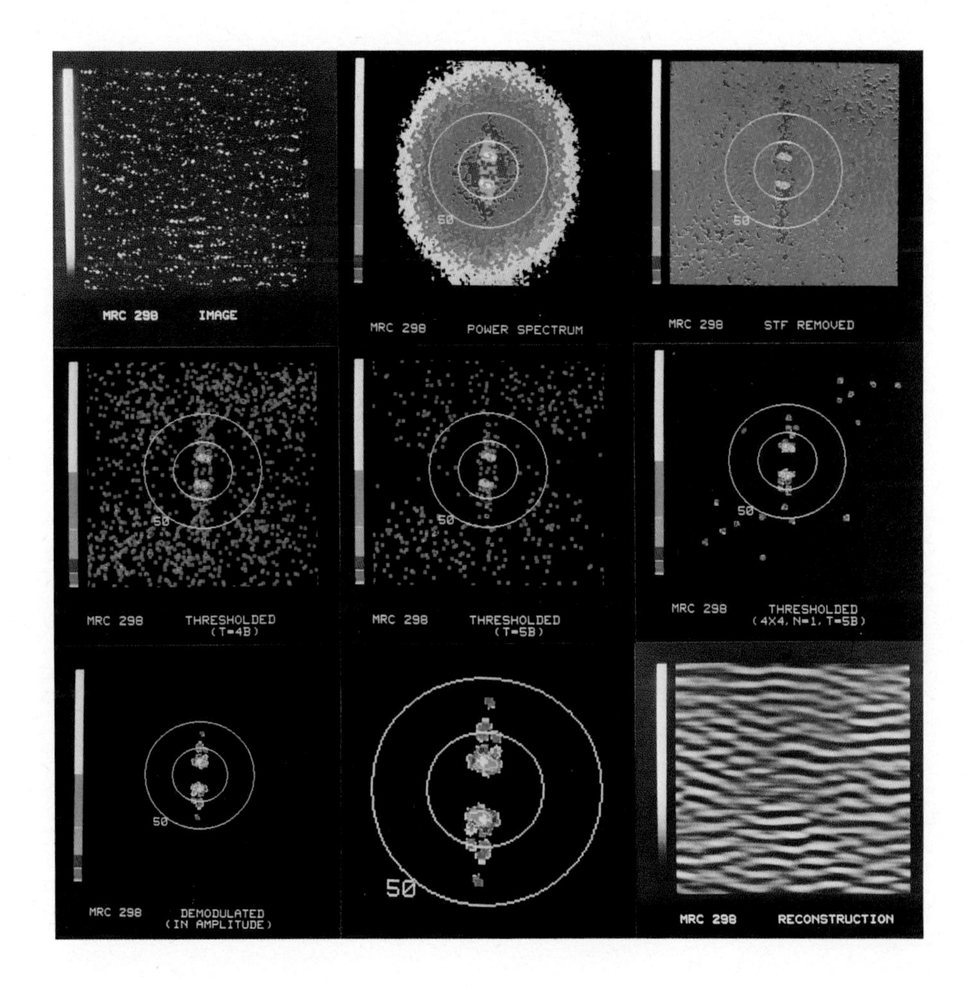

Fig. 10 SEASAT image and spectra. Orbit 762, latitude 60° 11'N, longitude 6° 41'W

4.2 Deriving the Sea-Surface Energy Spectrum

The image power spectrum consists of three main components;

a) a spike at the origin, $\underline{k} = 0$, given by (30),

b) a background contribution from speckle, proportional to $\tilde{Q}_0(k)^2$ and given by (25), and

c) a term containing information about the wavenumber energy spectrum $\tilde{E}_\ell(\underline{k})$ and given by (33).

An elementary inverse method, for reconstructing the sea-surface energy spectrum involves removing the terms (a) and (b) and demodulating the term (c). The term (a) is easily removed by setting the central pixel in the spectrum to zero.

The removal of the background due to speckle requires a determination of the static system transfer function $\tilde{Q}_0(\underline{k})$ using a method described by Beal (1982). The Gaussian form in (24) is based on a simple model of a SAR system; Rotheram and Macklin (1982) found that in practice this gave a poor fit to the background at high wavenumbers, so the more sophisticated form:

$$\tilde{Q}_0(k_x, k_y) = \exp(-\alpha k_x^2 - \beta k_y^2 - \gamma k_x^4 - \delta k_y^4 - \varepsilon k_y^2\, k_y^2) \quad \ldots \quad (38)$$

was fitted here. The constants, α, β, γ, δ and ε were determined separately for each image using least-squares fitting; the values obtained are given in Table 2.

Table 2

Fitted Static System Transfer Functions

Image	Figure	α	β	γ	δ	ε
300	8	38.9	39.8	54.2	- 83.0	- 121
002	9	38.8	31.8	- 11.1	- 87.9	- 41.4
298	10	42.0	35.7	13.5	- 181	- 123

Dividing the image power spectra by the appropriate $\tilde{Q}_0(\underline{k})^2$ then converts the speckle component into a background of constant expectation value. The results of this operation may be seen in the top right of Figs. 8-10; it is clear that no residual variations are present in the backgrounds of these spectra.

It is difficult to assess the significance of the difference between the fitted functions $\tilde{Q}_0(\underline{k})$ in Table 2 because there are five different parameters involved. Variations in $\tilde{Q}_0(\underline{k})$ may be determined approximately, however, by fitting the Gaussian function:

$$\tilde{Q}_0(k_x, k_y) = \exp(-\alpha k_x^2 - \beta k_y^2) \tag{39}$$

The method of Rotheram and Macklin (1982) was used, because this allows α and β to be determined independently, and the results obtained are given in Table 3.

Table 3

Fitted Gaussian System Transfer Functions

Image	α	β	converted to range position of image 300
300	42.0 ± 0.2	36.0 ± 0.3	
002	38.1 ± 0.4	27.4 ± 0.3	38.0 ± 0.4
298	43.0 ± 0.4	27.0 ± 0.4	40.5 ± 0.6

The errors in these values of α and β are due to speckle and the neglect of terms in k_x^4, k_y^4 and $k_x^2\, k_y^2$. There are clearly significant differences between the fits in range and azimuth, even for images 300 and 298 which come from the same SEASAT pass (Table 1). Some of these differences are due to the variation of range resolution across the swath; range resolution on the ground scales as $1/\sin\theta_i$, where θ_i is the angle of incidence at the target, so β in (39) scales as $1/(\sin\theta_i)^2$. This effect may be corrected by converting the values of β in Table 3 to a constant range position in the swath; the last column in Table 3 shows the result of doing this. The agreement between the fitted transfer is improved, but there are still significant differences in the fitted values of α and β. The residual variations which remain to be explained are approximately the same in both range and azimuth (about 6 per cent, corresponding to 3 per cent variations in resolution).

The next stage in the processing of an image power spectrum is the removal of the background due to speckle. This component has constant expectation value and obeys an exponential distribution. The mean value B of this background may be determined from a histogram of spectral intensities, as described by Rotheram and Macklin (1983): the probability density f(x) and mean value <x> of an exponential distribution are

$$f(x) = \frac{1}{B} e^{-x/B}, \text{ and}$$
$$\langle x \rangle = B, \tag{40}$$

so the slope of the histogram (on a log-linear plot) is simply related to the mean value. The estimate of B obtained is not significantly biased by the presence of long-wave structure in the image; such structure is present mainly at high spectral intensities, whereas the slope is estimated from the low-intensity part of the histogram. This is shown in Fig. 11 for images 300 and 298 as shown in Figures 8 and 10.

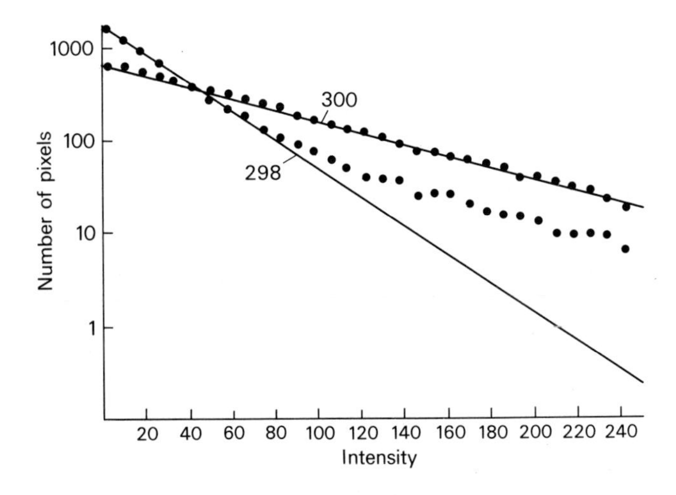

Fig. 11 Intensity distributions of image power spectra

Having determined B, the speckle background can be reduced using a thresholding technique. If the sampled image power spectrum (the middle picture in the top row of Figs. 8-10) is $\tilde{P}_s(\underline{k})$, the spectrum after removal of the static system transfer function (the top right picture in Figs. 8-10) will be

$$\tilde{P}'_s(\underline{k}) = \tilde{P}_s(\underline{k})\ \tilde{Q}_0(\underline{k})^{-2} \tag{41}$$

After thresholding, the spectrum becomes

$$\begin{aligned} \tilde{P}''_s(\underline{k}) &= \tilde{P}'_s(\underline{k}) - B \quad \text{if } \tilde{P}'_s(\underline{k}) > T \\ &= 0 \qquad\qquad\quad\ \text{if } \tilde{P}'_s(\underline{k}) \leqslant T \end{aligned} \tag{42}$$

where T is the threshold level. In other words, the background is subtracted from the spectrum when the spectral intensity exceeds the threshold T, but spectrum intensities below the threshold T are set to zero. Data above the threshold are deemed to come from the information-carrying term in (33), whereas data below the threshold are deemed to come from the exponentially-distributed background. It remains to fix T. The cumulative probability distribution for the background is

$$\begin{aligned} p(x) &= 1 - e^{-x/B}, \quad \text{so} \\ x &= -B\ \ell n\ [1 - p(x)]. \end{aligned} \tag{43}$$

A value of p close to unity must be used to remove most of the background. We have used $T = 4B$ and $T = 5B$ here (for which $p = 0.9813$ and $p = 0.9933$, using (43) with $x = T$); the results, displayed in the left and middle pictures in the middle row of Figs. 8-10, show the information-carrying terms plus randomly-distributed false alarms.

However, the above decision-procedure for thresholding (42) is done on a pixel-by-pixel basis; no attempt is made to allow for the fact that the information in the spectrum is correlated, in the sense that the peaks in the spectra in Figs. 8-10 are spread over several neighbouring pixels. The pictures on the right in the middle row of Figs. 8-10 show an attempt to allow for this correlation of signal. The thresholding procedure of (42) was applied with $T = 5B$ ($p = 0.9933$).

The resulting spectrum was then divided into blocks of size 4-by-4 pixels, and each block considered in turn as follows, depending on whether the number of pixels N in each block for which $\tilde{P}'(\underline{k}) > T$ was less than or greater than 1.

(a) If <u>no</u> spectral intensities in a given block exceeded T, the entire block was deemed to be background, so the thresholded spectrum was set to zero throughout that block:

$$\tilde{P}''_s(\underline{k}) = 0 \tag{44a}$$

The probability that 16 points drawn from the background all fall below T is $p_0 = (0.9933)^{16} = 0.898$.

(b) If <u>only one</u> spectral intensity in a given block exceeded T, this pixel was deemed to be an isolated background point, so again the thresholded spectrum was set to zero throughout that block:

$$\tilde{P}''_s(\underline{k}) = 0 \tag{44b}$$

The probability that only one of 16 points drawn from the background will exceed T is $p_1 = 16 \times 0.0077 \times (0.9933)^{15} = 0.097$.

(c) If two or more spectral intensities in a given block exceeded T, the block was deemed to contain signal and the thresholded spectrum in that block was set to

$$\begin{aligned} \tilde{P}''_s(\underline{k}) &= \tilde{P}'_s(\underline{k}) - B \quad \text{for } \tilde{P}'_s(\underline{k}) > B \\ &= 0 \qquad\qquad\quad \text{for } \tilde{P}'_s(\underline{k}) < B \end{aligned} \tag{44c}$$

The probability that a block of background data will be misclassed as signal is $1 - p_0 - p_1 = 0.005$. In terms of residual false alarms, this is a significant improvement over (42), for which the misclassification probability is $1 - 0.9933 = 0.0077$.

The results of Figs. 8-10 show that the method of (44) seems to be more effective than the method of (42) in separating the information term from the background term in the spectrum. However, the method of (44) is ad hoc. It also destroys the 180° rotational symmetry of the power spectra, since the zero wavenumber component (at the origin) is not at the exact centre of the spectra here, which have dimensions of an even number of pixels. Thus the zero wavenumber component is included in one 4-by-4 block but not in its diagonally-opposite counterpart. However, the 180° - rotational symmetry may be recovered by averaging each point in the top half of the spectrum with its mirror image in the bottom half.

The expected value of $\tilde{P}''_s(\underline{k})$, which is displayed in the middle right of Figures 8 to 10, is found from (31) and (33) to be

$$\overline{<\tilde{P}''_s(\underline{k})>} = <I_0>^2 \; \frac{L\;J}{4u^2w^2}\,\sigma_\ell^2 \; \exp\left[-\frac{X^2k_x^2\;\nu^2}{16}\right] |\tilde{M}(\underline{k},\omega)|^2[\tilde{E}_\ell(\underline{k}) + \tilde{E}_\ell(-\underline{k})] \tag{45}$$

This was obtained by (a) replacing $\tilde{Q}(\underline{k},\omega)$ by $\tilde{Q}_0(\underline{k})$ in (31) as the final term in the exponential in (15) is small for $N = 4$, (b) replacing the term between the modulus signs in (31) by $\tilde{M}(\underline{k},\omega)$, and (c) neglecting terms beyond $n = 1$ in (26). Note that it is a linear functional of $E_\ell(\underline{k})$.

The next stage is to demodulate this expression by dividing out the term $|\tilde{M}(\underline{k},\omega)|^2$, where $\tilde{M}$ is the scattering transfer function in (11). The form given in Figure 3 is used. This function is zero at $k = 0$ and so its inverse is singular at the origin. This tends to magnify any residual speckle, noise, or large scale structure for small k. To avoid this a regularised version, with $\tilde{M}$ replaced by $\tilde{M} + 1/(4\tilde{M})$, is used based on the methods of Tikhonov (1977). The result is

$$\tilde{P}'''_s(k) = \frac{\tilde{P}''_s(\underline{k})}{|\tilde{M} + 1/(4\tilde{M})|^2} \tag{46}$$

This is displayed in the bottom left of Figures 8 to 10 and in magnified form in the bottom centre of Figures 8 to 10. Its expected value is given by

$$\overline{\langle \tilde{P}_s'''(\underline{k}) \rangle} = \langle I_o \rangle^2 \frac{L\,J}{4u^2 w^2} \sigma_\ell^2 \exp\left[-\frac{X^2 k_x^2 \nu^2}{16}\right] [\tilde{E}_\ell(\underline{k}) + \tilde{E}_\ell(-\underline{k})] \tag{47}$$

No attempt has yet been made to remove the exponential factor which, through (27), depends upon the energy spectrum $\tilde{E}_\ell(k)$. Apart from this factor, $\tilde{P}'''(\underline{k})$ is a linear functional of $\tilde{E}_\ell(\underline{k})$ through the last factor in (47). Note that only the symmetric part $[\tilde{E}_\ell(\underline{k}) + \tilde{E}_\ell(-k)]$ can be found. $\tilde{P}_s'''(\underline{k})$ is our best estimate to date of the reconstructed sea surface spectrum $\tilde{E}_\ell(\underline{k})$ and is shown in the bottom centre of Figures 8 to 10. Figure 8 consists entirely of false alarms. Figure 9 appears to be quite a good extraction of all the useful information from the SAR spectrum. In Figure 10 some of the detail of the spectrum has been lost by the thresholding operation.

4.3 Inverting the image

An inverse theory may be constructed for the SAR image based on the linear part of the imaging theory given in section 2. It is quite analogous to the inversion of the spectrum and has the following steps:-

a) Form the DFT of the sampled image $I_{\ell j}$, as in (16). This gives the sampled image complex-amplitude spectrum $\tilde{I}_s(\underline{k})$; this is complex but Hermitian.

b) The zeroth component is removed by setting the central pixel to zero.

c) The complex spectrum is linearly demodulated by dividing by the linear modulation transfer function $T(\underline{k},\omega)$ in (10). The factor $\tilde{Q}(\underline{k},\omega)$ has been taken as $\tilde{Q}_0(\underline{k})$ and $\tilde{M}(\underline{k},\omega)$ as $\tilde{M}+1/(4\tilde{M})$, as explained in section 4.2.

d) The speckle has been reduced by setting to zero all those pixels that were set to zero in the power spectrum using (44) and the Hermitian symmetry of the spectrum has been recovered by combining it with its conjugated mirror image. The resulting complex spectrum is difficult to display but its power spectrum is indistinguishable from the bottom centre of Figures 8 to 10.

e) The inverse DFT of this complex spectrum is now formed to give the inverse image shown in the bottom right of Figures 8 to 10. These are our attempted reconstructions of the sea surface. In Figure 8 this is just the inverse of the residual false alarms and is quite meaningless. However Figures 9 and 10 are quite realistic reconstructions. They are much clearer than the originals due to the speckle-removal procedure.

5. DISCUSSION

No claim is made that the inverse methods described here are in any sense complete or optimum. They form the first stages in the investigation of such methods. By means of the step-by-step process described here, not only are inverse methods being developed but the imaging theories in sections 2 and 3 are being further explored and, in some respects, confirmed. Some parts of the imaging theory, particularly for the power spectrum, remain to be completed. Some further refinement in the step by step inverse methods can confidently be expected in the short term. If this gives confidence in the overall theory, the next stage should be the development of optimum methods. SAR images are stochastic through both the short and long wave fields. Resolution and speckle are traded off by varying the number of looks. A framework for the development of an optimum method is probably Bacchus - Gilbert theory which provides for such a trade-off, and this is the subject of our current research.

6. ACKNOWLEDGEMENT

Part of this work was supported by ESA.

7. REFERENCES

Alpers, W.R. and Hasselmann, K. 1978, The two frequency microwave technique for measuring ocean wave spectra from an airplane or satellite. *Boundary Layer Meteorology*, **13**, 215-230.

Alpers, W.R. and Hasselman, K. 1982, Spectral signal to clutter and thermal noise properties of ocean wave imaging synthetic aperture radars. *Int. J. Rem. Sens.*, **3**, 423-446.

Alpers, W.R., Ross, D.B. and Rufenach, C.L. 1981, On the detectability of ocean surface waves by real and synthetic aperture radar. *J. Geoph. Res.*, **86**, 6481-6498.

Alpers, W.R. and Rufenach, C.L. 1979, The effect of orbital motions on synthetic aperture radar imagery of ocean waves. *IEEE Trans. Ant. Prop. AP*-**27**, 685-690.

Beal, R.C. 1982, The value of spaceborne SAR in the study of wind wave spectra. Submitted to Reviews of Geophysics and Space Physics.

Ivanov, A.V. 1982, On the synthetic aperture radar imaging of ocean surface waves. *IEEE J. Oceanic Eng., OE*-**7**, 96-103.

Keller, W.C. and Wright, J.R. 1975, Microwave scattering and straining of wind generated waves. *Radio Science,* **10**, 139-147.

Larson, R., Moskowitz, L.I. and Wright, J.W. 1976, A Note on SAR imagery of the ocean. *IEEE Trans. Ant. Prop., AP*-**24**, 393-394.

Raney, R.K. 1981, Wave orbital velocity, fade and SAR response to azimuth waves. *IEEE J. Oceanic Eng., OE*-**6**, 140-146.

Rotheram, S. 1983, Ocean wave imaging by SAR. AGARD Conference on "Propagation factors affecting remote sensing by radio waves". Oberammergau, 24-28 May, 1983.

Rotheram, S. and Macklin, J.T. 1983, Inverse methods for ocean wave imaging by SAR. NATO Advanced Research Workshop on "Inverse methods in electromagnetic imaging". Bad Windsheim, West Germany, Sept. 18-24, 1983.

Rufenach, C.L. and Alpers, W.R. 1981, Imaging ocean waves by synthetic aperture radars with long integration times. *IEEE Trans. Ant. Prop., AP*-**29**, 422-428.

Swift, C.T. and Wilson, L.R. 1979, Synthetic aperture radar imagery of moving ocean waves. *IEEE Trans. Ant. Prop., AP*-**27**, 727-729.

Tikhonov, A.N. and Arsenin, V.Y. 1977, Solutions of ill-posed problems. V.H. Winston and Sons, Washington D.C.

Valenzuela, G.R. 1980, An asymptotic formulation for SAR images of the dynamical ocean surface. *Radio Science,* **15**, 105-114.

Vesecky, J.F. and Stewart, R.H. 1982, The observation of ocean surface phenomena using imagery from the SEASAT synthetic aperture radar - an assessment. *J. Geoph. Res.*, **87**, 3397-3430.

Wright, J.W. 1968, A new model for sea clutter. *IEEE Trans. Ant. Prop., AP*-**16**, 217-223.

INDEX